BRIGHT STAR

BRIGHT STAR

BEATRICE HILL TINSLEY ASTRONOMER

Christine Cole Catley

CAPE CATLEY LTD

For my three families

Sarah and Rob Beck and Anna
Nicola and Gavin Scott and Rebecca, Laura and Chloe
Martin and Jenny Cole and John and Annie

And for Michael King, colleague and friend

This book has been written with the assistance of a grant from Creative New Zealand.

First published September 2006

Cape Catley Ltd
Ngataringa Road
P O Box 32-622
Devonport
Auckland
New Zealand.

Email: cape.catley@xtra.co.nz
Website: www.capecatleybooks.co.nz

Copyright © Christine Cole Catley 2006

The author asserts her moral rights. This book is copyright under the Berne Convention. All rights reserved. No reproduction without permission. Enquiries should be made to the publishers.

Typeset in Sabon 10/13 pt
Designed and typeset by Kate Greenaway, Beach Haven, Auckland
Cover design by Think Red, Devonport, Auckland
Printed by Publishing Press, Albany, Auckland

ISBN: 1-877340-01-4

Prologue

A woman stood in a swirling snow storm and knocked at the door of a house in Chester, England. She asked for a piece of coal. It was needed, she said, for a baby born almost two months too soon. The baby's name was Beatrice.

A German air raid aimed at the Liverpool docks in World War II had brought on her premature birth. Without enough heat in her family's house, she would probably not last the night.

Coal was rationed in war-time England. In the depths of winter people shivered beside tiny fires made up of the few lumps of coal allocated to them by the government.

And yet, when the woman asked for coal for the baby, it was given freely. In an almost biblical scene, people came to their doors or to the baby's house and gave their mite of coal, or contributed anything else which would burn in the open fireplace in the infant's room. The fire was built up. Gradually the temperature rose. Despite all the odds, the baby survived.

Thirty-six years later she stood at the entrance to a Yale University auditorium in New Haven, USA, greeting those attending a conference. It marked a major turning point in the world's understanding of the universe. She had planned the conference, organised it and succeeded in bringing together the cream of the world's cosmologists to take part.

Neither before nor since, astronomers say, has there been such a meeting of minds on this particular subject as at the Yale International Conference on the Evolution of Galaxies and Stellar Populations, and it was due to her efforts. Almost single-handedly, Beatrice had created a watershed in cosmology.

The years that had brought her to this point were arduous and extraordinary.

As the conference began, she had less than four years to live.

Contents

Prologue V

Author's Note and Acknowledgements 9

Introduction 15

1 Beginnings 19

2 A Sea-Change 40

3 Towards a Life's Work 54

4 Freedom to Learn 78

5 What If 92

6 A Different Language 103

7 Mr & Mrs Tinsley 114

8 To a New World 132

9 The Doctorate 145

10 The Biggest Change 155

11 The Family of Four 169

12 Family Years 181

13 Rocky Years 195

14 The Unravelling Family 209

15 Decision 222

16 To a New Life 236

17 Going Somewhere 261

18 Year of the Galaxies 288

19 Time of Crisis 306

20 Face to Face 323

21 Acceptance 345

22 Letting Go 364

23 And Afterwards 385

Epilogue 404

References 406

Publications of Beatrice M. Tinsley 422

Bibliography 428

Appendix I. Edward, Poem by Rowena Hill 430

Appendix II. Obituary 432

Index 437

Author's Note and Acknowledgements

I said no when I was first asked to write this book, and for good reason. My background is in the arts, not the sciences. Beatrice Hill Tinsley is such a towering figure among world astronomers that I thought only another astronomer could begin to do her story justice.

Her father, Edward Hill, had approached me after I had written a book, *Springboard for Women*, the centennial history of New Plymouth Girls' High School.[1] Beatrice and I had both attended this school and both of us went on to Canterbury University, she being some 18 years my junior. I am a fifth-generation New Zealander, a writer and publisher. Beatrice was five years old when she came with her family from England to New Zealand in 1946. She soon felt herself to be a New Zealander and continued to love this country above all others, although her professional life was in the United States.

Beatrice Hill Tinsley was a professor of astronomy at Yale University when she died, aged 40, of melanoma in 1981. Until she came on the scene, people believed that galaxies were fixed, immobile and unchanging in the universe. She discovered (among very many other things) that galaxies are both changing and interacting with one another. She proved that the universe is still evolving. Yet, when I came to research the lives of her school's former students and teachers, I had never heard of her. The more I discovered, the more I was astounded and angered to find that virtually no one else in New Zealand knew about this extraordinarily appealing and altogether remarkable young woman.

In 1985 Edward Hill asked if I would go to Austin in Texas, where its university was about to honour his daughter at a big gathering of astronomers – after all, he pointed out, I could make a detour from my planned visit to England. I said yes. Sooner or later someone would be her biographer. I could take notes and pass them to this future writer.

My time in Austin was a revelation. Beatrice became alive. It was as if she were present with us when fellow scientists, her former professors and students told story after story, and honoured her. The professorial chair named for her was inaugurated, with her friend, Gillian Knapp – Jill – the first to hold it. Rowena Hill came from Venezuela. After the ceremonies and celebrations, Rowena, Edward Hill and I met up with the husband she had divorced, Brian Tinsley. Here was another dimension to the story.

I made notes and put them away but could not forget Beatrice. The next year Edward Hill selected a number of her letters, placed them in a family context and wrote a connecting narrative. His book, *My Daughter Beatrice*, a personal memoir, was then published by the American Physical Society. With a wise and illuminating introduction by Sandra Faber and including the resonating obituary by Richard Larson and Linda Stryker reprinted from the *Quarterly Journal of the Royal Astronomical Society*,[2] the small book made a considerable impression in the world of astronomy.

Edward Hill came back to me again. Would I not reconsider, and become Beatrice's biographer? By now I had realised that a full understanding of her work would be difficult for most scientists, including some astronomers. Most readers of such a biography would probably not have a scientific background. Perhaps if I could write a preliminary story and convey at least some of her work as well as her life and the blazing impression she left as teacher and mentor, if I could have help from her astronomer friends, if they would check what I wrote … Regardless of my other commitments, I said yes. I could not begin to imagine, then, the multi-layered task that was in front of me, and the many obstacles which illness and accident would place in my way.

What I did not discover until later was that Beatrice's letters home conveyed only a part of what she was really thinking, feeling and doing. There was so much she felt she could not say to her parents because it would cause them pain. Largely they are 'promise fulfilled' letters, in response to her father's 'Promise you will write home regularly.' She omits a great deal, so that some friends have called the letters a sham. That is unduly harsh. She does not lie, but she certainly obscures and gives a partial picture. Quite often her views of family relationships are not what others saw.

But Beatrice, the young woman for whom 'science was riotous living', as astronomer James Gunn has said, leaps from her pages. She cannot help exposing her joy in discovery and her immense pleasure in her work, to an extent rarely captured in the self-revelations of scientists. In her letters to her sister Rowena, too, a Beatrice-in-the-round begins to emerge, a more balanced picture of a young woman who was remarkable by any standards.

My researches revealed no real villains. Most of the protagonists deliberately tried to be good, and to do good, as they saw it. But key people, including Beatrice herself, made mistakes grave enough to distort lives and change the shape of her story.

As I made further discoveries I was glad I was not confined by having to write an authorised biography, in the sense that I had not been commissioned nor paid, so that I felt free to write as I saw fit, to be as true to Beatrice and her work as I could be. I did feel that Edward Hill, by lending me all the material in his possession, had trusted me to do what he once called 'the right thing' by his daughter. Father and daughter both in their own ways had a passion for truth. They were, however, on opposite sides of a philosophical divide.

Biographers owe their subjects the truth. Those who were or are close to them have feelings to be respected, too. How to deal squarely with all this, in situations likely to cause pain? When I began, neither Beatrice's father nor I could know what I would discover. I believe that it did not even cross the mind of Edward Hill – that benevolent, un-worldly man – that there was anything *to* discover except more testimonies to his daughter's brilliance, that daughter through whom, unconsciously, he sought in part to live his own incompletely fulfilled life.

I did find such testimonies, more than he could have imagined. I also found events and situations which would have distressed this deeply moral man and former practising clergyman: distorted and disturbed relationships, an understanding of which I believe crucial for an understanding of his daughter.

When he died in 2001, aged 94, I mourned him but then felt free to begin to complete and publish this book.

Far more than most biographers, I owe heartfelt thanks to the many people who have helped me since I began. I was venturing into unknown seas, a voyage impossible without the generous guidance of Beatrice's fellow astronomers, family and friends. Generosity in every sense, and a willingness to talk regardless of personal pain as the past was recalled and analysed and reconstructed, mark all the major contributors to this biography.

First I am grateful to Edward Hill, whose persistence led me through long years to the bright star Beatrice, the only member of the Hill family whom I have never met but who will always remain a powerful presence in my life.

To Beatrice's sister Rowena Hill, poet and professor in Venezuela, and to her musician and teacher sister Theodora Lee-Smith and her husband David in New Zealand, I am most grateful for discussions, letters, memories and considerations of the manuscript. With her particularly close relationship with Beatrice, Rowena has given me essential insights stemming from their childhood, and alerted me to aspects of her sister's nature and character. Because Beatrice destroyed her diaries not long before her death, her candid letters to Rowena are especially important.

Beatrice's children, Alan and Teresa (Terry) Tinsley, talked to me frankly and perceptively.

Brian Tinsley, whom Beatrice married in 1962 and divorced in 1974, said to me, 'This is partly my biography too.' I have tried to reflect this fairly, and am grateful to him for introducing me to Texas neighbours and colleagues, for sharing family memories, scrapbooks, photographs and letters, for elucidating some scientific matters, and for discussing the past as he saw it, and what he saw as his own part in it. Because Beatrice always regretted not keeping her maiden name of Hill, I have incorporated it in this book's title.

Five astronomers in particular, the five closest to Beatrice, have made this biography possible: Richard Larson of Yale University, Sandra Faber of Lick Observatory at the University of California, Santa Cruz, James Gunn and Gillian Knapp of Princeton University, and Linda Stryker of Arizona State University West.

Some if not all of them had mixed feelings about the project. How could a non-scientist from New Zealand begin to comprehend Beatrice the astronomer? How could such a person begin to convey Beatrice in the round, fairly and honestly, without tarnishing her? Naturally, before really talking openly, all had first to meet with me, question, assess, and be at least partly reassured, before agreeing to my requests for help.

That they all then so generously gave their time and many of their most intimate memories is something which will leave me always in their debt. Nothing could have spoken more strongly of their feelings for Beatrice than their efforts to help me understand her and her work. They have read and commented on my manuscript, in whole or in part, at different times and at different stages. Any errors of fact or perception, and inadequacies of explanation, nuance or context, are of course mine.

I thank most warmly the many others whose lives and Beatrice's touched, and who have contributed to this book. Usually I have interviewed people in person. Because of e-mail, and when I felt able to quote the support of Beatrice's closest friends in astronomy, I have been able to reach many of her peers, together with others who once taught her and those whom she herself taught.

In my country, I was led to discover Beatrice by two teachers at New Plymouth Girls' High School, Mary Heward and Helen Thomson, who persuaded me to research and write the school's history, and then encouraged me with this present book. Many former teachers and students of this school, Joyce Jarrold in particular, have been most helpful. I have also had assistance from Creative New Zealand. This body is more accustomed to giving supporting grants to novelists, poets and literary biographers than to anyone writing the life of a scientist. It nevertheless gave me a small travel grant, invaluable when I found it essential, first, to travel to universities in the United States of America in the early years of research so that Beatrice's colleagues could meet and assess me before deciding whether they could talk to me frankly. Since then, whenever it became financially possible, and if I could leave my book publishing business, Cape Catley, in the hands of others, I have returned for more research, in particular to Texas, and to Yale, Princeton and Lick Observatory, as well as visiting sources in Venezuela, England, Europe, Australia and of course New Zealand.

Writers long for uninterrupted time away from the distractions and claims of home and an ordinary working life. For two Easters, each of six days, I had what I called my Marlborough Cottage Fellowship, the use of a Picton cottage belonging to Triska Blumenfeld and Ernest Berry. This precious, focused time was the springboard for writing the whole of this book. Similarly, Pat and Roger Hindle of Whangarei lent me their peaceful flat. A week in Rarotonga in the Cook Islands in 2005 enabled me to stand back and look at my manuscript as a whole. These four occasions showed me, beyond doubt, how vital fellowships are for writers.

Marlborough people who helped me in my publishing business, thus giving me more time for research and writing, and who sustained me in the years of work on this book, are Pauline Summerville, Marie Perano, Coral Orsman and Shirley McEwan. In Devonport, Auckland, where I moved to live and work in April 2000, similar helpers have been Ruth Jones, Kate Greenaway, Garry Tee, Jo Birks, Geraldine Skelton, Kate Robinson, Jessica Fargher, Claire Smith, David Barton, Chris Tubb and Cynthia McKenzie, who has also worked on the index.

My sister Judith MacKenzie and my niece Elizabeth Aitken Rose have particularly supported and encouraged me over many years, especially through our 'laughing lunches', while Rod MacKenzie and Alex Cull have given practical help by giving me more free time to finish this book.

For comment and discussion on the family dynamics and personal relationships in *Bright Star*, I owe a great deal to my lifelong New Zealand friend, Polly (Evelyn) Lind, psychiatrist and psychoanalyst. My older daughter, Sarah Sullivan Beck, counselling psychologist of Melbourne, has spent much time sharing her insights and discussing relationships and likely outcomes, this while joining her husband Rob in supporting me in so many other ways.

I am grateful as ever to my younger daughter, Nicola, and her husband

Gavin Scott, of Santa Monica, and to my son Martin Reece Cole and his wife Jenny, of London, for their open homes and their discussions on parts of this book, and the financial and other assistance they gave me during the many years of my research in the United States, England and Europe.

For hours of the most intensive tuition in my life I particularly thank astronomer Sandra Faber. Similarly, I am most grateful to Richard Easther, a New Zealand physicist on the faculty at Yale. As the son of an old family friend, he volunteered to help me with more scientific explanations. He has been joined by mathematician Warwick Kissling and science writer Marilyn Head, both of Wellington, who have similarly helped with a number of explanatory notes for non-scientists. All these seek to throw light on some of Beatrice Hill Tinsley's main papers, and aspects of her particular fields. To the authors and publishers of the many books and papers quoted or referred to, I also owe much of what I have managed to glean of astronomy. One of my hopes for this book is that others will be led to read and study, and go further along the path.

I will always be indebted to my colleague and friend the biographer Michael King, who appointed himself my preliminary editor. He saw in Beatrice 'one of our great stories', and commented constructively and encouragingly on many chapters. Anne Tauté, my London editor of a previous book, again offered her editing skills and suggestions. Finally, Andrew Mason of Wellington has given me editorial advice unprecedented in my experience, and goals that I have tried to reach. The conclusions reached in this book are of course my own.

Beatrice was celebrated for her work as a synthesiser, the bringing together of apparently unrelated and individual scraps and strands of knowledge and theory, to help create a new whole. A biographer, too, must attempt to synthesise; hence the detailed attention given here to Beatrice's family and forebears – her genetic inheritance – and to her nurture, her far from usual upbringing. From these two factors, within the happenstance of daily living, came a woman whom nobody, anywhere, could forget.

Christine Cole Catley
Devonport, Auckland.
New Zealand.
May 2006.

Introduction

When I consider how my light is spent
Ere half my life in this dark world and wide,
And that one talent which is death to hide
Lodged in me useless ...
MILTON ON HIS BLINDNESS

People have been tantalised by the stars, and what they tell us about our place in the scheme of things, since the first light of awareness.

The search for the origins of the universe and answers about its ultimate fate will continue as long as humans are around for 'the last light'. The young Beatrice Hill Tinsley made discoveries that are recognised as the key to much of what is known today.

She pointed the way, and asked the big questions. How did it all begin? Where is it all going? She provided many answers, brought together many strands of astronomical theory and observations, and was one of the first people to grasp that galaxies evolve with time. Using diverse data, she was able to give astronomy what astronomer James Gunn has called 'an exquisitely powerful set of tools' for measuring the universe and testing rival cosmological models.

She looked at things in a new way. Her questions, her answers, and in particular her ways of looking, advanced the world's knowledge of the cosmos. Many details of her work have been superseded. It is the pathway that she pioneered for cosmologists, and the inspiration of her teaching, which remain.

Beatrice Hill Tinsley is one of the very few women to be honoured by the American Astronomical Society with an award and a medal. A chair in her name has been endowed by the University of Texas in Austin. She has been called the outstanding woman scientist of her age, although she would have said it is impossible to make any such claim in a field so diverse. Her attention to detail, her high standards and her personal concern for her students, her colleagues and others working in her field also made her a great teacher and model for other scientists. Yet she was nearly overwhelmed by the conflicting demands of family responsibilities and the emergence of her great talent. Beatrice was driven, as Milton said, by 'that one talent which is death to hide'.

Both her parents came from unusual and talented families, and the circumstances of her own upbringing were far from usual. In her case, genetics and environment interact in memorable fashion. As psychoanalyst Dr Evelyn Lind wrote, 'Apart from anything else in her life, it seems that enough has come to light about her forebears, on both sides, and about her upbringing to make her a classic study.'[1]

From an early age it was obvious that Beatrice's intellect was outstanding, and without doubt the way she was brought up had profound effects on the directions she took throughout her life. Her own nature inclined her to be a good person, intent on doing good in the world. Her parents imbued her with a strong sense of duty and of destiny. She was also unusual in that she combined intellect and great energy with remarkable self-discipline, so that it was often said of her that she did not waste a minute of her life.

In some ways Beatrice had an extended childhood. As she grew up it took time for her to sense the extent of her abilities and what she could and must do. By then she was married, at 20. She soon had two realities. On the one hand she found herself locked into an alien location which was barren scientifically, as a wife and eventually the mother of adopted children, with the full raft of family duties and responsibilities. On the other hand was her place in science – what she came to believe she owed the world of science, and what was owed to her, her rightful place where she could flourish and contribute among her peers.

Beatrice finally understood that for her it would be little short of living death if she denied her talent. Yet when she did at last step fully out into the scientific world it was to be, eventually, at the greatest possible cost. She herself believed her emotional conflict triggered the particularly virulent form of cancer that brought about her early death.

As she lay dying, comforting her family and friends, she could take some comfort herself from knowing she had pushed out the boundaries of the world's knowledge of perhaps the greatest mystery of all, the origins and future of the universe.

Scientifically speaking, Beatrice had the good fortune to be born at a turning point in the history of astronomy. At the beginning of the 20th century, astronomers knew that the sun was one star among billions in the Milky Way galaxy. They did not know, however, whether the Milky Way was a singular, isolated object in an otherwise empty and infinite universe, or if our galaxy was one of many 'island universes'. In 1900 this dispute was one of the foremost questions in astronomy. By 1923 it had been settled, with the realisation that our galaxy was one of many.

The challenge for astronomy then was not to find the other galaxies – they had long been observed and categorised as 'nebulae' because of their cloud-like appearance in optical telescopes. The problem, rather, was to determine how far away these nebulae were. If they were nearby, then it followed that they were very much smaller than our galaxy – but if they were distant, then they would be comparable in size to, or even larger than, the Milky Way.

By 1930, Edwin Hubble had established not only that galaxies were very distant but that they were all receding from us, and that more distant galaxies were receding more rapidly: the universe was expanding. Earlier, Einstein had handed the scientific community the theoretical tools to make sense of this discovery when he published his general theory of relativity in 1915. General relativity tells us how matter interacts with space and time and, in doing so, gives rise to gravity. In Newtonian physics, space and time are static – they label when and where events take place. In relativity, however, they are merged into a single entity, 'spacetime', and the solid wooden stage of space and time that is implicit in Newtonian physics is transformed into a magic carpet that rearranges itself as the actors move about the stage, becoming an active participant in the drama. From a mathematical perspective, the Einstein field-equations for gravity (the mathematical description of general relativity) explain how the 'shape' of spacetime changes as the matter and energy contained within it move around.

In the world of general relativity, an expanding (or contracting, for that matter) universe is far more 'natural' than one that is static and where the galaxies are not moving relative to one another. Just a few years after Einstein's original work on general relativity, the first papers describing what we now know as the Big Bang were written. Einstein, who was convinced the universe should not be expanding, added an extra term to his equations – the cosmological constant – which could hold the whole assembly in a delicate balance between expansion and contraction. Hubble, however, rendered this extra term moot when he showed that the universe was, in fact, expanding.

When Beatrice was a girl, the detailed theory of the Big Bang was not widely accepted. This was mainly because the measured age of the universe, which is derived from its expansion rate, had been seriously underestimated. Observations appeared to show that the universe was younger than the stars within it. In the 1960s the afterglow of the Big Bang – the cosmic microwave background – was serendipitously detected, and that observation proved fatal to competing steady state theories expounded by Hoyle and others.

Following our expanding universe backwards in time, we see that everything within it must have met at a point at some finite time in the past – the Big Bang. As the universe expands, the stars and galaxies coalesce from tiny ripples in the initial universe. Looking at distant galaxies gives us a window into the past – the light from a very distant galaxy takes billions of years to reach us, and so we are seeing it not as it is today, but as it was when the universe was a half or even a quarter of its present age.

When Beatrice entered the field, it was widely assumed (either tacitly or explicitly) that galaxies did not change radically as they aged. This may seem plausible – galaxies are so vast that it is hard to imagine them changing over time. It is now firmly established, however, that big galaxies are not born big, but grow from the mergers of many smaller galaxies; they interact and collide, and the stars within them create different mixtures of heavy elements, changing a galaxy's chemical composition, and the mixture of light they emit.

Beatrice Hill Tinsley's importance to astronomy includes the fact that she was one of the first to grasp that galaxies evolve with time, and to hammer this point home to her colleagues.

At the simplest level, failing to take the evolution of galaxies into account when exploring the distant universe will bias results if it is assumed that distant galaxies are similar to their nearby cousins. More importantly, any detailed theoretical account of the Big Bang does not just tell that galaxies have changed, but *how* they have changed – providing that 'exquisitely powerful set of tools' for experimentally testing rival cosmological models.

Beatrice was to absorb all this, revel in it, and bring her unique 'ways of looking' to cosmology.

CHAPTER ONE

BEGINNINGS

'To be a good person, and contribute to mankind.'
EDWARD HILL IN HIS UNPUBLISHED AUTOBIOGRAPHY, 'HALF A LIFE'.

The tiny infant Beatrice was born at home in England seven weeks prematurely, in the wake of a German air raid. The discoveries of scientists were being put to the worst possible use, and the most ferocious war the world had ever known was raging. It was 27 January 1941, and World War II was accelerating in Britain and Europe. Nobody expected the baby to live.

Her parents, Edward and Jean Hill, already had one daughter, two-year-old Rowena. Two years after Beatrice's birth they were to have their third daughter, Theodora. All were gifted.

What makes a child develop in one way, the next child in another? Nature and nurture, genes and upbringing. The Hill parents' genetic inheritance, how they were themselves brought up and particularly the convictions they applied to raising their daughters – much more than in most families, these are the keys to so much in Beatrice's life and career. What was said about her forebears was to be said, again, about her: it was assumed that the Hill daughters too would become illustrious people. And when their parents left their known and assured place in English society to live in a New Zealand which cared little for past glories, this roll-call of expectation became part of the daily fabric of their lives.

Edward Owen Eustace Hill was born in 1907, the first-born son of Eustace and Muriel Hill. His mother came from a long line of Welsh landowners, the Bowens, in recent generations more English than Welsh. His father was from a family of ship owners and builders who sent their sons (but not their daughters) to be educated at English public schools and universities.

Edward's paternal grandfather, Sir Edward Hill, was Member of Parliament for Bristol South while also taking an active part in the family shipping line based in Cardiff. These grandparents, with a large staff of indoor and outdoor servants, seem to have ruled their children's lives as they did their servants'. Edward's own father, Eustace, had left Sandhurst for a cavalry regiment, where he thrived. But he was required by Sir Edward to leave the army and join the Hills' shipping business. He claimed not to understand finance and business life, and seems never to have been happy.

The blight of not finding, or not being able to remain in, the career they wanted affected both Beatrice's grandfather and father, and helped alter the course of their families' lives. In Beatrice's childhood in distant New Zealand, Edward Hill told many stories of the Hill and Bowen families and their position in the world. He sincerely believed in an egalitarian society, yet wanted to make sure his daughters knew who they were and where they came from. The dark underside of these oft-told tales was frustration and unused talent.

Edward did have some happy memories of childhood. When he visited his grandmother Florence Bowen, she cuddled and hugged him, giving him what he describes in his unpublished autobiography, 'Half a Life'[1], as 'memories of unadulterated joy'. There was one happy week when his own father took him away trout fishing. For the rest he was brought up by nannies and governesses before the shock of prep school and then public school.

When he became a father Edward could remember these few golden times, but ordinary demonstrative love did not come easily to him. His own father was always in a world apart. Edward spoke respectfully of his mother, Muriel, as 'a little remote'[2]. She studied the piano at Dresden, intending a musical career, but she married young. Married women did not have careers or become serious artists. Muriel Hill got up early each day, organising her time so that she could read deeply (Hegelian philosophy, for instance, while she waited for Edward's birth) and keep up with her music. Her children in the nursery upstairs would wake to the sound of her distant piano.

She became a Justice of the Peace, and chaired the local bench, a figure remote from those who were habitually hauled in front of her. From this Bowen grandmother, Beatrice may well have inherited what Edward Hill described as a most orderly mind, and great powers of organisation. One of his cousins, a colonel in the Guards, said of Muriel during World War I, in the ultimate compliment, that she should have been a staff officer.

Edward Hill had two sisters, Rosamond and Patsy, and two brothers, Christopher and, much later, Francis. The children had ponies, trout streams and a world of woods and parklands to play in. Edward responded to nature with a delight he was to pass on to his own family. To outsiders it was an enviable life of wealth and privilege. For Edward, distanced from his parents whom he seemed never able wholly to please, it was a life riddled with anxiety and guilt. 'The English cuddle their dogs and leave their children to howl,' observed Mattie Hill many years later.[3] A family story has an anxious young Edward running away from his Hill grandparents' home (he did not recall why). A footman was pursuing him, and shouting 'Danger! Danger!' The apprehension of danger remained with him.

At his prep school, Hurst Court, which he hated and where he 'felt totally abandoned',[4] the nine-year-old Edward was made to write home every week. He was to inculcate this habit in his daughter Beatrice so successfully that it became her solemn duty to write regular letters home to the very end of her life, letters that her father carefully preserved. These windows on to her life and times are only part-open, however, in that they mostly contain only what she thought her family could absorb or what would interest, please or placate them.

The Hill and Bowen sons habitually went to Winchester, one of England's great public schools. Edward did well there, gaining a history scholarship to Magdalen. At Oxford, however, away from the discipline of his school and with all the temptations of high society in the new jazz age of the 1920s, he did little or no work in his first terms. And work towards what? He was uncertain of his future. Business life, the army, a life on the land – none of these appealed.

This tall, handsome, outwardly confident young man, with an appealing naivety and seriousness, and a genuine sweetness which he could suddenly reveal, found himself in a clique of those from similar backgrounds, wealthy and well connected, who were there mainly to enjoy the social life. He grew increasingly depressed. His submerged exuberance and zest for life did find expression in occasional revues and comedy sketches. Not knowing what he was looking for, he joined several university societies. Nothing gave him answers, nor did he know what questions he could be asking, or how to formulate them. His old feelings of never quite being able to measure up came to the fore. He was ripe for conversion to one cause or another.

For some of those in the 1920s who were idealists, the newly formed charismatic religious movement, the Oxford Group, sang a siren song. It supplied all the answers and kept its members in a close cocoon of certainty. Edward was more than ready for this. He was to meet his future wife within this movement, and it was to have far-reaching effects on Beatrice and on all members of their family.

The Oxford Group was formed and driven by an American, Dr Frank Buchman. It offered certainties and close personal contact in a new age of shifting standards, anxiety and doubt.

In this the Oxford Group had many parallels with Christian fundamentalist and charismatic religious groupings, but it had one specific and particularly potent characteristic. It sought its adherents in the ranks of the wealthy and the well-connected, so that exclusiveness if not downright social snobbery was a powerful if unacknowledged force.

The 1920s and 30s were a time of absolutes, of exclusive memberships, whether in religious or high society groupings or in the new communist and fascist parties. If a vein of idealism permeated the grouping, all the better, there being nothing like a dose of moral superiority to make one feel an insider.

Oxford Group adherents were instructed to have regular 'quiet times' in which they waited for 'guidance' from God. They were to strive to live by the 'Four Absolutes': Absolute Love, Absolute Purity, Absolute Honesty and Absolute Unselfishness. With the best will in the world, however, in a 'quiet time' one is apt to be strongly 'guided' to do precisely what one most wants to do. Lovely permission! Not only the blessing of the Lord but the enthusiastic support of other converts. It was irresistible. Or it was to many. Others retreated from the regular sessions of public confessions or soul-barings within The Group. One did not have to be particularly cynical to observe that these confessions were listened to with appetite, particularly when lapses from Absolute Purity were involved. Converts grew apace.

From feeling an outsider, a wavering agnostic bemused by the metaphysics that he was reading in philosophy, Edward was swept up by The Group. Here he found a direct, non-intellectual approach to faith – 'the insistence that something really happened to change a man if he invited God (or, as they usually said, Christ) to take over his life.' [5]

With his time and thoughts increasingly taken up by the Oxford Group, Edward graduated with a second in Modern Greats. He still did not know what he should do for a career except that he wanted 'to be a good person and to contribute to mankind', as he wrote in his autobiography.[6] The search for a path in life took Edward over the next 10 years to countless house parties and journeys abroad and to fashionable hotels and mansions, all places to which the Oxford Group's campaign teams had secured invitations for their missionary work

A small allowance from his father, (more than the wage of the average working man at that time), together with the hospitality arranged by the Group, carried him along. There was also settlement work with the poor in New York, where he came face to face with enough shocking reality to make him know that social justice must be his lifelong philosophy. In 1930 he enrolled at a Church of England theological college in Cambridge, breaking off for a time to travel to South Africa with an Oxford Group team. But he returned to pass the general ordination exam, which led him to become an assistant to an Anglican vicar in one of London's poorest areas, Lambeth. The experiment was not a success. Involvement in church work with the outgoing open-hearted New Yorkers was one thing, but he had had no training – nor shown any aptitude – for the kind of conventional parish work now required of him. Increasingly he again gravitated to the Group.

The family shipping business had been declining for years. This was not fully realised until Edward's father died in 1932, a frustrated man who had hidden his financial problems even from his wife. As for any close exchanges there might have been between father and son, the father had been 'no more willing to discuss money than he had sex',[7] two matters concerning which Edward continued to have small knowledge and, apparently, little interest. Religion, particularly questions of conduct, remained his obsession.

With the family home and land sold, and his mother and younger brothers and sisters settled in Brynderi, another large and comfortable house in Gwent,[8] Edward Hill was free to join the Oxford Group's campaign to Canada, which travelled via New York. It was during this journey that he got to know the woman he was eventually to marry. Jean Morton, four years older than Edward, was also a team member.

Dr Buchman had arranged for his team to be housed in the refurbished Waldorf-Astoria, known as The Palace of New York. Significantly, Edward in 'Half a Life' writes at length about this palatial hotel, thousands of miles of Canadian scenery and many Group meetings before he even mentions Jean Morton. Then he praises her performance in a revue which some of the Group put on privately for their own amusement. She was conspicuous as a character she named Mrs Total Recall, who said she believed, 'like dear Dr Buchman, that beautiful thoughts will change the world'.

Both Jean and Edward came alive with an audience. Some of their happiest times together were to be when they devised and appeared in satirical amateur revues. And they were good. In other circumstances, such as less wealth and different expectations surrounding them at birth, they could have become actors, she suggesting perhaps another Joyce Grenfell in one of her satirical monologues, he one of the 'I-say-old-chaps' who played such essential roles in the comedies of England's post-war Ealing Film Studios. But such an idea never surfaced.

The Group's campaign teams journeyed to Ireland and Scandinavia and throughout Europe. During those years Edward sometimes thought of asking Jean to marry him, but the 'guidance' of the Group or his own 'quiet times' did not point him in that direction. It seems that it was only when Dr Buchman took his campaign team to the Olympic Games in

Germany in 1936, where they had contact with high-ranking Nazis and where Buchman organised big public meetings increasingly in the style of Nazi rallies, that Edward began to see, and face up to, what was happening with the Group and his own life.

He came back from Germany in great distress as he struggled with his new criticisms of aspects of Buchman's evangelism. Home at Brynderi again, he had what he describes as an illness amounting almost to a nervous breakdown. But he kept his idealism, and out of this time came the decision, at last, to ask Jean Morton to marry him.

Her father would not consent to the marriage until Edward had a job or was training for a career. Still unsure about committing himself to the church, Edward became articled to a solicitor and began to study law. The church and the law have room for acting ability, too.

Edward had first met Jean in 1930, at an Oxford Group house party attended by all six of the Morton children. Three of them ardently espoused the Group's philosophies, with the others giving only temporary commitment, if that, although one son-in-law later became a Group leader. The Morton parents were not enamoured of this new kind of religion that devoured so much of their children's time and energies and money. They were quietly suspicious.

Looking back over his life and Beatrice's from the vantage point of the 1990s, Edward Hill said it seemed likely that Beatrice inherited much of her mathematical and scientific talents from her mother's side of the family.[9] And, he might have added, a good measure of the sturdy strength of character and justifiable confidence that characterised many of the men. In the random mix of genes which Beatrice inherited from her parents, so many of her forebears' characteristics were to emerge as she developed that the Hills saw them as unmistakable. They often commented on them, making a game of what they called 'ancestor-spotting'.[10]

The Morton family's lives are well documented. From Beatrice's great-grandfather, Alexander Morton, down through the generations, a number of the Mortons showed remarkable talent. They were weavers, practical and hard-working with a drive to experiment and find something new, and with a vision of the beauty that could be woven into carpets and fabrics of all kinds.

Alexander Morton was born in Darvel, Scotland, in 1844. Alex G. McLeod's *The Book of Old Darvel, Some of Its Famous Sons*, says of Alexander Morton: 'He came of grand old stock.' Alexander's father was Gavin, known as Guy, a hand weaver 'of the same independent character as his son'.[11] Guy died when Alexander was six. His mother, facing poverty, had to take the boy away from school when he was nine years old, not so very long after he had learned to read and write and do basic arithmetic.

To earn a pittance that would help his family, the wee lad was sent herding on a moorland farm. Far from becoming embittered by the life in the bleak outdoors, Alexander seems to have read what books he could, dreamed of a different future, and reacted to the farming experience by gaining a lifelong love of nature. At 12 he came home to learn the family craft of weaving.

This lad had visions of improving traditional methods, and he observed, experimented and persevered. All his life he worked very long hours and organised his time. He was thorough, thoughtful and responsible, taking thought for the welfare of others, and, says *Old Darvel*, he was 'ever on the outlook for new ideas'.

Alexander Morton was among the first to predict the fate of the traditional handloom. As the industrial revolution took hold in the English Midlands, he brought home the first weaving machine that Darvel had seen. Many of his visions became reality. He became a pioneer in the manufacture of lace and other fabrics, and the diverse businesses he and his family established did extremely well. Carpets, he saw, could still be handmade if they were of obvious quality – and made where labour could still compete with the new machines. He set up four large factories in Ireland, and his famous tufted Donegal carpets were held to be among the most beautiful examples of commercial art in Britain. They were admired and bought by the highest in the land, beginning with Queen Victoria in 1900. King Edward VII had Morton's Donegal carpets ordered for Buckingham Palace and the Royal Yacht. Later, King George V acquired one for the drawing room of Marlborough House. There are still Donegal carpets at 10 Downing Street and in the Houses of Parliament at Westminster. He delighted in finding ways to mass-produce beauty, and in this he was backed by the vision, taste and scientific talents of his second son, James, Beatrice's grandfather.

James Morton, born in 1867, studied the work of artists and writers such as John Ruskin and William Morris. It was not just the art of these men but their social idealism that inspired him: their creed that the arts should be brought into the lives and homes of ordinary people. He set to work to ensure that the beautiful designs of this new age, and their rich and striking colours, could be transferred into the firm's woven fabrics and eventually become widely available.

In many ways James built on his father's textile tradition, carrying it forward on lines of his own. He was an individualist, confident in his ability to think for himself and to disregard convention if it seemed to stand in the way of good sense. (An admiring doctor once told the young Jean Morton that her father had 'the best brain in Britain'.) Even in dress he opted for sense over fashion. He disliked the confining effect of ties and usually wore a cravat of Liberty silk. Like true inventors he went for the simple, the basic. An example was the design of his shoes. He had them made with vents or exhausts, so his feet would not become hot. Never mind if others laughed. He went his own way.

James Morton had great driving power and concentration. To a quite unusual degree, too, he combined business energy with an enriching involvement in both the arts and practical sciences. When World War I cut off his supplies of dyes from Germany, he personally worked with his own team of chemists to produce the dyes he lacked, as well as creating important new colours. His research and the subsequent production made him and his firm of Morton Sundour Fabrics pre-eminent in the development of new and strikingly beautiful fabrics.[12] Sundour furnishing fabrics were the first to be guaranteed truly fadeless, and were stunningly beautiful in colour and design. Here was art for the people indeed, and, if prices were too high for Morris's 'ordinary homes', Sundour standards had an immense effect in raising the quality of competing products.

James Morton received a number of honours for his work. He was knighted in 1936, and awarded an honorary doctorate of law by St Andrews University. What probably meant most to him was becoming the first recipient, in 1929, of the Faraday Centennial Medal, 'in special recognition of the signal service rendered to chemical science and industry in this country in the last 10 years'. This award was made on the joint recommendation of the Royal Society, the Royal Society of Arts and the Royal Institution. The Royal Society of Arts also awarded him its medal for a lecture he gave in 1929 on the history and significance of fast dyes.

His wife, Lady Morton, Beatrice Hill Tinsley's maternal grandmother, was born Beatrice Emily Fagan, the daughter of a major-general in the Bengal Staff Corps and from a prominent Irish landowning family that had long associations with India in both the army and the civil service. Beatrice's elder brother, Sir Patrick Fagan, was Financial Commissioner for the Punjab, and a respected mathematician. Particularly interested in education, she trained as a teacher and met James Morton while conducting adult education classes in art appreciation. It was said of her that she had a bright intellect and held strong views about the contribution women could make to the nation.

The Mortons were to have daughters Guenevere, Jean, twins Helen and Beatrice, and sons Alastair and Jocelyn. Music and art were important to everyone in the family. All six Morton children were educated at the co-educational St George's School, Harpenden, at a time when co-education was promoted only by those holding advanced educational views. But only the two sons went on to university.

The older son, Alastair, while still in his teens married the equally young woman who had been head girl to his head boy at St George's School. This marriage ended in divorce, an event that so shocked his family that the repercussions, down the years, were to have a decisive effect on his niece Beatrice. Alastair studied mathematics at Edinburgh University, and then at Oxford. A talented artist, he was an early disciple of painter Ben Nicholson, and became an outstandingly original textile designer and creator of highly original decorative fabrics. He also commissioned designs from some of the leading contemporary artists, such as Barbara Hepworth and Ben Nicholson. His firm, established by James Morton to produce designs and fabrics for the avant-garde, was Edinburgh Weavers, working with the parent company of Morton Sundour Fabrics Ltd

The *Times*, in a signed obituary of Alastair Morton, spoke of qualities that can also be recognised in his niece Beatrice. It used words such as capable and conscientious, sensitive and imaginative, adding 'with a true humility, friendship and generosity which made it possible – I almost said easy – for him to recognise, employ and encourage other designers, all of whom liked working for one whose ability in their sphere they acknowledged, and whose integrity was never in doubt ... and he spoke with conviction what was in his mind.' Similar things were to be said about Beatrice as teacher, mentor and appreciator of other astronomers.

The second son, Jocelyn Morton, took a first at Oxford, joined the family firm and made a long and important contribution to the business, becoming chairman of Morton Sundour Fabrics for its last 19 years. This firm was taken over by the giant Courtaulds in 1963.

Jean Morton had a closer relationship with her father than with her mother. Lady Morton abhorred public displays of emotion of any kind, and in private she does not seem to have reacted very differently. Apparently she had not had much mothering herself.

Jean as a schoolgirl had spoken of medicine as a career but in her mid-teens she became seriously ill with pneumonia, an illness that was then often fatal. Her health was affected, she became more self-absorbed, and from then on she was perceived by the family as 'delicate', an omnibus term which in those days covered a multitude of ailments, real and otherwise.

When she and Edward became engaged they visited a London doctor together to confirm that she would be able to have children, as she had had one ovary removed some years earlier. She also suffered another bout of pneumonia not long before her marriage, but it was her first severe illness, in her impressionable teenage years, that was Jean Morton's great misfortune. A miasma of concern and anxiety from her family, from her father particularly, constantly hovered about her. This changed her perception of herself; she came to believe she was someone special, deserving special care and consideration. She was not to be crossed. Beatrice from quite early childhood learned to be wary of how 'poor Mummy' was feeling. Anything triggering an outburst of hysteria was to be avoided at all costs. Edward appears to have dealt with this by absenting himself in spirit if not in body.

Jean Morton was also intelligent and creative, and forceful even when she was being languid. But, in spite of her parents' advanced attitudes, she was launched into life with inadequate education and no training for a career. True, she had been educated to the age of 16, unlike her own grandfather, Alexander Morton, plucked from school and so many of his hopes at the age of nine. In many ways, however, she had been brought up as a princess with her own retinue, and she had the very considerable Morton fortune to fall back on. Like the princesses of fairy tales, she was to find her private income both a blessing and a curse.

By temperament she was reserved, seemingly remote, although she could be generous and a delight to her few intimates. Where she flowered was when taking part in amateur plays and revues, particularly in comedy roles, and from childhood she had studied music seriously, both the piano and the cello. She became an excellent accompanist but the cello was her principal instrument. This informed delight in music was perhaps her greatest gift to her three daughters. Beatrice was to become an outstanding violinist, and to find music and music-making her solace.

The young Jean Morton was at a loss in her twenties. Women of her background were not expected or encouraged to work. Without training of any kind, no household skills, and with no obvious career path to follow, she was filled with uncertainties as well as undirected idealism. All of this was made to measure for one undertaking. She became another obvious recruit for the Oxford Group.

Edward Hill and Jean Morton were to give their future children a remarkable inheritance. The parents brought with them the attitudes and assumptions of their own unusual families, the Hills from the assurance of stately homes in park-like settings, the Mortons with their background of science applied through glowing colours in treasured carpets and furnishings. The characteristics and aptitudes of the two families were shot through with brilliance, but also with flaws.

A fairy-tale wedding, everyone said of their marriage. How right it was. They were a golden couple who would live happily ever after. England's late summer sun streamed down on Carlisle Cathedral and then on the grounds of the bride's parents' stately home, Dalston Hall. The champagne flowed. Voices rose in assured counterpoint. The wedding guests milled from the marquee to the hall itself, the tide of champagne helping transform the grim lines of this former castle.

As happens when the bride and the groom are older than usual, and when family and friends may have wondered if marriage were going to come the way of either of them (Jean's younger sisters had long since married), the good-hearted feel a special pleasure at seeing the happiness of the two people on their wedding day. On this day, 31 August 1937, no bad fairies seemed to lurk.

But in fairy tales there is usually one small cloud, a portent. In this case it was Edward's seeming inability to settle on a career, to feel completely at home within himself, to see a settled path ahead. It was not a cloud that loomed. In that circle few would have paid it attention or seen it as cause for unease, although those few almost certainly included the purposeful Morton men. On this fine day, however, any such cause for concern, and where it might eventually take the bridal couple, would not have been foremost in anyone's mind.

Carlisle Cathedral with its bishop and choir in full attendance had been filled with their families and friends. Edward's old nanny was in the cathedral, together with the Mortons' many business acquaintances and connections, and a large party from the Oxford Group. The latter wanted to catch a train back to London that afternoon so the wedding was timed for their convenience, at 1pm. A motorcade of the kind rarely seen in Carlisle took guests to the reception at Dalston Hall, some four miles from their business headquarters.

More like a castle than a home, Dalston Hall was built around a medieval Peel tower originally used as a refuge from Scots marauding across the nearby border. At its centre was a huge sitting hall in what had been the cattle yard of the Peel tower, a vast, draughty and uncomfortable space in spite of being adorned with the Mortons' fine carpets and fabrics. Edward privately thought the draughts had contributed to the second bout of pneumonia from which Jean had recently recovered. But on this day the English sun was unseasonably warm, and there were no omens of ill health or anything else untoward.

Edward and Jean Hill began married life in Llandaff House on the outskirts of Cardiff in Wales, within easy driving distance of Brynderi. The Mortons presented them with curtaining and carpets, and even that sedate and impressive early model car, a Lanchester. Edward had been shocked when he discovered how big Jean's private income was.[13] But it must have been an agreeable shock for someone who had not yet settled into a career. They were able to live comfortably at Llandaff House, and employ a married couple to look after them, the woman as cook – Jean had never cooked – and the man as gardener-chauffeur.

The heart of the house was the billiard room, but it was not used for anything as frivolous as billiards. This room was large enough to seat up to 70 people and they came weekly, many recruited from the mining valleys, to attend the Oxford Group meetings which the Hills initiated, canvassed for, and led. Reinforcing each other, Edward and Jean were more than ever intent on obeying the Group's requirements. Besides, to do this in such style was satisfying. Satisfying to Dr Buchman, too. Upper-class converts were his aim.

Edward took up his law studies again, in Carlisle. He had already passed the first of the three main examinations then required, and became articled to a large legal office dealing mostly with the business of a major coal combine. Soon he began to feel frustrated and alienated from the concerns of this demanding client, wanting human contact but with men of his own kind. Jean, trained for no career, stayed at home and before long became pregnant.

Their first child, Rowena, was born on 26 September 1938. This was the week of the Munich crisis, when Britain's Prime Minister, Neville Chamberlain, returned from his meeting with Hitler, waving a piece of paper and declaring 'Peace in our time.' Jean of course had a nanny for her baby. Edward was a proud new father. The 'Social Merry-Go-Round' column of the Cardiff newspaper, the *Western Mail*,[14] in its account of Rowena's christening described the parents as 'ardent Buchmanites'.

In most parts of the Western world by then, the air was heavy with apprehension. Winston Churchill led those who were warning and fulminating about Germany's intentions. Even those most reluctant to believe that Britain should re-arm, and there were many, were seized with foreboding. When Hitler invaded Czechoslovakia – so much for Chamberlain's piece of paper securing its security – guilt over Britain's betrayal of the Czechs, and a facing up to reality, sent many men to enrol in the Territorial Army. Edward was one. He was at once made a second lieutenant because he was of the upper classes and because of the training he had received in the officer training corps at his public school. Either in Britain would have sufficed to make him an officer.[15]

While Europe moved closer to war, important events were occurring in the Hills' private lives. They now began to believe that the Oxford Group, which had been centre-stage for both of them for some 12 years, was changing its character. Edward had been perplexed and distressed by the propensity of the Group's leaders to fawn on prominent Nazis at the time of the Olympic Games in Berlin. They noted, too, that the Group was becoming more concerned with propaganda, with worldly affairs, than with the quiet and exclusive evangelism they had always known. Why, they were even required to distribute pamphlets outside public places. To stand around outside a cinema, offering leaflets on the Oxford Group? This was too much. In their quiet times neither felt any 'guidance' to do so, and therefore did not.

In Edward's words, 'We still felt that conversion and personal experience [of Jesus Christ] were vital to a Christian life, and the only true basis for ultimate change in society.'[16] Dr Frank Buchman and other Group leaders, however, appeared to be cashing in on the growing clamour throughout the country for more and more rapid re-arming. This was when they re-named their movement Moral Re-Armament.

Representatives from the movement's London headquarters began visiting the Hills to spread the new vision, and their place in it. They even took over the running of the weekly meetings in the Hills' home. Adherents split into factions. Feelings ran high, as they are wont to do among true believers of whatever persuasion. The Hills found themselves on the outer. Before long, when meetings were under way, the Hills were reduced to hanging around in their own home, denied the sanctuary of their own billiard room.

To the outsider, and from this distance in time, the goings-on seem comical, a plot twist from Gilbert and Sullivan. But both Edward and Jean Hill seriously and sincerely believed in the tenets and structures of the Oxford Group and tried to live by them. These comprised their vision of how to be, and remain, good people. Betrayal by those whom you trust must always wound even the worldly wise, and worldly wise the Hills were not, regardless of their place in society. So, although they continued to try to live by the same principles as before, they became divorced from the mainstream of the movement. Possibly because they felt themselves alone as the true flag-bearers, in Jean Hill's mind in particular the 'Four Absolutes' were to take on a weird and sometimes almost savage intensity, out of place in everyday family life.

War became ever more imminent. Edward's autobiography shows he found aspects of those early army days 'a jolly good lark'. He was occupied, useful and needed at last. When Jean again became pregnant (although she was to have a miscarriage), it was thought best that she and Rowena and the nursemaid should go for a time to the safety of her mother-in-law's establishment at Brynderi. Llandaff House was too close to the docks of Cardiff should there be air raids, as indeed happened later. Edward was called up to the regular army. He was outfitted with heavy boots and put to drilling and marching, but then the arches of both feet gave way. His feet, in medical parlance, were 'unreliable' as far as marching was concerned. It was not a heroic disability, but it was real.

There was a significant consequence of his call-up. Had it occurred one month later, he could have sat his second main law examination. This would have seen him on his way to a law degree so that he could have picked up his study again later. In that case his life, and his family's, could have taken a very different direction. But because he missed this exam he was to discover, after the war, that he would have to begin his law studies all over again, from scratch.

Britain declared war on Germany on 3 September 1939 after Hitler moved his forces into Poland. The six bitter years of World War II had begun.

Edward Hill might have been posted anywhere, probably to a semi-sedentary position in view of his medical history. But his new colonel was a close friend of his mother's. He knew there was a need for junior administrative officers, 'so he put the typically British "old boy" system to work and I found myself most unexpectedly posted to Western Command in Chester as a staff captain dealing with accommodation.'[17]

Then, in spite of the war and in part because of it, a period of family stability began for the Hills in Richmond Lodge, an attractive house and garden near Chester in north-west England, where the family was based. Edward was energised by a new sense of purpose. His duties were to drive around the countryside, assessing and requisitioning stately homes. The government wanted these as possible hospitals or billets for the armed services, for those evacuated from particular enemy targets and, much later, for servicemen arriving from the United States. Friendships he made at this time remained important to him, notably that of the then Marquis and Marchioness of Anglesey and their son Henry.

Jean, pregnant again, had a married couple to run Richmond Lodge and to queue for extras to eke out the already meagre food rations. The couple's daughter was nursemaid to two-year-old Rowena. Air raid alarms occurred quite often. Chester itself was not a prime target but, if the German planes did not penetrate nearby Liverpool's anti-aircraft barrage, they were likely to drop their bombs on Chester as they flew back to Germany. Alarms therefore sounded whenever Liverpool was raided.

One night the Hills heard a bomb whistle overhead. They threw themselves on to the floor. The house was shaken by a tremendous explosion, the attic ceiling bulged ominously and the latch flew off the back door. They heard the whistles of two bombs but only one explosion. Everyone in the neighbourhood knew what that meant. Somewhere close by there was an unexploded bomb that could go off at any minute. The people of Britain were almost to get used to this, in so far as anyone gets used to the prospect of sudden death or maiming, but that was later in the war. Tension grew in this part of Chester until the second and unexploded bomb was discovered lying near the back of the Hills' house. A local bomb disposal squad came and defused it.

Soon after this, Jean Hill went into labour, seven weeks prematurely, in the depths of winter. Edward was summoned home from Western Command headquarters. The maternity nurse told him matter of factly that he should not expect the baby to be born alive, or to survive if it was.

Edward, waiting outside the bedroom on 27 January 1941, about 8pm at last heard a thin cry from its adjoining bathroom. He learned later that the doctor had had trouble getting the tiny child, Beatrice, to begin breathing.

The household's energies swung into keeping the bedroom warm enough to keep the baby alive; 70 degrees or the temperature of a hot summer's day, the doctor had said. The nurse stayed on to care for the baby but an incubator of any kind was out of the question. The devoted servants, George and Cissie Folkard, and their daughter canvassed the neighbourhood for any donations of rationed coal. In an almost biblical scene, people came to their doors or to the baby's house with offerings of coal to keep the room at nearly hothouse temperature until Beatrice was out of danger and could be weaned to temperatures more normal for Britain – still very cold by today's standards. Mother and baby could then go downstairs; just as well, as spring brought a change in the weather so that air raids began again. These were all the more terrifying if people could not get to the comparative safety of the ground floor or cellar.

Beatrice Muriel Hill, named for her grandmothers, was dressed for her christening in the same robe which had been worn before her by Rowena, her father, her grandfather and great-grandfather, and which was also to be worn by the youngest of the Hills' three children, Theodora, born two years later.

Beatrice's two grandmothers, both living in the countryside but staying in Chester for the christening, had their first personal experience of what an air raid could mean. A bomb went off in mid-air above the town the night before the ceremony. Every pane of glass over a wide area was shattered. 'The grandmothers, so far from feeling nervous, were glad to have shared something of what much of the nation was enduring,' Edward Hill notes in his book, *My Daughter Beatrice*.[18] Among strong-minded people of conscience who were living remote from likely danger areas, guilt at their good fortune tinged their natural feelings of relief. Some even visited targeted areas as a matter of principle, so that they did not feel unfairly privileged and cut off from others' suffering. Some went to the other extreme, leaving Britain for the United States or Canada. Beatrice's grandmothers were strong-minded.

Soon after the 'christening air raid', and perhaps in some way connected with it, Jean Hill was asked to bring the two children to stay for the summer in the peaceful atmosphere of Brynderi. It was remote from air raid territory, although nowhere in Britain could count itself wholly safe.

Jean felt she had not fully recovered after Beatrice's birth and it seems that nearly all the care of the baby up to then had been left to the maternity nurse.

From that time on, the small Beatrice received even less of her mother's attention. For a while, as the maternity nurse also went to Brynderi, Beatrice was cared for by no fewer than three nurses. Edward's old nanny, Nanny Jenkins, still held what he called 'a somewhat undefined but vital role' in his mother's household. And Jean had found for the children a permanent nanny, Nanny Gullidge.

Constance Gullidge's arrival on the scene was one of the most important events in Beatrice's life. By the greatest good fortune she was a woman of warmth, sense and reliability, for she was to play a central role in the child's early years. She became, in fact if not in name, Beatrice's real parent, always there and seemingly with bottomless love.

Nanny Gullidge was an experienced children's nurse who came to the Hills well recommended by one of the Marks families of Marks and Spencer. She had not married until her late forties. Then, a few months after her marriage, she became a widow during the war. Edward Hill says of her: 'She came to us trying to pick up the threads of her life again, and with pent-up love to bestow on her new charge. It was returned in full measure.'[19]

Jean Hill, like her own mother not noticeably maternal, had a keen sense of the ridiculous and could laugh at the situation she found herself in. Here she was, she wrote to Edward, lying on a chaise longue under a cedar tree overlooking a stretch of the most beautiful and peaceful countryside, and reading a book called *The Single-Handed Mother*.

It was probably around now that she posted off the date and time of Beatrice's birth to an astrologer in Edinburgh, and received an astrological chart showing the child's aptitudes to be music and mathematics. Jean Hill, again like her mother, was interested in astrology, the occult, extra-sensory perception and alternative methods of healing. When she was recovering from her teenage bout of pneumonia, her mother had sent her away to stay with a spiritualist medium who supposedly could hasten her recovery, although there is no evidence that anything of the sort occurred. In her youth Jean for a time also studied various natural healing methods in Edinburgh.

When Beatrice was a few months old and they were still staying at Brynderi, the household and guests were sitting out on the lawn when a cousin said dreamily, 'Isn't that smoke coming from the baby's room?' It was. Jean raced upstairs to the rescue. Clothes arranged to air had caught fire from a heater, but Beatrice was unharmed by the smoke.

By the time Beatrice was about 18 months old and thriving, it was

Rowena's health that concerned the Hills. Rowena developed pains in her legs, and the parents began to dread rheumatic fever, a scourge in Jean Hill's family. In Rowena's recollection this was the only time in her life when her mother used to come into her bedroom at night to tuck her in or to say goodnight: many years later she was to say,'Psychologically, I didn't have a mother.'[20] Nanny Gullidge played an important part in her life, too, however.

Prompted by their doctor, who thought the damp from the nearby river Dee could be affecting Rowena, the Hills discovered The Astburys, which Edward described as 'a perfect miniature of a large country house set in idyllic grounds', but with a rather daunting rent. In their 'quiet times', which they both kept up, they felt guided to take the house. No sooner had they made this decision than Edward – providentially, they agreed – was promoted to major with an increase in pay of 10 shillings a day, a considerable amount.

The children flourished at The Astburys. Theodora was born there, also prematurely, on 22 October 1943, and during her first 18 months she was not even taken out of the grounds. They had more than enough land, with space and interest for all the children.

Beatrice from the beginning was exceptionally bright-eyed and alert. As she grew into a toddler, her red cheeks and dark curly hair reminded people of a Dutch doll. All the children loved Nanny Gullidge, but Beatrice was the nanny's special baby and grew sturdy and confident, a child who enjoyed life and purposeful play, and who was able to keep herself happy and busy on her own.

Of the family stories from this time, one was reported by Nanny Gullidge who was watching from an upstairs window. The Astburys had extensive grounds for a comparatively small country house – half a million crocus bulbs had been planted in the lawns, and many flowers flourished – but Jean Hill like most mothers had forbidden the children to pick flowers indiscriminately. All the Hills loved nature so the children were told they could pick any flowers that had fallen over. Four-year-old Rowena, always a rebel and precociously intelligent, was observed sending Beatrice toddling down a garden path ahead of her. She had told her sister to tread on the bordering flowers so they would indeed have fallen over by the time Rowena reached them.

Like all the family, Beatrice had a deep love of music from when she was very young. In her teenage years she became such a good violinist that her teacher hoped she would become a professional. Playing informally in chamber music groups became her great relaxation and joy in later years. As a three- or four-year-old, Beatrice played games in which she was a musician. The family heard her being a BBC announcer: 'The next item will be a violin solo. The violinist will be Beatrice Hill. The pianist will be someone no one has ever heard of.'

This anecdote, one of her father's favourites, is included in the book, *Springboard for Women*, in the chapter about Beatrice. It is followed there by the comment: 'As the recollections of her teachers and friends will show, Beatrice was a singularly modest and unassuming person, so this early utterance must have been simply a statement of fact.'

NO – in large capitals – is scrawled in the margin beside this remark in the copy of the book owned by Richard Larson, professor of astronomy at Yale University. He was to become the most important person in Beatrice's life in her last years. *Not* modest and unassuming when he knew her, he meant. By then she had endured a life of being housebound and frustrated, feeling that her talent was evaporating and in danger of drying up for ever, and her ideas being misappropriated. Those years had left her determined to make her mark in astronomy, and to be seen and acknowledged as doing so.

The assertion of modesty is true of her childhood and student years, however. Beatrice may in fact have been overly modest, so that the extent of her talent was not fully perceived by teachers who may have been able to open up new vistas for her, although advanced mathematics was not taught in her years at her high school. In any case it would have been hard for her to comprehend her abilities; young people do not have the experience to evaluate their own potential. Nevertheless, granting her modesty, in hindsight it is possible to see in this young child her belief that she was going to become someone special, someone who *would* be heard of. Her father's expectations of her bit into her bones from the beginning, as she was to tell him in one of her very last letters.

Playing at BBC announcers was one of her favourite games. In the years of the war, people stopped what they were doing so they could listen intently to the BBC at news time. To the child, here was a distant, assured voice, a presence out of thin air, bringing the revealed truth. She identified with this voice, and learned by heart the BBC's frequent admonitions, such as the need to carry one's gas mask with one always, and to carry around one's name and address, 'clearly written', she would enunciate. She also mimicked the BBC's exhortations such as making the most of whatever food was available, a vital matter during wartime food rationing. Beatrice, at no more than four, was heard earnestly broadcasting: 'People must eat up all their potato peelings. The guv'mint will take away the rest.'

For a time Beatrice alarmed her family by sleep walking. She would get out of bed, arms outstretched, and walk downstairs for no reason that they could discover. They knew enough not to disturb her but to hold back and watch until she walked herself back into bed.

'She was always different,' Edward Hill was to say. 'Rowena was so obviously our child, and Theodora too, but Jean and I thought Beatrice sometimes seemed a throwback to one or more of her forebears.'[21]

Edward meanwhile was once again considering his future. By 1944 he was 37, and he found himself increasingly reluctant to think of returning to training as a solicitor after the war. As an army accommodation officer, he had been particularly busy trying to find billets for American troops. They had been brought over to Britain, thousands at a time, in the great pre-war luxury liners, *Queen Mary* and *Queen Elizabeth*. With the Allied landings in Europe, this work was coming to an end. By VE Day, on 8 May 1945, he was again thinking about becoming a minister in the Church of England, and began having discussions and tutorials with the warden of a theological institution. The boundaries of the Anglican Church, in England at any rate, are liberal. The warden helped Edward become convinced that he was orthodox enough in his beliefs to be ordained, although not all clergymen welcomed those who had the Oxford Group in their backgrounds.

But his uneasiness and uncertainty continued. The life of a country squire had never appealed, particularly if this meant a continuation of his mother's dominance. (She was eventually to make over Brynderi, the large house and farmlands, to her second son, Christopher.) Did his future, his family's future, lie outside Britain? Jean reluctantly agreed to think and pray about the idea, mainly, she said, for the children's sake.

When he was demobilised, not long before World War II ended with the capitulation of Japan on 15 August 1945, Edward took his family back to Wales, to Llandaff House in Cardiff. He quickly decided against going back to his legal studies when he discovered he would have to begin all over again. That at least was one certainty. By then Jean had been doing the cooking herself for some time. Household staff were hard to find. The war, with what seemed glamorous possibilities of jobs in factories and new if still limited vistas for the working class, had put paid for ever to the traditional servant class. Nanny Gullidge was the only resident helper at Llandaff House, although others came in by day. Without a household staff, several rooms were unused.

The Hills decided to find tenants for these. In one of those happenings that can completely change the course of a family's life, their tenants had just arrived from New Zealand. The Revd Merlin Davies, Welsh-born, had been the Student Christian Movement chaplain at Canterbury University College. His wife, Kathleen, had inherited a large collection of books from the noted New Zealand poet Ursula Bethell. They included the remarkable series of volumes and periodicals produced to celebrate the country's first centennial, in 1940, making the most of its past, present and promise. Edward, who held an intellectual belief in being equal while knowing he was not, was fascinated. 'Both of us wanted to bring up our children in a less class-ridden society than the Britain we knew, and that wish was fulfilled. I believe too that they received as good an education in New Zealand as they could have had in Britain, and possibly had more chance to be themselves.' [22]

Meanwhile the winter of 1946 was cold, wet and miserable. The roof of their house had been damaged by flying stone when the nearby cathedral was hit by a bomb, and no plumbers were available to fix the leaks. Food was still rationed. All the children got whooping cough (no vaccinations then), and their father recalled small Beatrice carrying 'a throwing up' bowl and pursuing the smaller Theodora, in case the 'whoops' got out of hand. Then the children got chicken pox, to which Edward also succumbed.

Jean, looking after the four invalids with only Nanny Gullidge's help, began to think there could be something, after all, in the idea of emigration. Edward as he convalesced continued to read about distant New Zealand and to study the photographs, but he also considered other countries: Canada, South Africa and Australia in turn. Post-war currency and travel restrictions and long-term planning problems made decisions difficult. They quickly ruled out South Africa because of growing racial tensions and its Boer-dominated government.

Then Jean thought of a former neighbour and family friend in Carlisle, Canon Campbell West-Watson, who had gone to New Zealand to become Bishop of Christchurch, and eventually Archbishop of New Zealand. They wrote to him in a preliminary and general letter of enquiry about prospects for Anglican clergymen in this distant country. What they did not expect, and were not prepared for, was a reply offering an immediate and specific position, a curacy in the Christchurch diocese.

They were in a quandary. Events were moving too quickly.

Edward's mother declared, 'If you go to New Zealand, you burn your boats and Christopher will inherit Brynderi!' [23] Edward retorted that the Anglican Church in England was too stuffy. His mother asked what he imagined it was like in New Zealand. Behind this exchange was the mother-son tension which had existed since Edward was a baby.

Both Edward and Jean were romantics in the sense that their personal Utopias seemed always just within their grasp, if only they would look or move somewhere else. The Oxford Group had fed them visions and dreams of life as it ought to be. Now here was New Zealand, whose climate they soon persuaded themselves was much warmer than England's. It had progressive social legislation and great natural beauty, nobody could deny that. Besides, it produced seemingly unlimited supplies of health-giving food. In England the continuing wartime rationing now seemed more irksome than ever, particularly for somebody who had never learned basic cooking and how to make do with kitchen scraps. Above all – although this realisation probably came to them only in their personal 'quiet times', and apparently was not openly discussed – New Zealand was a long way from the scenes of the unhappiness and frustration in which Edward had lived much of his life, and which in a number of ways had also touched Jean.

Such were the waiting lists at shipping offices that plainly it would be very difficult to book passages out of England unless definite jobs were waiting. Now here was such a job. Edward had been told he was needed. And a family connection with the bishop, one of their own kind, helped swing the balance. They knew they would bring with them the trappings of a more gracious way of life, together with their familiarity with the arts and the *beau monde*. Here they must surely find their place in the world.

They accepted the offer, and selected their most cherished furnishings, books, paintings and music. Just how civilised was this country they were going to? To ward off possible colds and influenza, Jean packed enough hoarded mustard, for mustard baths, to last for years. She packed months' supplies of toilet paper. What could one be sure of finding in this far-distant, almost mythic country? It was not exactly pioneering, but for the Hills with their three small daughters it was the biggest adventure of their lives as they took their hopes and boarded RMS *Akaroa* in July 1946 for the voyage to the bottom of the world.

Most emigrants to New Zealand have set off with very much less.

CHAPTER TWO

A SEA-CHANGE

*'Nothing of him that doth fade
But doth suffer a sea-change
Into something rich and strange.'*
SHAKESPEARE, THE TEMPEST

Beatrice never forgot the voyage to New Zealand. She was suffering from the first great emotional shock of her life. Nanny Gullidge had been left behind in England. She was nanny to the other two girls, too, and much loved, but for Beatrice, for all of her five years, Nanny Gullidge had been the centre of her life.

Her parents, true to English upper-class mores of the time, saw nothing unusual in this, if indeed they realised what was happening. It would not have occurred to either of them that children primarily need their parents, and need more than conditional love and attention – conditional, that is, on good behaviour and obedience. Onlookers believed that Edward, when he was around the family, had always been drawn to Rowena, his bright and attractive first-born. Jean, distant with both Rowena and Beatrice, had focused her emotional life on Theodora in what became a process of keeping her tied to her side, the baby of the family. Beatrice had grown up with Nanny Gullidge.

Rowena in later years was astonished that onlookers should think she was ever her father's favourite. 'From the time I knew I was me, nobody ever approved of me.' And, of her situation in her family, 'Dear little Beatrice – nobody could have resented her. I did resent Theodora, always Mummy's favourite. Not Theo's fault, of course.' She and Beatrice, discussing their childhoods, had agreed that they had always felt different, insecure, with so much expected of them.[1]

Jean had relied on Nanny Gullidge to run the household but she did not care greatly for her, complaining to a relative that Nanny was ugly and that she was sick of her. In New Zealand there would be other servants she could engage instead of Nanny, she told her sister-in-law, Kate Morton: 'Then, when she was in egalitarian New Zealand, she pined for a nanny and servants.'[2]

On board the *Akaroa* was an Englishwoman, Teddy Fardell, who became an important part of Beatrice's life in her new country. Her husband, John, was a New Zealand engineer who later became a lawyer. Teddy, a warm-hearted and practical woman, had trained as a nurse. The Fardells were the kind of people the Hills met socially. They were pleased to see them 'take up' young Beatrice, and saw nothing strange in the amount of time the Fardells devoted to their middle daughter during the voyage.

As Teddy saw those weeks at sea, Rowena was with her father or off by herself, exploring. Theodora was 'almost always' in the cabin with her mother, who suffered from seasickness. That left Beatrice, aged five, 'that little, lonely figure'. Teddy was also concerned about Beatrice's physical health: 'Her legs were quite often cold, blotched with purple. I wondered if she had a bad circulation. Her mother seemed oblivious to this. Beatrice was a golden girl. She was so lonely and so lovely.'[3]

Edward Hill said in later years that the Fardells wanted to adopt Beatrice.[4] Asked about this, Teddy replied, 'We never actually said we wanted to adopt her, although oh how we loved her! I thought the Hills had more to offer her because they had money and we didn't, or not till later. Beatrice would have come to us, though!' Beatrice did come to them, alone and repeatedly, for her school holidays two or three times a year until her family moved from the South Island to the North in 1950.

In spite of reservations, Teddy Fardell remained fond of both the Hill parents. She stressed how Edward was always so charming, and how kind Jean was to her after she had had an operation, 'although it must be said that hysteria was often near the surface'. Her love, however, went to Beatrice. 'It must also be said that Edward spoilt Rowena, and Jean spoilt Theodora. Beatrice remained herself. She loved us to hug and cuddle her.'[5]

Soon after the family arrived in Christchurch, in September 1946, Edward Hill was ordained. He became a curate in the parish of Merivale, one of the most English suburbs in New Zealand's most English city. But life was still very different from that of home.

In letters to family and friends the Hills tried not to dwell on the rawness of New Zealand life – that houses were considered old after 50 years because they were built of wood, that the New Zealand accent was, well, not that of the educated English, and that orchestras and concerts were scarcely abundant. Christchurch did happen to be better off with theatre than other New Zealand cities at that time. Ngaio Marsh was already world-famous for her detective novels featuring Inspector Alleyn. In her home city of Christchurch she may have been equally known and honoured for her role as a Shakespearian producer and remarkable teacher of drama to university students since 1942. The Hills met Ngaio Marsh but were not involved in either of her worlds.

Rowena and Beatrice attended St Margaret's, the diocesan school for girls, which charged lower fees for the daughters of clergymen. In spite of Jean's private income, the lower fees were welcome as Edward's salary as a curate was very small. Rowena and Beatrice must have heard this discussed. One day their parents found them by the front gate of their home at 72 Chapter Street. They were selling bunches of flowers they had picked from the garden, to help redeem the family fortunes.

Then a joyful reunion occurred. Nanny Gullidge was brought out in time for their first mid-summer Christmas. In January 1947, when Beatrice was not quite six, the nanny took the children to Banks Peninsula to stay with a farm family near the sea. The first of the very many letters which Beatrice wrote home throughout her life was written on this occasion:

> Theodora and I can go nice and deep in the sea with Rowena . . . Some nights Rowena and I watch the calves being fed, they are sweet . . . Rowena Theodora and I see the cows being milked by machines.
> Love from Beatrice.[6]

Part of that summer holiday she spent with the Fardells, the first of many visits. Teddy remembered her as both merry and grave, full of fun but serious and intent in her play, too. Her hazel eyes sparkled and her dark curls bounced as she walked. She would play happily by herself. A year later her standard one report contained an A for arithmetic with the comment, 'Mental excellent'. Her highest marks, however, were for reading, spelling and grammar.

In mid-1948 the family left Christchurch, full of optimism, for the village of Southbridge, some 50 kilometres south, where Edward was to be vicar. He was also to take services at three small outlying churches. For the children, Southbridge meant open countryside and ponies.

The Merivale experience had not always been the happiest. Edward was progressive, left-leaning, attracted to this new country at least in part because of its Labour Government's record of working for social justice, yet the parish was wealthy and conservative. Southbridge offered new territory and new opportunities. As it is also an excellent fishing area for both salmon and trout, Edward's spirits rose. In any case he intended to do his best, in line with the family motto of *Perseverantia omnia vincit*, or 'Perseverance overcomes all'. (Theodora remembers Beatrice, as a small child, thinking it meant 'Percy sitting on the verandah, blinking.')

The two older girls, with Theodora to join them in 1949, were enrolled in the primary wing of the district high school in Southbridge. Nanny Gullidge was still the mainstay of the family. Jean continued teaching Rowena the cello, and Beatrice and Theodora began learning the piano.

Beatrice was happy and busy, which for her were synonymous. When her mother was away for some reason, on 16 December 1948 she wrote her a letter of eight small pages, describing the Southbridge school excursion to Lyttelton harbour. It ended:

> On Sunday night dec 12 there were 16 children after carols. We had lovely fun IN BED 10 o'clock.
>   Food we had chocolate biscuits
>   plane   "
>   cake
>   raspberry drink.
> Love from Beatrice.'

This letter is an unusual effort from a not yet eight-year-old. Obviously she could already write well and fluently with very few errors, and in a home where languages were prized the letter must have seemed early confirmation that here, like Rowena, was another daughter who would make languages her field. As her parents did, the principal of her high school was to take this for granted when Beatrice's choice of university scholarship subjects was questioned. There is also a clue to the way her mind was already working. She saw that those 'plane biscuits', as compared with the chocolate ones, need not be written out in full. It is quite a sophisticated thing for a child to use ditto marks. It is also a small sign of Beatrice's instinctive feeling for economy, an ability to express something quickly, in the shortest possible way, which was to astound her science teachers at high school.

In mid-1949, Beatrice was promoted to the next year up, to standard four. From that time on, she was at least a year younger than the next youngest in any of her classes. To some extent this hindered her development in that she seems always to have been the 'class pet', something like a special little sister who is protected, even babied. Her formidable intellect was not really noticed by her classmates until her second year at high school. By then the attitude to 'little Beetle' had become general. Former classmates were most surprised to learn later that Beatrice was the first among them to become engaged, an engagement she broke; and then the first to marry.

As a child Beatrice seems to have been bored until she was promoted to the higher class. Her school report for 1949, the first half covering the time she was in standard three, and the second after she was placed in standard four, shows much higher grades once she had the challenges of the higher class. 'The best in the class' reported her teacher.

In Christchurch John and Teddy Fardell, still without children of their own, continued to enjoy having Beatrice to stay during school holidays. 'We gave her family warmth and fun,' Teddy said. 'At home she had Nanny, and in Christchurch she had me.' They would take her to play in the Botanical Gardens, and John took her rowing on the Avon. Most mornings he would leave her a crossword puzzle he had made up the previous evening, and would check her solutions when he came home to lunch. 'She had a very good brain. "Too easy, Uncle John!" she'd say.'

Before her eighth birthday, which Beatrice elected to have with the Fardells, Teddy stayed up till 1am making clothes – 'underwear, a hat, everything' – for a doll she had bought for Beatrice: 'She loved that doll. And when our son Edward was born, she'd sit with a book by his pram for hours. It was when she was eight and nine, and she was so drawn to him.'[7]

A devastating event occurred when Beatrice was eight. Nanny Gullidge returned to England. For Beatrice it was like a death in the family. 'A black hole of grief opened up' she was to say in later life.[8] Rowena and Theodora were shocked by the departure, too. As Rowena recalled it, 'I'm sure Nanny's going was shattering to Beatrice. It was to me, too, but I don't remember ever talking about it.'[9] Theodora has good, small-child memories of the nanny as being 'lovely to sit on' (that ever-welcoming lap), and kind and reliable.[10] But these two girls may have understood that they each had a parent 'special' to them, even if this did not necessarily make them feel special.

It is not possible now to disentangle the reasons for Nanny Gullidge's departure. Edward Hill has written that she was not feeling well and had said she wanted to 'lay her bones in the old country'.[11] In fact she was to live on to a good age, doing her best to keep in touch with Beatrice. Constance Gullidge was an uneducated woman. This may not have mattered much to Jean Hill when the children were young, but her speech and possibly some of her mannerisms seem to have grated as the years went by. Above all it is never easy for a parent to see a child obviously preferring to be with somebody else; Jean told Teddy Fardell that Nanny was trying to take Beatrice away from her.[12] With their backgrounds, certainly, neither parent could be expected to imagine the consequences to Beatrice – to all the little girls – of having this central part of their lives cut away.

The decision was made during the winter of 1949. Three months before Nanny Gullidge was to leave, eight-year-old Beatrice began writing a book which somehow remained in the possession of the family instead of being given to the nanny as Beatrice intended. She called her book 'My Lovely Nanny', and filled it with songs and prayers for her.

In *My Daughter Beatrice*, Edward ends his account of the nanny's return to England by saying: 'Up to that time Nanny Gullidge had been possibly the most important person in Beatrice's life and she was to recall many years later the enormous sense of loss she felt on her departure.' Beatrice threw herself into play and into writing and schoolwork, as if by filling every minute she could also fill the void.

A few weeks later, towards the end of 1949, Edward wrote one of his periodic letters to friends and church supporters in Britain, a blend of amused and decently restrained sophistication and determinedly positive Christian outlook. Nanny's place, he said, had been taken by 'a strong and cheerful girl of 19, admirably trained by her mother, who is our bi-weekly "help" and the youngest of a family of 15. She does most of the cooking.' Locals said of Jean that this was the time she learned to use a can-opener.

Early the next year Edward learned that a friend, the Revd J. T. Holland, urgently needed assistance in his large New Plymouth parish. A new life in the warmer North Island seemed irresistible. And New Plymouth, that greenest of all New Zealand cities with its looming mythic mountain, had had a particular appeal for Edward since he first studied that collection of New Zealand books back in his home in Wales.

The Hills did not know, then, that New Plymouth Girls' High School was one of the great schools in the entire country – great for languages, music and 'school spirit', that is. Not great for mathematics and the sciences.

New Plymouth, with its space and its long views, its mountain, parks, gardens, river and sea, offered balm to a grieving child who was always quick to respond to the natural scene. To the end of her life, Beatrice's face lit up when she spoke of New Zealand, but particularly of New Plymouth.

The Hills in 1950 found it a reassuringly well-ordered town which, they were told, had not changed much in 30 years. The children were enrolled at Central Primary School. Beatrice in later years told her father she did not think it particularly distinguished, but at least she had an opportunity to absorb the beauty of the area, to concentrate on her music and, most important of all to her for a time, to fall in love with a pony, Jimmy. Nanny had gone, but here was a warm, responsive creature waiting to be loved. 'My Happiest Days' was the title of the book she wrote over several months at this time. Mostly it is about Jimmy the pony, in nine chapters and with step-by-step accounts of their adventures. She illustrated and printed it carefully.

Edward was writing again to their friends and supporters back in Britain. Everything for him and Jean seemed to be going very well, suggesting that these satisfactions were rubbing off on the children, too. Jean, who had already been called on to be an accompanist on the piano, had begun playing the cello in a trio, and with a small orchestra. This was run by Vinnie Ross, a notable musician who was quick to perceive Beatrice's talent, which she called 'quite unusual'. Beatrice continued to play the piano, too, but she had to drop these lessons as there was no time for them as well as her violin. Rowena still learned the cello, and Theodora already showed the aptitude for the piano which was to become her life's work and interest. In a family letter, Edward compared her musicianship with Beatrice's: 'Theodora seems to have an equally real gift for the piano and plays by ear in an amazing way for a child of six, as well as transposing both hands of her "pieces" into another key with no apparent difficulty.' Both parents took for granted that music and languages were to be the lives of all their children. Science was not thought of.

As exotics set down in a plain provincial town, the Hills saw that they were observed and talked about, but they could not have realised just how much. Their comparative wealth was obvious. Their accents, assumptions, manners and clothes were different. They delighted in private family sayings, a joke language that quite a lot of families enjoy, particularly those confident of their place in the world. Jean Hill used to say of herself in relation to money that she was an incomepoop. One could be deaded by examinations, be in utter suspenders, have a Norrible Cold, decide something was a Good Thing (capital letters were part of the fun), or observe something about 'uming nature, all phrases Beatrice used in letters home in later life. A certain self-consciousness imbued their daily lives. As Rowena was to say, looking back, 'It was very difficult to be yourself in our family.'[14] The parents, however, would not have countenanced any signs of boasting or showing off in their children. If there had been any 'skiting' at school – the most egalitarian section of New Zealand society – people would have remembered. Nobody saw this, ever, in Beatrice. Her contemporaries and teachers saw her as modest, unassuming and kind.

Jean Hill evoked Beatrice, at home in New Plymouth at the age of 10 years and two months, extraordinarily clearly in a description she wrote in March 1951 for her own mother in Carlisle:

> Beatrice, 10, short, burly, rosy-faced, with big sparkling hazel eyes and brown hair whose natural state is one of extreme dishevelment. Her nose is of the 'baby beak' variety, finely chiselled, and her mouth small and firmly set.
>
> Beatrice is always up or down, mostly up, fortunately for those with whom she lives. When clouds come up they gather, and a storm

is inevitable. It may be a long and noisy storm. Nearly always it happens when some creative project, whether it be one of her many 'Heath Robinson' contraptions or only some elaborate 'pretend', has been frustrated or interrupted.

From the moment she wakes at 6am or earlier she is busily involved. Seven is the earliest she is allowed to get up and dress, but that doesn't mean she can't put in a good hour's concentrated work on something she has on hand. Music comes first. Sometimes it seems as though she thinks in music.

No need for Beatrice to be actually playing one of her instruments, piano or violin, to be making music. All she knows of theory, note values, chords, phrases, rhythm has to be passed on to Bruce.

Meet Bruce. He is a woolly koala, rather dull to the uninitiated observer but mildly benevolent and patient in expression, as indeed he needs to be because he endures much, much tuition, much imparting of every kind of knowledge, at all times of day not spent by his owner at school. 'Bruce Brown' is taught by 'Muriel Green'; Beatrice's room is littered with papers which testify to the thoroughness with which he is taught. Her mother and father feel they might learn quite a lot of the rudiments of musical theory if they could only steal Bruce's ms book from time to time and study it. Some of the directions are: 'Write the scales, chords and inversions as directed' – 'Put in bar lines' – 'Also write the time signature' – 'Name the key of the following' – 'Write the following in the same position on the treble staff. Prefix the new key signature.' There are numerous headings and sub-headings. 'Theory. First Examination' – 'Theory. 2nd Examination' – 'Hints and Rules' – 'Memoranda'. Nothing is copied. It is all out of her own head.

Now meet the Rubapluc, an instrument made and invented by Beatrice and played by Bruce.

While Beatrice is in bed, before or after lying-down hours, he can be seen perched up with music in front of him on the folding music stand, clutching his harp-like instrument and being patiently instructed in its use.

It took hours to make the Rubapluc. It is kept in a cardboard box, neatly covered with Christmas paper. On the outside is a label saying, 'This is a box containing a WONDERFUL RUBAPLUC instrument and Bupluk.' The latter is the metal top of a coat-hanger. It lives in a slot in the lid and is used to twang the strings. The idea is a wooden frame, shaped most ingeniously, through which go wooden posts, a rubber band going above and below each so that there are six strings, and each makes a definite note on the piano, which is marked on the cardboard beside each. Bruce has his own music. Chords can be played by twanging two strings at once and are written in the music provided, or single lines of nursery rhymes can be picked out.

On a brief tour around Beatrice's room the following were discovered:

    1. A carefully written double sheet of a school notebook, 'Ballet Grade 1.' Full description of different positions and dances (all original) with pin-figures to illustrate it.

    2. A large sheet with much coloured lettering, 'Second Examination! Give the meaning of the following' – with many musical signs and marks allocated accordingly.

    3. A piece of cardboard with a design of triangles – 'How many are there?' Very elaborate calculations and diagrams to find the result.

    4. A sheet of paper about 4ft by 2ft covered with divisions, each containing written out instructions in large childish printing (not her normal writing) for the music lessons of other members of the koala family, Big Rookie, Ann, Stripie, Teddy and Wee Rookie, with comments – 'Fair work' – 'Good' – 'Count out loud' – etc.

    5. An odd sheet of music ms with coloured notes and huge gold stars at intervals. (Purpose – who knows?)

    6. A small box containing two bottles of home-made scent, with 'Shake a drop. Grand Double rose and lavender' printed on the outside.

    7. Several cardboard lids with shells held down by transparent cellophane paper, neatly numbered and named.

Her room is generally in a normally untidy state, but the things she has actually made are characterised by neatness, originality and ingenuity.

To Beatrice, engaged in some of her private operations, meals and requests to wash her hands or get ready for bed are obviously nothing more nor less than tiresome interruptions. Grown-ups are intruders and the ordinary grown-up world another country.

She goes into protracted day-dreams and is interestingly absent-minded. One evening when she seemed to be undressing more slowly than usual, with long dreamy intervals of doing nothing, her mother was delighted to see that at last she at least had on her pyjama tops. Not long afterwards Beatrice wandered into her mother's room without them, and only the trousers on. 'Mummy,' she said in a rather worried way, 'I know I had my tops on, but now they're off again.'

The first occasion on which she was allowed to walk home alone from her violin lesson (not more than 15 minutes' walk) she turned up after 45 minutes in the dark, 20 of them having been spent in anxious surmise by both her family and her violin teacher. When she wandered in, blissfully happy and dreaming, she couldn't understand what all the fuss was about and said she hadn't done anything on the way. She 'just came'.

A very familiar sight is Beatrice at the piano, with Bruce alongside high on a cushion on another chair, being given a lesson. Original little themes and phrases are used as she goes along. Occasionally one of them seems to please her. It is repeated over and over again, and now a large musical ms book she has been given is being used to write out these little musical ideas. They are fully harmonised, and generally illuminated with coloured pictures and figures round the edge.

Non-musical 'pretends' are always elaborate and thorough – picnics, cinema shows, school, everything for the 'creatures', with Theodora as faithful collaborator. If the news gets round that it's Bruce's birthday today, or Teddy is to have his tonsils out, everyone knows that at least half a day is going to be blissfully happily spent. All that is asked of mere grown-ups is to keep out of the way.

Beatrice was both consolidating what she herself had been taught and, going further, devising her own examples to extend her lessons. Rowena had played inventively with Beatrice in earlier years, but was now well beyond such play. She was highly intelligent, quick to sum up a situation, and already clashing with her mother and what she had begun to consider were her mother's unreal expectations. Beatrice, from the time she lost Nanny Gullidge, began to adopt an unconscious strategy of going along with the family mores – actually being a dutiful daughter as well as appearing to be – while living an increasingly rich life in her own enquiring imagination. That way she could avoid difficulties with her mother. Edward seems rarely to have intervened. He stayed aloof from such matters.

While Rowena was exploring forms of rebellion, Theodora, younger and content to follow, was delighted when she was included in Beatrice's pretend-games. 'She was endlessly inventive, the best possible playmate.' Beatrice was the most important person to Theodora, apart from their mother, during her childhood:

> Clear in my mind is the image of her face with its high, broad forehead, large hazel eyes and rosy cheeks. She was full of ideas for what we could do together. I can hear her voice saying, 'Theodora, I've got an idea ...'

She was always a very distinctive personality, not an imitator of either parent. Her activities had their own personal stamp. She had enthusiasm, energy and determination. She was also warm and affectionate, closely bonded to her family and deeply attached to her friends.

Theodora describes 'a certain intensity' in the way Beatrice lived, including speaking very rapidly. She could get excited and competitive, too, such as over a family game of cards.[15]

Not all girls of 10 still play at tea parties or hospitals with their toys. Beatrice was not sophisticated. It seems that at this stage she did not often visit the homes of other children or have many friends come to her; Jean Hill had definite if usually unspoken ideas about those who might be invited to her home. The family rarely visited the cinema, television was not to reach New Zealand until 1961, and then only in Auckland. Apart from Girl Guides and Brownies, which the Hill children did not join, there was no tradition of holiday camps or after-school activities. Beatrice, alone or with Theodora, played little-girl games with a significant intensity. It is worth noting, too, that she played teachers rather than mothers and fathers.

For Edward's birthday in 1951 Beatrice made him a decorated card with a three-verse poem. She inscribed it to 'Poppikins from your pet Beetle Bug', and wrote on the back, 'I am the only beetle who can draw and make up poems.' Her childhood poems were no more distinguished than the usual kind; it was Rowena who was to become the poet, writing in Spanish as well as English. But from about this time Beatrice began to be called Beetle by her family and close friends, and throughout her life she often used the Beetle signature in letters home.

Edward in *My Daughter Beatrice* describes her as a busy and happy child, but 'by no means altogether placid. When frustrated she could roar!' He gives as an example the time when one of the elastic bands on her Wonderful Rubapluc had perished, and she could not re-tune it. 'Her anguish was prolonged, loud and wholehearted.' He also remembers her coming home from school in a seething rage, one winter's day when she was 10 and had joined her older classmates for a cooking lesson. The girls had a new teacher. An old-fashioned coal-burning stove, of a design that opened on the top, was used both for the children's cooking and to heat the room. The teacher opened this top, jeered at the efforts of one child, and scraped them into the stove. Then she turned to the class and said, 'Let's see whose work will be next for the flames.' Edward Hill commented, 'This sarcasm and bullying absolutely enraged Beatrice. She really exploded. She couldn't let an injustice slip past.'

Outbursts of rage for whatever reason did not often happen. A more typical memory of her father's is of Beatrice deep in conversation and oblivious of anything else while up in the branches of a tree with a friend, Molly Hurst. 'There they were, these children, having a peaceful talk up out of the range of adults.' [16]

Her violin teacher, Vinnie Ross, saw her as a charming little girl, very polite, very intelligent and very musical. 'Sometimes in music your intelligence takes you a fair way but you lack that essential intuition that can't be taught. She had this, too.'

This teacher saw Beatrice as dignified, poised and composed, but slightly remote. Six years later, when Beatrice and some of her other one-time pupils were members of the National Youth Orchestra, Vinnie Ross went backstage after a performance in Wellington. The others hailed her but Beatrice, though welcoming, remained a little distant. 'That prominent, bulging forehead – I don't think one ever knew what she was really thinking.' [17]

Edward's parish and the girls' school agreed that the family could visit their English relatives during the northern summer of 1952. They had received 'an unexpected windfall', as Edward called it, when one of the Morton family trusts was wound up, and they decided to fly from New Zealand to England. There the children stayed mainly with Edward's mother at Brynderi, with her horses and dogs. Beatrice went alone by train to London to see Nanny Gullidge. She was to do this, or write, throughout her life, and the Hill parents remembered the nanny on visits to London. She was living in a room in a half-sister's house and was feeling the cold, so they paid for a gas heater to be installed. They also hired a car and took her on a drive through Richmond Park. Her thoughts were centred on Beatrice, but she was also concerned for the other girls. Edward remembered her saying of Rowena, years later, 'What is she doing with her life? That girl could have done anything!' [18]

Back in New Plymouth, Edward was put in charge of the three small churches at the western end of the parish. The Hills bought a bigger house, with separate bedrooms for all the girls, in Peace Avenue on a hillside site overlooking the harbour to the west. Jean arranged for a large room suitable for entertaining to be added to the house. Thus extended, long, low and painted dazzling white, the house in its prominent position became known to locals as the *Mauretania*, after the luxury passenger liner of the day.

The move took the family some distance from where the ponies were kept and, as so often happens with young riders, the demands of homework and other school activities meant that Rowena in particular had little time for riding. After a while riding was abandoned. The family, always fond of animals, acquired a corgi called Russet, which Beatrice and Theodora tried to train to do circus tricks.

Going to the Hills' house was always different, old friends of the girls have said. A neighbouring child, Jean Ford (now Jean Ross), remembers Theodora's 10th birthday party:

> We played the traditional English nursery games but really they were for younger children. When it was the game called pass the parcel, the parcel was most elegantly wrapped. And when we played musical chairs, the chairs were Chippendale, or that's what people thought.[19]

What was talked about at the house was different, too: world affairs, philosophy and religion. New Plymouth people who knew the Hills would smile (in some cases with flashes of envy or derision) and say that, where others would talk at mealtimes about the small events of the day, the Hill parents as a matter of course would introduce lofty concepts or aspects of international affairs as suitable topics for all the family at the dining table. The big questions of life, death and the hereafter were often discussed around that table, with Beatrice participating from an early age.

The Hills found New Plymouth society difficult to come to terms with. Their hopes had been largely unformulated, but finding people with whom they felt they had something in common was a preoccupation. Both enjoyed the idea of a salon of the intellectual and the artistic. To find even a nucleus was not easy.

Two who became friends and to some extent mentors, and whose views and interests influenced the Hills – and, later, Beatrice – were Alec and Fanny Niblock. They too were English, but much older than the Hills. Although world travellers, they had cheerfully come to terms with New Zealand provincial society. The Revd Alexander Moncur Niblock, who preached at New Plymouth's principal Anglican church of St Mary's, had been a missionary in China and later in India, where he was influenced by the poet and mystic Rabinath Tagore, among other philosophers. The Niblocks lived not far from the Hills, and the four were soon sharing discussions and experiences. Alec Niblock was a forceful man, well-read. Fanny Niblock was assiduous in researching their current interests, and sought out books from the public library which all four read and discussed.

Besides philosophy and religion, the Hills' main interests were the arts, and in particular languages and music. To these, for a time, the Niblocks added the sciences.

Fanny Niblock's diary for the 1950s [20] records her researches, at her husband's behest, into the first astronomers, and Aristotle's theory of the universe. She writes with some excitement of Galileo, with the newly invented telescope, as the founder of observational astronomy. There are respectful entries about Halley and Herschel, and Fanny Niblock writes: 'In this century astrophysics will give a lead, and scientific knowledge will grow among all nations. Man may even know what the moon is like and will certainly land on Mars.'

Edward Hill agreed that discussions instigated by the Niblocks and books they lent the Hills would have introduced an astronomical view of the universe to Beatrice, who was then still at primary school. New Plymouth had an observatory. When Beatrice became a pupil at New Plymouth Girls' High School she joined the school's astronomy club and made visits to this observatory.

She herself used to say it was the *Junior Oxford Encyclopaedia* and then Fred Hoyle's books, particularly *The Nature of the Universe*, which turned her attention to astronomy. Through her family, however, ideas, discussions and books were introduced at the right time for a young scientist. Beatrice was ready.

CHAPTER THREE

TOWARDS A LIFE'S WORK

*'Language is primary, and science would be inarticulate
without the tool of language.'*
CHARLES FLEMING

All the Hill daughters were achievers. In later life Beatrice was to say of herself that early on she felt intensely driven, with her parents expecting her to become famous. She grew up believing that her parents were distinguished and superior, and the family atmosphere was one of expectation. She did not see herself as exceptional. Rowena, in contrast, seemed to her to be dazzlingly gifted, achieving while having to do little work. Beatrice felt she had to work extra hard to get anywhere.[1]

In February 1953 she entered a structured and competitive world for the first time, the top stream among the third forms at New Plymouth Girls' High School. It was to be a decisive year. After her primary schools, where little seemed expected of her, this was something quite different. The staff was led by the headmistress, the stylish and academic Rose Allum, who streaked like a meteor across New Plymouth's staid society. In their graduates' gowns the staff swept up on to the dais for assembly at the beginning of each day. Girls stood to attention. Not a murmur could be heard in the hall. This discipline, and the school's focus on the beauty of its surroundings and its linguistic and music traditions, were the legacy of past generations but particularly of the previous headmistress, Doris Napier Allan.

The school had been established and had developed very much along the lines of girls' grammar schools in England. It focused on academically minded girls rather than on developing the potential of everyone in the school. The Hill sisters attended at the height of this emphasis, but it was not academia in the round. English, Latin, French, history and music were prized far above mathematics and the branches of science, in so far as they were taught at all. This was just what the Hill parents were looking for.

Beatrice on her first day was the youngest in the entire school. She rejoiced in the work and the teaching. 'I can see her to this day, so quiet and well disciplined and so very young,' said her French teacher, Sylvia Philips. 'She sat in the front row, lapping up everything – an immaculate student with a protruding forehead with delicate veins. I could almost see her brain working.'[2]

Her violin teacher, Vinnie Ross, having given her an excellent foundation, had gone to England, and she became a pupil of Willi Komlos, the school's part-time music teacher. Komlos was a fiery Hungarian refugee who had been in a Japanese prisoner-of-war camp. He was an excellent violinist and an even greater teacher, exacting and demanding a student's very best. Most of the first violins in the soon to be formed National Youth Orchestra were taught by Komlos. Often they had originally been Vinnie Ross's pupils, too.

Komlos saw Beatrice's talent immediately. He was to award her the school's prize for strings in this, her first year, and from the beginning saw her becoming a professional violinist. For her part, just as her father had had times with his Bowen grandmother which were his joy to look back on, Beatrice, even in her darkest days, was to keep fresh the absorbing passion and pure joy of listening to music and of playing the violin herself.

In charge of the school's music, its singing and major musical occasions was Hilda Veale. She became a member of the Hills' social circle, and on one occasion Jean Hill drew her into Beatrice's room and showed her the big timetable on the wall above the bed. Every hour of every day of the week had been given a square, often subdivided: '6.30am, get up; 6.40am, make bed ... ' Hours for sleeping and for each school subject were marked in, together with mealtimes and hours for music practice, homework and playing. 'And she sticks to it,' her mother said.[3]

The girls took part in the charades and word games that were a feature of the Hills' social evenings. Hilda Veale recalled Beatrice as 'so sensible', Theodora as 'little and overpowered', and Rowena as 'excitable and highly strung like Edward' and, in any case, often somewhere else. There was always music, sometimes billiards or table tennis, and often pen-and-paper games, the latter intimidating for the less sophisticated or less well read. An invitation to the Hills' home could be daunting to the uninitiated. 'The Hills were aware of social background. They invited people they thought would make a contribution to their life.'[4] When the three sisters had friends to stay, everyone joined in wild games of cards, and in the world history or geography quizzes at the dining table. Newcomers were amazed, but they never forgot learning that there were other ways of living.

Beatrice was poor at gymnastics and sport, although she joined in cheerfully. A girl two years behind her, Claire Campbell, (who as Claire Stewart was mayor of New Plymouth 1992–2001), remembers her 'giving hockey a go', just as she was prepared to take part in any aspect of school life.

Her co-ordination was poor, but she and a close friend from primary school, Julienne Lobb,[5] listened to the appeal for girls to learn life-saving. New Plymouth Girls' High School has always had a strong tradition of life-saving; Taranaki has long stretches of treacherous coast and many mountain-fed rivers. 'Beetle was always such fun and she gave me confidence,' Julienne recalled. 'We went together and said we'd practise hard.' One of the school's most stalwart teachers, Helen Thomson, recalled 'two tiny scraps of girls saying they wanted to try for the intermediate life-saving certificate. We found they could barely struggle across the baths, and they were probably too young for it anyway, but did they work!'

Julienne remembered the patience of their instructor, Dorothy Geddes, and how much she helped them:

> We completed the part of the test where the 'rescuer' had to tow the 'victim' to safety. I (the rescuer) said to Beetle (the victim), 'Did I duck you much?' 'Oh no,' she said, 'only once.' She told me later that this was quite true; I'd apparently ducked her once, but then held her under most of the rest of the way. She hadn't liked to say so too loudly, however, because she'd noticed one of the examiners standing nearby.

Dorothy Geddes was also the instructor when Beatrice went for a higher life-saving award. 'She had to steel herself to dive from the high board, a requirement for the silver medallion or award of merit. It took a lot of courage, for a girl who didn't like diving, to do this part of the test.'

Meanwhile Jean Hill had been finishing a religious novel called *Wind May Blow*. It does not read as if it is by the same person who wrote that engrossing, carefully factual account of Beatrice and her Wonderful Rubapluc. Jean had the novel printed by a local firm. Such was the aura surrounding the Hills that the novel sold and went into two more small printings.

Edward Hill was not as happy. Being a clergyman was not what he had thought it would be. He was not at ease with ordinary people and visiting the sick was something of a strain, on both sides. His sermons were not striking home. A friend,[6] although appreciative of Edward, heard him preach and said the congregation did not understand a word. He could not see advancement, and aspects of church services and attitudes were not to his taste. Once again he felt frustrated, his hopes dashed. In particular he did not feel fulfilled by the work he was given to do within the church, and he genuinely wanted to be of service, to do some good in the world.

If his career did not lie in the law, and now not in the church either, perhaps his future lay in politics? If he needed more experience before standing for Parliament, then how about local politics? The local body elections would be held on 31 October 1953, and by then the sitting mayor would have been in office for 21 years. Jean, bored with being a clergyman's wife – 'I had never wanted this'[7] – felt suddenly stimulated again. The talk at the Hills' dining table became focused on politics, and the family became the focal point of New Plymouth gossip.

From this time on, Beatrice began to be protective of her parents. Almost certainly she did not realise at first what she was doing; it was instinctive for her to shield people from pain if she could. But the gibes and derision, the envious jokes and snide comments which very soon began to swirl around her parents, and to which both she and Rowena were exposed at school, were sharp and painful. They were sufficiently pointed to set her off on what she came to believe was her life-long duty: protecting 'poor Mummy and Daddy' from anything that might upset them. What they did not know about, what they could be prevented from discovering, could not send them into the excitements of distress. She, Beatrice, would be the buffer.

On a deeper level, here was a child who had lost her dearest person, her nanny. Here in this new situation – her unconscious could have been telling her – was something she could do which might make things right: protect them, be the good daughter, do the proper thing, and then she herself might feel part of a whole again. Be that as it may, the criticism and ridicule of her parents' ambitions helped set Beatrice on a path which eventually helped change the course of her life.

The outward story of that time is told very well in press clippings which Jean Hill cut from New Plymouth newspapers and pasted in a scrapbook. Jean enjoyed making up scrapbooks, a habit Beatrice was to adopt when a university student and in her early years in the United States. Jean labelled her book 'The Mare's Tail'. She could laugh at the goings on while feeling exhilarated by the spotlight suddenly beamed on the family as Edward's political ambitions took root. The clippings also throw light on the life and times of a small New Zealand city in 1953.

New Plymouth then had two daily newspapers, the *Taranaki Herald*, the oldest in the country, and the *Daily News*, which alone survives. The *Herald* provides most of the clippings in this scrapbook, beginning with a provocative letter to the editor on 15 June from Edward Hill, a letter arising from the coronation of Queen Elizabeth II.

There was much talk of another Elizabethan age. Schools had enjoyed a whole day's holiday, and the country felt it had made its new queen a personal gift when Edmund Hillary became the first person ever, together with Sherpa Tensing, to conquer the world's highest peak, Mt Everest. New Zealand towns rejoiced and outdid themselves with special events, carnivals and loyal decorations. All, it seemed, except New Plymouth. For days its papers had carried letters from citizens criticising the mayor and council for the meagre and unimaginative ways in which the city had marked the occasion. And not everyone thought it 'screamingly funny' that the ageing mayor had solemnly announced on Coronation Day that mountaineer Edmund Hillary had just performed the great feat of climbing Taranaki's Mt Egmont.[8]

The letters to the papers were nearly all written under pseudonyms: Outraged Ratepayer, Pro Bono Publico, Mother of Five and John Citizen were among the favourite cloaks sheltering most correspondents to New Zealand newspapers at that time, and indeed into the 1980s. Here was a small country with a population of just over two million[9] where everyone knew everyone else, more or less, so it was often awkward to be forthright. In fact, speaking out under one's own name was not necessarily considered one of the virtues; it smacked of egotism.

But the Revd Edward Hill was accustomed to the correspondence columns of the *Times* and the *Daily Telegraph*, and in any case was ready to stir this New Zealand pot. His first letter to the *Herald* editor appealed for 'a bit more moral courage and forthrightness all round'. Correspondents should stop writing anonymous letters of attack. If he were mayor, he wrote in an ingenuous piece of kite-flying, he would want to know who his critics were, and also who was thinking of standing for the mayoralty and the council. 'It would be most unsatisfactory all round if this election should go by default after this spate of anonymous abuse.' Edward ended with the hope that in the election campaign 'a vision may emerge of how this overgrown struggling township is going to grow up into a real city'.

The *Herald* that same day, 15 June 1953, used this phrase from his letter as the peg for its long editorial. Without directly pointing the finger at those who had held office for so long, unchallenged, while the district's problems grew, it managed to endorse and elaborate on the points Edward had made: 'there is vast and continuous scope for vision.' The elections, it said, were 'of vital and even historic significance to this district'. The right persons in the right jobs were 'fervently' wanted, 'so that the district can remove the reproach of drift, apathy, bungle and delay'. Fighting words. Nobody missed the connection with the Hills.

Nine days later, Edward announced publicly that he would no longer be a clergyman, although he remained in holy orders. 'You do, unless you're defrocked,' he later explained.[10] He also said he had been thinking seriously about contesting the city's mayoralty, although this was only one of the possibilities he was entertaining.

Edward had never joined a political party, although he had had some engagement with the Labour Party; his interest in getting better conditions for the poor was sincere and of long standing. Jean had attended one Labour Party meeting with him and said later that it had made her absolutely exasperated. Never would she attend another.[11] She had also told a relative before they left England that she could not see herself as a clergyman's wife.[12] She could, however, see herself as the wife of a mayor, and encouraged Edward in his ambitions.

He was receptive when the New Plymouth Junior Chamber of Commerce, not known for leftist leanings, asked him to address one of its luncheons. Then a delegation of businessmen called to ask if he would stand for the mayoralty in the forthcoming elections. Mayor E.R.C. Gilmour had been in office for so long. That he was the Hills' next-door neighbour in the street named Peace Avenue, and that there had been some pointed discussion over misbehaving dogs, were facts which titillated the small city. Its population had reached 23,000 but in many ways it seemed to be standing still. Loans had been secured for a number of essential works and services, but nothing had been done about them. Not only the mayor but the town clerk had grown old in the job. Altogether, said the businessmen's delegation, it was time for a change. Edward let it be known that he would think about it.

Mayor Everard Gilmour came out fighting. Yes, he agreed that he had said, before the previous election, that he might not stand again, but it was different now. The new Queen and the Duke of Edinburgh were to visit New Zealand in December and January, and – said His Worship the Mayor with touching candour – he would like the honour of representing the city during their visit.

Then Edward issued a statement containing a remarkable series of 'ifs'. If he should decide to stand for the mayoralty, he said, and if he should win, then it would be his 'genuine wish' that the official reception to Queen Elizabeth II and the Duke of Edinburgh should be conducted by the present mayor – if that were possible. He had not yet made a decision about standing, he said.

But by the very next day, 26 June, he had. The papers featured his announcement that he was now in the mayoral race. Long news stories and editorials about the now-revitalised campaign began appearing. Letters, mostly anonymous and mostly condemning Edward's temerity, were printed on most days. Especially hurtful were those signed Anglican, and even Staunch Anglican. A correspondent called Fair Play took the opposite stance. Letter upon letter took up the banner for one side or the other. 'There is a rumour afloat', one would begin, with another trumpeting, 'The hour draws near.' Then began a campaign unique in New Zealand, to have the local body elections postponed in New Plymouth until after the Queen's visit.

The announcement of the royal tour had excited nearly everyone. The young Elizabeth II would be the first reigning monarch to visit New Zealand. Months in advance people began to plan how they would travel to catch a glimpse of her.[13] In New Plymouth the royal visit and the election of the new mayor somehow became intertwined.

Edward had never before stood for any local body. Indeed, he had had no relevant experience whatsoever, except perhaps his army years as an accommodation and requisitioning officer. His friend John Fardell warned that this was bound to bring problems. Edward did not see it this way. His sincerity was unquestioned. He yearned to be a force for good, but at the same time both he and Jean saw being mayor – and mayoress – as something of a lark as well as a privilege to take seriously.

Their submerged acting ability came to the fore. Here were roles worth playing. They held public meetings and composed slogans. Edward had a van decorated with billboards: 'Coming to New Plymouth – a New Era Featuring Edward Hill', and 'A Vote for Hill Is a Vote for Dignified and Progressive Leadership'. He drove this van around the city. It was not the same as being a clergyman.

It was impossible to be in New Plymouth and not hear about the Hills. Beatrice's high school class in particular heard most of what was going on. Edward was always in the news, by no means always agreeably. The fathers of two other girls in her class were deeply involved. Mary Honnor was the daughter of the senior councillor who later became deputy mayor, and Lesley Aderman's father was the district's frequently publicised Member of Parliament. That they were good-hearted young girls, fond of Beatrice and sensitive to her situation, did not stop all the talk and speculation.

Beatrice's loyalty to her parents solidified. If they had no need to know something, why tell them if it would distress them? She must often have been distressed herself. Theodora, still at primary school, was young enough to be insulated from most of the barbed talk, but for Rowena the gossip was every bit as painful – and the gossip flowed.

Nevertheless, in the biggest turn-out to the polls that New Plymouth had ever known, on 31 October 1953 Edward was elected mayor for a three-year term. What is more, in every single polling booth, in every suburb, he polled far more votes than any of the other three candidates. PhD students investigating a former British colony's attitudes to the Crown and Britain's ruling classes will find a thesis in microcosm in what went on in New Plymouth that year.

Edward was installed in office, and revived a traditional custom: he arranged to be presented with a pair of white gloves to signify that he began his term 'without stains of discredit or dishonour, and as a reminder that the city's affairs must be handed over to his successor in as pure and unsullied a condition as when they were taken over'.[14]

The voters had certainly spoken. The glamorous, wealthy and upper-class Hills would point the city towards a new era.

Beatrice, in everyone's recollection, remained self-contained, modest and unassuming. This was the persona she seems to have unconsciously adopted, that of the good dependable daughter. In spite of the publicity turned on her family, she worked on quietly and steadily, winning both piano and violin prizes, passing her violin Grade 4 exam with distinction, and – as did Rowena, who was outstanding at languages – winning a French reading prize. She also won the form prize that year, as she was to do every year of her school life. One girl, Jackie Bryant, was fated to come second to Beatrice throughout their school lives. She has said:

> Beatrice was a sweet-tempered, friendly classmate, so obliging and unassuming that none of us minded how much more clever she was than the rest of us. She joined in even at the things she wasn't very good at, such as sport, but we always knew she would excel academically. For one thing she worked. We all dramatically set our alarm clocks for 5am the week before exams, but she put hers on the opposite side of the room so she actually had to get up.[15]

That 'obliging' included unobtrusively helping anyone with subjects where they were lagging behind. Julienne Lobb remembers receiving this help. 'Beetle and I used to stay together in the holidays and swot for part of each day, but only part. We used to laugh a lot. We had tremendous fun.'[16] The two girls went off to Crusaders Camp together that summer holiday. This was run along evangelical and fundamentalist lines, and both signed a Decision for Christ pledge. Jean Hill did not approve of this narrow approach to religion, saying it was without an intellectual basis. Similarly she had not tolerated some of the things her daughters were being taught in Sunday school at the little Anglican church at the port, Moturoa. She had taken the girls away.

In both her third and fourth form years, when she was 12 and 13, Beatrice often sat next to Lesley Aderman, who lived nearby, for the bus rides to school and, on Saturday evenings, to the Boys' High School for dancing lessons. Lesley Aderman, who became Lesley Powell, said of Beatrice:

The dancing lessons were hilarious. We had great fun communicating in a sort of pidgin Latin. The boys we admired acquired names like Mucus Spurious Supercilius, or Gaius Postumous Rusticus.

It seemed that Beetle was within our range in the third form but with each year the intellectual gap became wider and wider. Yet there was no condescension in Beatrice. I never heard Beatrice put anyone down. There was always great courtesy, a respect for the individual and individual conscience which came with her home background.

'And Beatrice never became separated from us socially. She was always interested, and eager to participate. Her eyes twinkled and her words bubbled out – always so quickly that we often had to ask her to repeat what she'd said. She had a very quick wit.' [17]

As New Plymouth's new mayor and mayoress, Edward and Jean Hill presided over the huge gathering and civic welcome at Pukekura Park when Queen Elizabeth and the Duke of Edinburgh visited the city on 8–9 January 1954. (The suggestion that the previous mayor should do this had not been revived.) The country was still trying to recover from the shock of the disaster at Tangiwai in the central North Island on Christmas Eve. The railway bridge over a normally peaceful small river had been torn away by a torrent of muddy water and boulders that erupted from the crater lake of Mt Ruapehu. Six carriages of the Wellington–Auckland express train, filled with holidaying and home-coming families, plunged into the river, bringing death to 151 passengers and crew.

The disaster had visibly distressed the young Queen, and both she and the hundreds of thousands of flag-waving New Zealanders wherever she went[18] seemed to make a special effort to surmount the tragedy. No crowd could have been more loyal than that which assembled in determined carnival mood in New Plymouth, to cheer the Queen as she and the Duke were slowly driven in a Land Rover around Pukekura Park. Everyone was there. Hospital patients, even some from the tuberculosis ward, were brought to a special enclosure, to benefit from the tonic effect of seeing the Queen.

As the city's hosts, the Hills soon saw there was no special rapport between the royal visitors and Prime Minister Sid Holland, who was accompanying them. When a venerable Maori, chosen because of his lineage to speak for the local Maori tribes, had finished a long address of welcome at the park, he whipped off his head-band and offered it to the Queen. She was nonplussed and looked to Edward, who, as ignorant at that time as she was of Maori protocol, stepped forward and accepted it on her behalf. 'What should we do with it?' she asked him quietly later. 'Let's give it to the Prime Minister,' Edward replied. They looked at each other in tacit understanding.[19]

Although the Hills had cautioned their daughters not to repeat household conversations about aspects of mayoral duties, sometimes a story leaked (not necessarily from the children) to an appreciative audience.

Thus, after the public welcome, the Queen invited the Hills and the Prime Minister back to their hotel, the Criterion, for lunch. When biscuits and cheese were offered at the end of the meal, the Queen lightly asked the Prime Minister if New Zealand had any other varieties of cheese. 'Or are you going to live on eternal mousetrap?'

It was the Prime Minister's turn to look non-plussed. It was obviously a subject to which he had never given thought. 'There's only one kind of cheese as far as I know,' he said in some bewilderment.[20]

Family and friends were to enjoy Jean's account of part of the lunchtime conversation. Largely removed from everyday domesticity, she had recently had that very new appliance, a dishwasher, installed. She mentioned this to the Duke. 'Really?' he said. 'We've been thinking of having one put in, in our place.'

Later, and 'quite against the rules', as Edward wrote in a letter to his mother, he and Jean gave Queen Elizabeth a small oil painting of Mt Egmont. It belonged to Beatrice and she and the other Hill children wanted it to be a gift for four-year-old Prince Charles, left in the care of his nanny and nursery staff in England. 'The Queen was very pleased. Her eyes lit up when her children were mentioned.'

As for the public, they reportedly descended on the Criterion Hotel in droves, taking sheets of toilet paper as souvenirs.

For Beatrice, 1954 was the year she entered her teens, the traditional time for challenging one's parents' philosophies or at least beginning to evaluate them. She does not seem in any way to have rebelled openly against their constant talk of the Oxford Group's rules for behaviour, its four Absolutes and its view of human destiny. Instead she came more and more to think about the big questions for herself, the origins of the universe, where it was going and why. She wanted the scientific theories, not her parents' version, and began to build her private world of science, where they could not follow – or rather, as she was to say sadly in her adult life, where they would not follow.

Rowena, thinking back to those early times, and not wanting to criticise her parents because of the way they themselves had been conditioned, much later put into words something she had been thinking about for years:

> I think we must thank our parents for sensitising us and opening us out to the big questions, this pre-occupation with the depths

– though the pressures that accompanied it all and the difficulty of keeping in contact with 'reality' were hurtful. For Beatrice, Daddy's specific ideas and attitudes never ceased to be a point of reference and rebellion almost to the end of her life.[21]

For the present, the girls were wrapped in the world of school. Pupils had been visiting the local observatory for some time. One of the teachers at the girls' school, Fay Dunlop, organised an astronomy club that Beatrice joined. No records remain except for a short note in one issue of the school magazine:

> Mr Moorshead made available to us the library at the observatory, and gave us many interesting talks on the moon, and Jupiter and other planets and stars. We at least know the difference now.
> Some of us gave talks about our solar system.

Mary Heward taught general science to the 36 top-stream fourth form girls. Known as core science, its syllabus contained physics, chemistry and biology sections. She remembers their mid-year two-hour examination well. The biology section gave a possible escape route to girls not scientifically minded. Some questions could be answered with mini-essays, so 'great wads' of pages were handed up, she said:

> My heart sank when I looked at Beatrice's chemistry section – only three-quarters of a page. When I got it home and could study it, I saw she had given the most economical of answers. Wherever it was appropriate she'd referred me back. For example she'd written, say, 'as in 3B-1' if there were two questions with some similarities, such as about getting solids from liquids. I had to do all the work, and check her previous answer. It was correct, of course.
> That sort of economy is genius.

Already, at the age of 13, Beatrice knew about energy, and force. She had a gift for perceiving and linking things which went well beyond anything her teacher had ever encountered, and she did this while going directly to the heart of a matter and expressing herself in the simplest and most precise way. Her command of language was exceptional. At 13 she was already a synthesiser, Mary Heward believed:

> A synthesiser puts together the relevant facts. When she was a junior, Beatrice didn't have to swot science as all the others did. She put all the facts together but not in the way she'd been taught them.

Teachers get so used to having their own notes and textbook definitions regurgitated at exam time – from the top students, that is. I never got my own notes back from Beatrice, but rather a synthesis of what she'd learned from me and learned by herself. She'd listened, read, observed and reasoned it all out for herself.

This teacher saw Beatrice as self-contained although eager and fun-loving. She was younger socially than most of her classmates, not interested in such things as 'boy-spotting' or casually being where boys were likely to pass by. 'Many young ones in a group try to get in big with the others. Beatrice didn't.' [22]

That 'fun-loving' comes through in many memories. Josephine Leigh, later Dodd, wrote this account for *Springboard for Women:*

> Beetle was an exceptional girl, seen as 'different' by her classmates yet accepted totally.
>
> There were her educated and cultured English parents, her skill in music, her intelligence and a quality I find hard to pinpoint – not naivete but a rather innocent straightforwardness. She was always cooperative and law-abiding which is why this trifling incident is worth recounting.
>
> I can still vividly picture the scene in 1954 when we were in the fourth form. Beatrice came to school with a perfectly regular-looking fountain pen, but when you unscrewed the top it emitted a loud bang. We girls were delighted by the possibility of entertainment, and decided that our target would be one of our youngest, and gentlest, teachers. We were agog when she came to mark the form roll which was checked at the beginning of every period. Beatrice promptly offered her fountain pen which the teacher accepted without hesitation. (Beetle was the least likely in the form to be suspected of being up to mischief.) I clearly remember being very impressed that she was prepared to get herself involved in a scrape quite willingly.

Beatrice was extremely contrite at the young teacher's distress. The trick pen was a trophy Rowena had brought back from Canada. With five other girls from New Plymouth Girls' High School, in July 1954 she went on a school exchange visit. Among her own trophies was a pair of pedal-pushers, the very latest American fashion in pants for girls, tight and to mid-calf (unseen in New Zealand before, and remembered in New Plymouth for decades).

Family friend Teddy Fardell from Christchurch, always particularly happy to see Beatrice again, had a birthday on one occasion while she was staying with the Hills. Jean had organised a birthday meal that evening, a Friday, but Fridays were when Rowena had won permission to go off to town to meet her friends. Teddy Fardell said:

> There was a dingdong row, with Jean saying Rowena wasn't to insult her Aunt Teddy by going out, and Rowena claiming, in truth, that I'd be glad to let her go. The argument raged, with Edward not saying a word and Beatrice and Theodora silently getting on with things. Jean won, but Rowena was both fiery and remote all evening.
>
> Jean used to agonise over the time Rowena came home in the evenings. She made such unfortunate issues over things like that.[23]

She was not the only person to notice the tight control Jean Hill attempted to keep over all her daughters – 'Who was that on the phone for you? And who are you going to see?' The Oxford Group's commandment of Absolute Purity (except within marriage, where sex was assumed to be decorous too) took on ever-increasing importance. Jean's incipient hysteria flared when Rowena was suspected of talking to local boys.

Rowena was the only one of the three sisters to rebel, at school as well as at home.

Beatrice, less headstrong and more ruled by logic, could see ways of avoiding these family battles. To Rowena's natural resentment, Beatrice and Theodora were held up as the 'good' daughters, a manoeuvre of their mother's which resulted in the two older girls becoming not quite as close as they had been as young children, not really close in fact until later in their adult lives.

Beatrice increasingly became a dependable, 'good' daughter. She most probably also had an unconscious fear of risking the precarious hold she had on her mother's affections. She was a warm and loving child. She loved Rowena, but her concern for her parents was real, too. She wanted to shield them if she could, as well as keeping close to them. The more she worked at this, the closer she came to a mind-set that was to have far-reaching and eventually tragic consequences.

The children were not often involved in Edward's mayoral duties but he always remembered their distress when a down-and-out man, not long out of prison, turned up at their home one night. He asked for help from the Mayor's Relief Fund, the discretionary fund operated by all New Zealand mayors at that time. Edward produced money, and the children made up a parcel of food and clothes.

He involved everyone in mealtime discussions about civic problems and possibilities. The new town plan for New Plymouth needed particular study, and his interest in the welfare of the aged led to an old persons' welfare council being formed. His became the first centre to take advantage of the new government subsidies for councils that were building special accommodation for the aged. He also chaired the state house allocation committee, taking very seriously the circumstances of each applicant family so that the most deserving got these low-rental houses.

This was not always plain sailing. Beatrice's bedroom was nearest the telephone. One Christmas morning, just after midnight, the phone woke her. It was a woman, an unsuccessful applicant and drunk, who wanted to wish the mayor a very unhappy Christmas.

Because of his role, Edward found himself involved in situations he could never have envisaged back home in England. On one occasion he found he was expected to visit a Maori marae in Waitara for the tangi of a much-respected old man, a blind tohunga. With no knowledge of Maori protocol but eager to learn, he discovered he was also expected to address the body, which was lying in state in the meeting house.

All around him crowded hundreds of Maori mourners, whole families and sub-tribes. Oratory came naturally to Edward, but what to say here? He remembered that the tohunga had been blind and this gave him inspiration. Walking out into the cleared centre of the meeting house he declaimed in his beautiful voice that great sonnet of Milton's on his blindness, the sonnet which he had learned at school and which his daughters in turn learned after him.

> When I consider how my light is spent
> Ere half my days in this dark world and wide,
> And that one talent which is death to hide
> Lodged with me useless, though my soul more bent
> To serve therewith my maker, and present
> My true account, lest he returning chide,
> Doth God exact day-labour, light denied,
> I fondly ask; but patience to prevent
> That murmur soon replies, God doth not need
> Either man's work or his own gifts. Who best
> Bear his mild yoke, they serve him best. His state
> Is kingly. Thousands at his bidding speed
> And post o'er land and ocean without rest;
> They also serve who only stand and wait.

Edward, the Anglican cleric, could declaim these noble words in total belief – and they would have been received in respectful silence. But Beatrice, in every fibre of her being, came to reject the religious passivity of 'They also serve who only stand and wait.' For her, waiting was to become intolerable frustration.

If those who taught English, French, Latin and music had known nothing of Beatrice's ability in mathematics and science, they would have been unanimous in seeing her advance unswervingly to a career in any or even all of these. After all, these were the subjects for which the school was famous, and Rowena, as dux in 1955 and winner of one of the country's 10 National University Scholarships, was considered a brilliant linguist.

These teachers, with others, gave their impressions of Beatrice for the centennial history of the school.[24] They and Beatrice's contemporaries – some of whom were learning for the first time of her death in 1981, and whose responses were loud with shock and grief – were without exception in agreement that she was a golden girl with whom they found no fault. The suspicion of course arises that nobody would offer the mildest of contrary views because of the tragedy of her early death, but persistent questioning has produced nothing at all in the way of criticism or shortcoming except the one incident of the mildly booby-trapped fountain pen.

Criticism, even condemnation, came later in her life, when she was faced with agonising life choices. As a schoolgirl, merry and bubbling with the possibilities in front of her, she lived within the framework of the way she had evolved to deal with the loss of the person who had been closest to her.

Helen Thomson at that stage taught her mathematics. 'She was streets ahead of anyone else but she was never bored and never intolerant of the class pace. She worked quietly and didn't irritate the others with a know-all attitude.'

One weekend Beatrice's mathematical insights surprised no less a person than the Governor-General, Sir Willoughby Norrie, who was lunching with Edward and his family. A kindly man with a love of card tricks, he called for a pack of cards and entertained the children with some simple magic. Beatrice then quietly picked up the cards and did a mathematical trick that left the Governor-General 'most surprised and admiring'.[25]

Billie Steuart, who taught Beatrice English, said: 'She was a pleasure to teach and to know, especially for her quick perception and her sense of humour; and especially because it is such a comfort when a gifted science student so thoroughly enjoys English literature and all its surrounding landscapes.'

Beatrice's ability to write well – clearly, quickly and to the point – was to play a part in her remarkable output of scientific papers, her own or in collaboration. Even today, people well versed in the humanities are frequently virtually illiterate in mathematics and the sciences. Those who turn to the sciences – doctors, nuclear physicists or whatever – often do not have the sound grasp of their native language needed to write lucid research papers. Beatrice was unusual in her rounded interests and abilities.

Billie Steuart also said of her: 'There was, I believe, another side too, even then; a very self-contained private little person who lived and studied, necessarily alone, in a small universe of her own – the mark of a real scholar. This was an impression at the time, not hindsight, for after she left school I never saw her again.'

Another teacher, Thomas Kardos, said of her French: 'Her linguistic ability was outstanding and it gave me a real pleasure to teach her. She was cheerful and most conscientious, a person to whom nothing was a trouble if it concerned comradeship or achievement, somebody whose encyclopaedic knowledge earned the admiration of all who were privileged to know her.' (He also told the Hills on an earlier occasion that Rowena had a greater gift for languages than anyone else he had ever encountered.)

Latin teacher Dorothy Geddes was definite: 'There is no doubt she was a genius and combined with all that ability a very attractive appearance, a very warm, outgoing personality and a willingness to join in all school activities.'

Beatrice was fortunate in her physics teacher, Joyce Jarrold, who was responsible for making physics available at senior level. She also had a particular sympathy for girls wishing to advance in mathematics and the sciences in general. She herself had been a pupil of New Plymouth Girls' High School in her senior years, 1937–39, and credits the school with opening her eyes to the possibility of university.

Joyce Jarrold won a Taranaki Scholarship.[26] After graduating she returned to the school for 10 years to teach sciences before becoming long-time principal of Waitaki Girls' High School. Mathematics had been her favourite subject, and as a schoolgirl she had been excited at the thought of making it her degree field. But the then headmistress, Doris Napier Allan, disabused her. 'Girls don't become mathematicians. You had better do home science.' So she did. This was only 14 years before Beatrice entered the same school. By then there was a new headmistress but the attitude of the school as to what was or was not suitable, or even possible, for its girls had not changed much.

The syllabus for both boys' and girls' high schools gave an option of physics or home science from fifth form level. Almost all New Zealand girls were taught home science. It was Beatrice's good fortune that Joyce Jarrold returned to her old school to teach. For Beatrice she opened the most important door of all by giving the physics option to her New Plymouth girls.

The idea of physics had been in Beatrice's mind since an epoch-making book fell into her hands through her parents' friendship with the Niblocks. This book, she was to say, changed her life. Fred Hoyle's *The Nature of the Universe* was based on his weekly radio talks on the BBC's Third Programme. They had had a huge following when they were first broadcast in 1950, and made Hoyle, after H. G. Wells, the voice of popular science throughout the Western world, and the inspiration for countless future scientists.

It was Hoyle in a radio talk who coined the term 'the Big Bang', standing for the belief that the universe was created in some unimaginably massive explosion at some definite point in the very distant past.[27] Hoyle, however, did not go along with this belief. He had his own 'steady state' theory which proposed that the universe had always existed in a form not so very different from what is known today. But his arresting name stuck, and the Big Bang was part of the excitement that made astronomy or cosmology intellectually attractive.

For Beatrice, Hoyle's book asked the most exciting questions possible. How was the universe created? Is the universe running down? What is the fate of the earth? Is life present on other planets?

And two sentences of Hoyle's came to possess her imagination:

> It is my view that man's unguided imagination could never have chanced on such a structure as I have put before you in these talks.
>
> No literary genius could have invented a story one-hundredth part as fantastic as the sober facts that have been unearthed by astronomical science.

Unearthed up to now, Hoyle meant. She, Beatrice, could be part of the discoveries to come. And physics seemed to be the key.

With *The Nature of the Universe* very much in mind, Beatrice Hill, aged 14 and in the fifth form, accordingly knocked on the door of Joyce Jarrold's laboratory one day in 1955. 'Such a pretty, neat little girl – I'll never forget her.' She impressed her teacher as being vivacious with many interests, not just narrowly devoted to academic success. Beatrice pointed to some physics textbooks and university reference books on Joyce Jarrold's bookshelves, and asked if she might borrow them. The teacher was momentarily sceptical. Beatrice had only just begun learning the elements of physics in class, and these were the most advanced physics books in the school. 'But I knew she was bright, and there was something about her that made me say yes, after only a moment's reflection.' From then on Beatrice began to work by herself in physics, sometimes coming to consult her teacher.

Some 30 years later, reflecting on this time, Joyce Jarrold said: 'When you teach you're mostly trying to din something in. ... Very occasionally you realise you're dealing with a mind that is infinitely superior to your own. Beatrice came into that category. I would call her a genius.'[28]

This is a subjective judgment, but intelligence alone does not a genius make. Michael J. A. Howe puts it this way in his book, *Genius Explained*: 'The exceptional talents of those we call geniuses are the result of a unique set of circumstances and opportunities, but in every case they are pursued and exploited with a characteristic drive, determination and focus which the rest of us rarely show.'

Such people, he says, question, and devote themselves to answering their questions. They find great joy and satisfaction in their own creativity. The answer or part-answer to what is genius may lie in their unique combination of genes inherited and individual upbringings, nature and nurture, and their reactions to these. Beatrice's unusual inheritance plainly tipped the odds a little in her case. Her upbringing inclined her to constant questioning. Her innate temperament, it seems, was yet another key.

By the time Beatrice was 14 and set on beginning to teach herself physics, Howe's criteria do seem to apply to her. Beetle, however, was never submerged in grinding study, and she did not narrow her interests. She wrote poetry and prose for the school magazine, kept up her languages and continued to win music prizes while passing with distinction the Royal Schools of Music exam for violin, Grade 6. Life was good.

It was not a good year for Rowena Hill.

Attractive, vivacious and daring, and always the top student in her class, Rowena was the centre of her own social set. 'Such fun to be with,' recalled Lynne Hall, formerly Lealand.[29] But it was as if Rowena were compelled to lead small rebellions and disobediences in her school years – although nothing in the least horrifying, for she was good-hearted and passionate in her enthusiasms. But Rowena was a Hill, and the daughter of the mayor. Her headmistress and her mother, part of the same close social set, colluded 'to teach madam a lesson'.[30]

Although it was obvious to them and to everyone else that at the end of that year Rowena would be dux, the school's top scholar, they decided to humble her. They would prevent her from becoming one of the girls' own rulers, the prefects. No doubt they told each other it was for Rowena's own good; it would teach her the kind of lesson she needed before going out into the world. If these two powerful women had had more insight, they surely would never have done such a thing. In the world of school, their decision was a wound that did not heal.

By the next year, 1956, Theodora had joined her middle sister at secondary school while Rowena was in Christchurch in her first year of studying languages at Canterbury University College. For generations there had been a tradition followed by many New Plymouth young people that they went there if they were university-bound. Rowena was only too happy to journey down to the South Island, leaving her family in the North.

Beatrice and Theodora played together, violin and piano, to win the top prize in that year's school music competitions. In the big music festival in which the orchestras of the girls' and boys' schools combined, Beatrice led the first violins while Theodora, very young but already an exceptional musician, was the pianist for the mass orchestra.

Beatrice also entered the senior speech contest, vying in the finals with her friend and classmate Lesley Aderman. One of those listening, Janice Griffin, has written of this contest:

> The whole school was fascinated. The mayor's daughter or the MP's daughter? There was nothing between them in the judges' count-up of marks in the final. Rather than leave it as a tie, it was decided they should each give an impromptu speech of five minutes before the whole school at assembly. They weren't permitted to hear each other, and they were given two minutes' notice of the topic, which was money.
>
> Lesley's speech was hilarious, not always very coherent or logical, but it had the whole assembly of girls in fits of laughter. Beatrice's speech was more precise, better structured and altogether more factual, and she seemed to have a greater knowledge of the topic. It was delivered in a more serious manner. She won, and deservedly so, but Lesley's speech was the more entertaining.[31]

Lesley Aderman herself said: 'My inspiration petered out very quickly but Beetle's clarity of mind brought forth a controlled and logical argument.' Beatrice's original speech, called 'This Brave New World, Mechanised', was printed in the school magazine. The next year the contest was won by Beatrice's old friend from primary school and life-saving days, Julienne Lobb, who has recalled, 'We both felt it was nice that our names would be engraved next to each other's, on the cup.'[32]

The school was divided into four houses for competitive purposes and somehow Beatrice, although she was not an athlete, found herself in the senior six-girl relay team of her house, Tainui, on sports day. Photos exist of these six senior girls with their regulation sports rompers pushed up to show a daring amount of leg. Janice Griffin again: 'With her high forehead and thin legs, Beetle looked totally out of place when dressed in the terrible rompers and white blouses we had to wear for phys ed. But she tried very hard to participate in all sports activities and though she received a bit of gentle ribbing she was always treated with respect.'

One of the photographed girls, Berrie Stronge, later Stewart, said of this picture that they certainly must have been seniors, the clue being that their voluminous rompers were tucked up to an abbreviated smoothness to show as much leg as possible, 'and it took us all at least four years to get to that daring stage'.

As conscientious parents the Hills were involved in the school's parent-teacher association fund-raising, which had the main aim of building a new assembly hall or auditorium. As frustrated actors they had fun staging popular revues of the kind they had enjoyed in their Oxford Group days in England, but this time scripted for New Plymouth's appreciation. They wrote much of the material themselves and vastly entertained their audiences.

In 1956, while Edward was mayor, they wrote skits for the 'Parents' Pep Show', one of them featuring Edward as Henry VIII and Jean as Catherine Parr. The word 'litter', as in 'Don't drop litter', had just come into public consciousness so there was a Litter Ballet of the kind long popular at students' capping revues, with men of rugby-playing build performing, in tutus, to the music of *Swan Lake*. This time the violinist Willi Komlos was the litterbug pursued by anti-litter fairies, who captured him and carried him off in a bin. The antics of Edward and the rotund police sergeant, Percy Smeaton, both wearing a tutu, were well remembered in New Plymouth some 40 years later. The long years of Mayor Gilmour's reign had never been like this. The school and whole community buzzed with talk.

Local body elections came around again that year. Edward had all his abilities stretched by the responsibilities of being mayor. Without the constraints of having to go to work elsewhere, Edward had found himself 'landed with everything – from state house allocation and the Roads Council to all those debutante balls. I had decided to treat the mayoralty as a fulltime job. I found out it was.'[33]

This election there were no denunciatory campaigns in the newspapers, and Edward as sitting mayor did not go to the lengths of canvassing with a van bedizened with billboards. He had gone beyond that. He was standing as an independent, but with his concern for the under-privileged he had joined the local branch of the Labour Party. Word of this got out. In the more conservative parts of New Plymouth, which were very conservative indeed, it was interpreted as Edward's having become that dread creature, a socialist. There were other reservations. And he also had a challenger. This opponent, Alf Honnor, the father of one of Beatrice's friends and classmates, Mary Honnor, was a builder, a practical and well-respected man who had been deputy mayor to Edward for the past three years. The situation could easily have led to unpleasant tensions and undercurrents in the classroom, but everyone agreed that neither Beatrice nor Mary let their fathers' rivalries affect the world of school.

When the votes were counted on the Saturday of the elections, Edward lost the mayoral race to Alf Honnor, although he was returned as a city councillor. Lesley Aderman wrote:

> The Monday after the elections, Mary and I came into the form-room. Beatrice made a quick, smiling, ice-breaking comment and the situation was eased.
>
> Later I asked Mary what Beetle had said, but – characteristic of her – she had spoken so quickly that Mary hadn't been able to distinguish the words. But it didn't matter. Her action achieved the effect. It was a hard time for Beetle, though.[34]

The Hills then bought a holiday house, little more than a bach, near the small fishing village of Hatepe, just off the main road running past Lake Taupo towards the town centre to the north. Everyone enjoyed it. Meal preparation was minimal, Edward could fish for trout all day, Jean could read and write, and the girls could scatter to their own amusements. Beatrice had a lot to think about that summer.

In February 1957, when the family returned to New Plymouth and she began her last school year, she took up a stand which for the first time put her in opposition to her parents, and to her headmistress. She was a prefect – that office everyone took very seriously – and she was still deeply immersed in music but she had made up her mind that this, her University Scholarship year, was to be devoted to mathematics, chemistry and physics, as well as to English which was compulsory.

Headmistress Rose Allum was far from happy. She was not at home with the sciences. The Hills were totally non-scientific. Armed with the evidence of Beatrice's quite brilliant record in languages, Miss Allum called on the Hills, to talk over this disturbing plan. It would be much better all round if Beatrice took advantage of the school's strengths, as Rowena had done, and concentrated on languages, she said. Rowena had been dux and had gained a National University Scholarship. Beatrice would certainly do the same – indeed, she might do even better and win a Junior University Scholarship, the very top ten on offer for the whole country – if she built on her obvious strengths. Miss Allum meant the strengths that were obvious to her and to the Hills.

Beatrice knew herself better. She put her case.

You had to be brave, in those years, to tell school principals you did not agree with them. Rose Allum was accustomed to getting her own way. Beatrice was accustomed to being obedient. But this was the most important thing of all, her very life as an astronomer or astrophysicist, and the right foundations had to be laid, now. She must have used all her debating skills, this still-tiny girl who had just had her sixteenth birthday. More than any specific arguments, it must have been her blazing conviction which won the day. Beatrice was not capable of fudging issues.

It was important to the school that its girls won scholarships. If Beatrice had given up a virtually certain scholarship by turning away from her languages, it was all the more essential that she should now be helped with her chosen field of study. But how? She wanted to do additional mathematics as well as the regular syllabus, and no girl within memory had tried to do this.

Rose Allum arranged for her to have extra mathematics at New Plymouth Boys' High School, and for her form's maths and chemistry teacher, Alma Davies, to give her additional mathematics, called admaths, by devoting her free time to working with her. The textbook was new to this teacher so they studied it together. On one occasion, Alma Davies recalled, she asked Beatrice where they were in the textbook:

> Beatrice replied, her eyes twinkling, 'Well, I'm at chapter nine, and you're at ... ?' She was a delightful child to teach, and had wide interests. One day she had a book on reading character from handwriting. She looked at my writing and read my character, including weaknesses as well as strengths. She was a very natural girl with a delightful sense of humour. And grateful. At the end of our year's work she gave me a gold brooch which I treasure still.

Alma Davies noted that Beatrice's pioneering study of additional mathematics opened the way 'for quite a number of girls' to take the subject at scholarship level, now that the teaching was available.[35]

A few weeks before Beatrice and her class were to sit the scholarship exams their physics teacher, Joyce Jarrold, was severely injured in a motor accident. Rose Allum permitted her class to visit her in hospital but they were expressly forbidden to raise the subjects of work and exams. This must have been frustrating for everyone, certainly for the teacher, who was keenly aware of what needed to be learned in those last vital weeks.

The senior physics class were sent to the Boys' High School but found they had already done the work the boys were doing, so back they came to work at their own school, with Beatrice in charge. When the results were available Joyce Jarrold said, 'In the scholarship physics exam all the girls got better marks than usual – and this without my final words of wisdom.'[36]

This was Beatrice's first experience of class teaching if one does not count her earlier, equally serious, teaching of her koalas. It was to lead on to her inspirational teaching of many of the new generation of astronomers.

She was dux of her school in 1957, and won a coveted Junior University Scholarship. Her marks were: English, 122; mathematics, 173; chemistry, 153; physics, 143; additional mathematics, 145; a total of 736 out of 1000. She was still only 16.

New Plymouth Girls' High School was given a holiday to mark her achievement. Everyone felt the whole school was honoured.

There was one area where most girls were to feel that their school had let them down. Many years later four women, three of them older than Beatrice, met to discuss their schooldays. Each had been the school's deputy head girl in her time and accordingly particularly close to the school's ethos. They discussed what the school had done for them, and how much it had taught them. But what had it omitted to teach them about? And the answer was unanimous. 'Men!' [37]

Nothing approaching family life education or sex education – facts, feelings, sensitivities, rights and responsibilities – was taught at the school. The frankness with which virtually any aspect of any hitherto intimate subject is now discussed in all the media came about in the 1990s, and then largely because of the urgent action needed to inform people about AIDS. In the 1950s, well-brought-up and obedient girls like Beatrice met boys at carefully chaperoned school dances, or talked with them, usually awkwardly, if they were brothers or cousins of school friends. Boys were almost a different species. After all, the Swinging Sixties did not hit London and New York, let alone New Zealand, until three years after Beatrice had left school. Most young people left school in an abyss of ignorance apart from such actual pieces of information as the more sensible and secure parents felt able to pass on.

It was against the law for teachers to tell pupils about contraception, and in any case few would have known enough about the subject to present a satisfactory lesson. Dr Margaret Sparrow, medical director of the Family Planning Association of New Zealand, was (as Margaret Muir) at New Plymouth Girls' High School for her two sixth form years in 1951–52, just before Beatrice entered the school. She has said that this schoolgirl ignorance was a trigger for her lobbying for many years to help bring the country's education system more in line with reality and what was needed by teenage children.

In her last year at school Beatrice was still not socially mature. She was not included in most of the talk about boys, nor did she become a regular participant in the 'down town' encounters on Friday evenings. The

Hill sisters were given information at home about sex, always with the underlined 'don't'. The Oxford Group with its Absolute Purity cast a long shadow. The mother left her daughters in no doubt that there was only one relationship approved between men and women, and that was marriage. 'If you feel any urges, put them in the bank for Mr Right,' was Jean Hill at her most emphatic.

When Rowena, as the eldest, rebelled against the restrictions and tight supervision, Jean's distress was real, but so was Rowena's. Other girls did not have to account for their every movement or come home early after being out with friends. One evening Rowena, pushing the limits, returned a little later than had been laid down. Her mother unlocked the door for her, distraught: 'Jesus Christ will punish you for this.'

Once she was away from this straitjacket of home and was a student at Canterbury, it did not take Rowena long to try to be her own person. Christchurch society was very easily shocked, and Rowena easily shocked it. Stories of what she allegedly said and did filtered back to New Plymouth from Hill friends and acquaintances, in letters of the 'I thought you should know ...' variety, although Teddy Fardell was careful not to distress Edward and Jean Hill by joining the tattling. Beatrice and Theodora, however, could not miss seeing and hearing their mother's reactions, which were dramatic and prolonged. 'Poor Mummy' became ill, her emotions running from bitterly reproachful through distraught to hysterical. Edward seems to have managed to ignore a lot of the gossip and to feel that he remained close to his oldest daughter while not deflecting Jean.

The effect on Beatrice was profound. Something seemingly dreadful must have been happening, but what? She both knew and did not know. The most socially daring thing she herself had done was to pull up her rompers to that unseemly height. She had the normal young-girl curiosity about sex and boys, but here was the quandary: however did one get to know a young man unless one were already engaged to be married? Again it was the familiar story. She wanted to spare her parents more distress, and she had no wish to be criticised herself.

The answer, without doubt come to unconsciously and without guile, was to do precisely what her parents seemed to approve of – become engaged to be married. In the long summer holidays after she left school, before she had had her 17th birthday, she did just that.

CHAPTER FOUR

FREEDOM TO LEARN

*'Many take to science out of a joyful sense of superior
intellectual power; science is their own special sport to which they
look for vivid experience and the satisfaction of ambition.'*
ALBERT EINSTEIN

Lake Taupo in the summertime sparkles in clear air under blue skies. Heat shimmers from the shore, and holiday-makers drowse over books if they are not sailing or fishing. Not Beatrice. She had her books for active reading, and she took her violin into the bush, with a clothes peg to clip her music to a manuka branch. There would be weeks of sun and the clean smells of bush and lake water. At the end of January she would have her seventeenth birthday, and a larger horizon would open up for her. Like Rowena before her, she would go away to university in the South Island, and begin to discover what she could become.

Perhaps only those who come from a close and conventionally loving family in a small enclosed town really know the particular excitement that comes from getting away, striking out on one's own, heading for a different world; and Beatrice already knew in outline what her world would be. One way or another, physics would be the centre of her life.

Physics had already brought her an unlikely friendship. In the 1950s many young people had pen-friends. Innocent, or mostly innocent, advertisements would appear in newspaper columns: 'Girl keen on horses wants to correspond with similar ...' Beatrice, lonely at being left out of some of her classmates' activities because she seemed so young, put this in her advertisement: 'Girl interested in wave technology ...' Surprisingly, it had brought her a reply.

He was a young man scarcely more than a year older than Beatrice but already experienced in the ways of the world. His interest in waves was true enough, but they were radio, not cosmic, waves. He was an amateur radio operator but his passion was the theatre. Six years later he was to star in New Zealand's first modern feature film, a romantic adventure produced by John O'Shea at his Pacific Films company in Wellington. *Runaway* made a big impact with its story of a young man who is in trouble with his girlfriend, and who embarks on a set of wild, self-seeking adventures through the more picturesque parts of New Zealand. Hitting the country just before the Beatles in 1964, *Runaway* had many thousands of young women at fever pitch over the star. His name was Colin Broadley.

In 1957, however, he was an ambitious unknown, discovering himself and trying out different personas. He was intrigued by Beatrice's advertisement, and more so when he learned that her father had been a vicar and the mayor of New Plymouth. In their letters she wrote about music, in which they shared an interest, as well as physics. He wrote about drama and art. He remembers being impressed by the catalogue of things she had arranged for each day. They wanted to meet.

One day Beatrice went by bus to Hamilton, where Colin Broadley had attached himself to a repertory company and was busy with backstage work. He remembers the meeting ruefully:

> I was so surprised. She looked so young – just a child. I hadn't expected this, from her letters. She came to the theatre where I was up on the stage, painting a set. I had to get on with it, get it done, so I got her to sit in the stalls and wait for me.
>
> When I could I went to her. There she was, sitting in the dark in the back row of the stalls, weeping.[1]

Colin Broadley was not insensitive. He realised Beatrice had felt rejected. They went for a walk, and talked, and she was comforted. The Hill family was accustomed to having friends to stay during their Taupo holidays, so she invited him to Hatepe that January. There was no room in the house, so he stayed in a tent in the garden with another guest. Jean Hill seems not to have paid him much attention, but he and Edward hit it off from the beginning.[2] In fact they were quite alike. Both were tall, good-looking and full of easy charm in the right company. Both enjoyed acting and were good at it.

Colin played the situation to the full, taking part in the drama of fighting a bush fire at the edge of the lake, and acting in charades with the family and friends. He and Beatrice arranged to get up early one morning and go out on the lake in Edward's dinghy, to see the sun rise on Mt Ruapehu. The lake water lapping, and the rosy beauty of the mountain set against dark bush and water, worked their magic. He asked her to marry him. She said yes.

The Hills were distinctly taken aback when he and Beatrice returned for breakfast and told them they were engaged to be married. Beatrice was still only 16. Asked, in later years, why they had gone along with this extraordinary announcement, Edward Hill said they did not take it seriously. In one way they were right; the two young people seemed to be playing parts in one of Colin's ongoing dramas. But it did not occur to the parents to look at the serious aspect – to wonder what deep need in Beatrice could have prompted her to look for closeness in an engagement. There is also the possibility that Jean, at least, was gratified: here was one of her daughters at last doing 'the right thing'. Rowena looked on sardonically, and with disbelief.

Colin Broadley looks back on this time also with some disbelief:

> We had a romance – but we weren't romantic. We didn't even kiss, just gave each other little pecks. There was no touching. The closest I ever got to Beatrice was when I was her partner later that year at her debutante ball, and we danced together. I never saw any welcoming sign that she would like even a cuddle. Perhaps she couldn't show how she felt ... she grew up in a bubble, a protective bubble.[3]

Any of the girls with whom she had gone through school would have said that Colin was a boyfriend, not her fiancé. But it seems likely that Beatrice, the family's little Beetle, felt that an engagement gave her more freedom to be herself. The formality of an engagement, too, made it more probable that her mother would accept that fraught situation – her daughter having time alone with a man.

Beatrice, however, mostly needed time to herself. Her new fiancé went back to his repertory company, and she returned fulltime to practising the violin for the Cambridge Music School. This annual summer school had a large and ongoing influence on many of New Zealand's most promising young musicians. Here Beatrice met a cellist, Philippa or Pip Harding, (later Jackson), who was to become a professional musician and a lifelong friend. Pip remembers their first meeting:

> This little, bright, vivacious person with the big smile – she was always direct and enthusiastic, and so very kind, kind in subtle ways. Very keen and a pretty good fiddle player – and she worked so very hard. She took so much joy in working hard and developing her wonderful talent – all her talents.

Pip was brought back to stay at Hatepe, but heard surprisingly little of Colin Broadley. She was three years older than Beatrice, and this was to be her second year at Canterbury. Now they took violin and cello out into the bush and played and talked. For Beatrice, who already loved Christchurch because of her happiness there with Teddy and John Fardell, whom she planned to visit often, it was a perfect way to learn more about what awaited her at university.

Pip for her part was fascinated by the Hills. Edward was usually away fishing or doing Jean's bidding. Jean, remote and engrossed in her thoughts, sat around reading or writing – 'reading in the daytime!' Pip, from her close and busy family, had never encountered anyone like her before. She recalled that Beatrice was eager to leave 'dear Mummy and Daddy' – not in the way that Rowena, distressed and openly rebelling, had been, but because of the wider opportunities for learning and discussion which university would offer. 'I could see that Beetle wanted to get the fullest possible measure of university life.'[4]

Her school had placed high expectations on its girls. In Beatrice's year, or in the years just above and below her, were young women who left school with deeply held ambitions to do something important in the world. Among them were two who, as Claire Stewart and Margaret Evans, became mayors of cities, New Plymouth and Hamilton respectively. Another, Philippa Black, became the first woman president of the Royal Society of New Zealand. Those leaving the school's top forms carried its banner with them. Beatrice had absorbed this same background of self-expectation, but in her case there was something else. What she openly expected of herself overlay a far deeper and older conviction, albeit unconscious, that she must, and she would, achieve something special.

Rowena had given her younger sister a vivid picture of what awaited her at Helen Connon Hall in Christchurch, where she herself had stayed as a first-year student. The city's pre-eminent hall of residence for young women attending university, it was named for the first woman to graduate with honours in the British Empire, in 1881. Across the road from the River Avon and beautiful Hagley Park,[5] Connon was only a short walk to the many Gothic-style stone buildings which made up the grouping of Canterbury University College, Christ's College and Canterbury Museum.

At Connon Hall Beatrice was allocated an upstairs room in the original brick building. Next door to her was a young woman, Nita Nitschke (later Hanna), who came from a smaller and even more provincial town, Marton, and who was revelling in being away from small-town constraints: 'Wearing slacks on Sunday was a moral issue in Marton.' Nita was to become another lifelong friend and, without realising it, was to have a decisive influence on the way Beatrice's life developed.

In appearance and character, Canterbury University College was the closest to Oxford and Cambridge of any New Zealand university; more important for Beatrice, it was the right distance away from home. The college had begun classes in 1874, and admitted women and granted them degrees well before any other university in Britain, the British Empire or the United States. It was proud of its traditions and its famous names. Pre-eminent for Beatrice – indeed for everyone – was the Canterbury graduate and Nobel Prize-winner Ernest Rutherford, Lord Rutherford of Nelson, known worldwide as the father of atomic science.

In 1889 Rutherford had won a Junior University Scholarship and studied at Canterbury for five years before graduating in 1894 with double first-class honours in mathematics and physical sciences. Even in his time the space at the young college was inadequate. His early research included experiments in very high frequency magnetisation of iron, devising a way of detecting short, fast, current pulses, so that he was able to measure and record minute changes in the particles as they became magnetised. He had problems in finding the right place to conduct these delicate experiments, but eventually gained permission to use a basement room, away from likely vibrations.[6]

Beatrice, attending classes in the same laboratories and lecture rooms as Rutherford had done, was soon well aware that in these rooms there had been someone who had altered, for ever, the world's view of nature. What is more, he had done so three times. Through his experiments he had explained naturally occurring radio-activity. He had determined the structure of the atom, becoming the first person ever to split it. And he had become the first successful alchemist – that dream since the Middle Ages – by transforming nitrogen into oxygen.

Not that great names, including Rutherford's, in any way intimidated Beatrice. Rather it was the reverse, as this exuberant letter of 1 March 1958, the first she wrote to her family after classes began, makes plain:

> I'm meant to be attending to a bloke who's lecturing on the trig we've all known for two years, so I'm writing this to save myself from going to sleep... Pure maths is rather a farce, and at present so is applied, but that'll improve as we advance.
>
> Physics is really super. Doc Gregory is a most interesting lecturer and goes frequently into the philosophical reasons for knowing anything; anyway it's miles more satisfactory than learning strings of facts in an orderly manner and never knowing why anyone knows it. His first lecture was a long discussion on 'What is science? What is real? What is length? What is being?' etc etc and going into the attitude towards learning being building everything up from experiment, not using the theories we've been taught to see if an experiment 'worked'. In other words doing the very opposite from the unbearable school attitude. Now we're starting on the subject

of measurement, going into how man perceives length and how it's measured, and today it was 'time' which was terrific. Wonderful training in taking nothing for granted. It's our job to try to prove a theory wrong rather than assuming it's right. Everything must be thought of from a logical and perfectly basic beginning.

In chemistry it's much the same; we begin from an historical point of view and learn how the theories were built up. At school we learnt up theories and saw how the experimental knowledge was explained by them, and then (to quote our lecturer) said 'That's nice!' Well here we see how the theories were thought of, then built up or discarded, starting with experimental evidence; which makes a stimulating atmosphere for working in. (End of lecture thank goodness.)

(Afternoon) The result is that I now can't look at any 'fact' in a textbook without thinking 'How did someone come to believe that? And if I had the same evidence would I believe it?' So I'm learning to question everything. I realise that all advances in science have been made by people who have thought along totally unconventional lines and haven't been misled by the authority of a Great Name having said it was true. Just to think that because Aristotle said things were true, so many centuries passed before someone said they weren't. Now we learn things that Newton and such Great Names have claimed as true, but why shouldn't we question them? Well the fact is we can and do, and we're constantly reminded that in science a theory is accepted only as long as it isn't disproved by experimental evidence. I doubt people would mind what theories or questions anyone would like to put forward here if they felt like it. It gives one the courage to think originally, and that is the beginning of research. Of course I've known in theory for a long time how all this is, but how marvellous to be in a position to think and experiment myself.

Almost certainly Beatrice had never before felt as happy, eager, and on the verge of great discoveries. Physics in particular excited and intrigued her. But that was not all. She ended the letter with an account of her first rock and roll dance – 'super fun' – and thoughts about being a woman in a lecture environment which was almost exclusively male. Neither in this nor in subsequent letters was there any mention of Colin Broadley.

In *My Daughter Beatrice* Edward Hill prefaces an excerpt from an April letter with the comment, 'By the following month music was absorbing a good deal of Beatrice's time.' This is wide of the mark. True, she was playing chamber music up to four evenings a week, often in a group with Pip, but this was still only a comparatively small allocation of time in her crowded days and nights. Her childhood habit of using every hour was carried over into university life. Instead of the times allotted to play in her early timetables, she now made sure she gave time to talk and debate, that most important part of education.

Her friends have said of this memoir by her father, which consists largely of extracts from Beatrice's letters home throughout her life, that it gives a lopsided, in fact unreal, picture of her university and later years. Taken on their own, these first letters home could suggest that music was unquestionably her predominant passion. But Beatrice enjoyed writing letters. She brought her particular gusto and immediacy of impression and experience into focus for each of her family and friends in turn, while always bearing in mind their individual interests and spheres of understanding. If others had kept all her letters as assiduously as did the Hills, a fuller picture of her interests might have emerged. By writing home at length about music, she could be sure that her parents and Theodora could participate in at least one of her experiences and delights. There was no point in trying to convey subjects beyond their ken, or activities they would not want to follow.

It was this innocent, even thoughtful, way of writing letters – thinking of how the recipients would react – which led her into carrying on an eventually dangerous correspondence with her parents. It was risky both for her and for them. The Hills were lulled into thinking they knew their daughter while seeing only a part of her. Without meaning to, Beatrice deceived them. It became harder and harder to write about her life and feelings as they actually were. When the time came that she had to give painful facts to her family, they were totally unprepared and reacted accordingly.

Presenting the part as if it were the whole had another danger. Beatrice's parents were unworldly, bound to the Four Absolutes of their Oxford Group. Beatrice herself was young in every way except for the power of her intellect. Quite unknowingly, she began to weave for herself a false persona, one she was bound to try to live up to. She wanted to experiment, to reach out and try everything she could. But even having a boyfriend was suspect. More and more, as the months and years went by, she contorted herself in her letters into what she hoped her parents wanted. But life has a habit of emulating art. As in Rowena's prescient words about their parents, Beatrice too, at times, 'could not see what is because of what ought to be'.

She and Rowena did not often see each other. There was the usual gulf between science and arts faculties. But Rowena took her along to perhaps the liveliest and certainly the most intellectual of the university's clubs, the Socratic Society. Fellow members of Soc Soc, as it was known, remember Rowena as dramatic and fiery in her arguments, and Beatrice as composed and logical.

Chamber music sessions at the home of her chemistry lecturer, Dr Wal Metcalf, were very important to her. He was a violinist, and his wife, Glen, a pianist. Pip had introduced her to this family and Beatrice was soon included in all their music-making. But it was science which absorbed her. So many doors and windows seemed to be opening that she could not stop herself trying to convey the excitement of what she was learning, in a way few young scientists even attempt. In an undated April letter home she wrote:

> Every day I get more and more thrilled with science. The other night at the Metcalfs I spent ages talking with the Doc about chemistry and physics. Gosh he's an interesting person. He was telling me about various living scientists of most outstanding insight and ingenuity, who are working on things now and the work they've done – all revolutionary ideas and results. Dr M is totally fascinated by it all himself – it sort of pours from him as he speaks – and he talks and talks about all the discoveries and says 'Isn't it thrilling?' I do agree! He sees science as it really is and not as a mere 'subject'.
>
> I told him I thought it was a pity the way stage one had to skim over all sorts of atomic things or learn half-accurate explanations for them, and he quite agreed and told me all sorts of things about the way he had to adjust his lecturing technique so that he can teach over-simplified things without letting us think that they are sufficient.
>
> I also let forth a lot of things about how maddening it was (a) to have to learn the dry sides of physics that would have no bearing on an atomic scientist's career,[7] and (b) to have to do such ghastly easy maths – all of which he explained why it had to be so. He doesn't care in the least about degrees and exams etc; he only cares about knowledge.
>
> I'm therefore feeling very revolutionary and I'm damned sick of working for exams. Consequently I'm going to put first things first and instead of listening to Mr – waffling through applied maths that I did last year, or – mumbling thru' calculus I did in 6b, I'm going to sit in the back in lectures and work on stuff that's going to get me somewhere, namely learning new maths. Also, instead of just sticking to the syllabus work in chemistry and physics, I'm reading and learning as much as I can about everything. Obviously there's no chance of doing any original work until one has a wide background of present knowledge and a very wide knowledge of maths.
>
> So here I go. Even if I never achieve any of my aims, I'll at least have had the thrill of knowing where I want to go – along those lines,

that is. And the same applies to all parts of life – it's the wanting that matters, and having a definite goal ahead. Even if one has about ten thrilling goals it hardly matters! One thing I certainly know, because I'm a woman my home must come before my science. That is right and it must be so and it would be unbearable otherwise. How bleak and bare life would be.

Just how true to herself Beatrice was being throughout this headlong letter is open to question. She says (and one can hear the joy in using a mild oath) that she's 'damned sick' of working for exams, but she is also careful to use the form of address her parents would expect her to use when talking about the Metcalfs. In her letters Wal is always Doc or Doc M, and Glen is Mrs Metcalf. In actuality, within a few meetings she was calling them Wal and Glen, and babysitting for them.

Her insistence that her home must come before her science, 'because I'm a woman', is worth examining. It may, of course, be a specific response to a letter from her mother. (Beatrice kept very few letters throughout her life, and conclusions drawn from a one-sided correspondence can be shaky.) The easy, loving and equal relationship of Wal and Glen Metcalf was now regularly before her as an example of a marriage which was both close and free. Glen had an MSc in organic chemistry and it was implicit in the Metcalfs' situation that, when their two children were older, she would pick up her science studies, gain a doctorate and continue her career.[8] There were certainly women in New Zealand at that time who took it for granted that they would have both a family and a career without having to choose, or to place one before the other. It is as if Beatrice is reassuring her parents that she will do the Right Thing, that her values are their values. The most likely explanation is simply that she was young – 17 years and three months – and afloat in a sea of heady ideas. She could express herself – that 'damned sick' – but still be the dutiful daughter. And her excitement with science keeps shining through above a row of kisses and her childhood signature, Beetle.

A couple of significant sentences are slipped in at the end of this lengthy letter, which began as an expression of sympathy and love after her grandmother's death: 'Don't worry about either of your two elder offspring. R's feet are more on the ground than mine are. She's fine.' The rift between Rowena and her mother explains why Beatrice rarely mentions her in letters home. Rowena had had a relationship which she herself later described as disastrous.[9] This had led – high drama – to their mother's demanding a promise of virginity from Beatrice. The sisters led different student lives but Beatrice kept in touch with Rowena more than her family mail indicates.

In all the letters from this time there is still no mention of Colin Broadley, the dashing but not quite real fiancé. It is possible that Jean Hill, unnoticed by Edward, lost or destroyed some letters, and almost certainly, despite their intentions, the parents failed to keep all of them. But it is not easy to read any specific passion or intention to marry into the context of Beatrice's 'my home must come before my science'.

Colin came to New Plymouth during the May university holidays in 1958, to partner Beatrice at her old school's debutante ball. This 'coming out' was already being dismissed by many bright young women as quaint and old-fashioned. In the world in which Beatrice had been brought up, however, being a deb was still *de rigueur*. She and 18 other former pupils of New Plymouth Girls' High School, many from her old class, posed for the traditional group photograph, all smiling, wearing white and carrying small posies. Becoming a debutante, having a white wedding and, after a proper interval, having two or three children and then living happily ever after were the expectations which Beatrice shared. Her fiancé, however, faded from her life over the next few months. There is only one passing reference to him in the letters which have survived. Although he came down to Christchurch, and Beatrice went to some trouble to try to find him a flat, he did not stay long. She happened to see him in a coffee bar with a woman, and, though she may not have attached importance to this, she did describe the incident to Nita. As Nita has said, 'She realised he wasn't for her.' [10]

So many new things were happening in this, her first university winter. Steeds Hut, which students had helped to build at Arthurs Pass in the Southern Alps during the 1940s, was the centre of winter activities off-campus for many students. Besides arguing and playing chess, it was traditional to sing one's self hoarse in the steam train puffing up the mountains, and on the return journey. Songs that well-brought-up young people had never heard before came from the *Students' Song Book* : 'The One-Eyed Riley', 'Weeping and Wailing', 'Twas on the Good Ship Venus', and the rollicking new words to 'John Brown's Body' – 'My father is the captain of the Lyttelton ferry boat'. This was all part of the release of being away from authority and in command of one's own life. The mountain air was exhilarating, adults were absent, the days were packed with climbing or other adventures in the snow, and the evenings were for talking and trying out ideas in point and counter-point.

With Nita Nitschke and seven other friends from the Socratic Society, Beatrice made the first of many journeys to Arthur's Pass. She was not a climber but had fun with others, careering around on a hired sledge. She also enjoyed long walks alone, miles over the pass road in the stillness of the bush and the snow.

Her letter of 20 July 1958 also describes how they decorated Helen Connon Hall for a ball. She was probably the only young woman resident who had not given thought to that central pre-occupation: who was going to the ball with whom. Not to have a partner was to be shown up by the dreaded spotlight of social inadequacy. But Beatrice cheerfully wrote: 'I in a typical fashion suddenly found myself on Friday afternoon with no partner.' It was a matter to be approached logically. She had already arranged for a young man to be Nita Nitschke's partner, so phoned him and asked if he could find her a partner, too. 'Well, one way and another, at 6pm I suddenly had three offers. My black stole inspired me to get a large black rose to decorate my deb frock, so went all in black and white. I had a wonderful time.'

To Jean Hill, 'wonderful times' were safe. The words probably suggested to her a parade of young men, well brought up, taking turns to dance decorously with her daughter. And Beatrice gave her mother no cause for alarm. She was engrossed in study and university life in general.

With Nita in the adjoining room at Connon Hall she shared the same subjects: chemistry, physics and maths. There were so many male students, however, that there were several classes and they never found themselves in the same one. Nita had a physics lecturer named Walter Roth, a Hungarian. He did not believe physics was for girls, and Nita could imitate him perfectly.

'Do Mr Roth,' Beatrice would say.

Nita would gather herself up, sweep into a pretend-lecture room and, ignoring the few female students, say, 'Gutt morning, chentlemen.'

Another stock piece was the student having a problem. 'Vell, you't not be doing physics – it shoult be music or zomzing.' Beatrice always laughed with great enjoyment at the absurdity.

Many years later, Nita said:

That's all it was to Beetle then, an absurdity. With her home and background she couldn't believe there were people in the world who thought like that. She'd learned better when she came back from Texas in 1972. But when we were students she took feminism as a matter of fact.[11]

An overwhelming body of both lecturers and students in the science faculty did not. It was common for her fellow women students to feel they were not always taken as seriously as the males. A student majoring in geology, Jan Heine, has said: 'Some lecturers looked on you as if you wouldn't last the distance, anyway. "You'll only get married" sort of syndrome, and "Once you get married you'll have a family so you aren't much use to train".'[12]

Both Beatrice and Nita, however, felt confident that they would remain themselves within eventual marriage. Both were full of the joy of life and its possibilities, but Nita, two years older and from a more typical New Zealand background, was much better prepared. She was also an efficient home-maker and certainly more worldly-wise. They soon decided they would share a flat in the coming year, 1959.

Beatrice undertook to return to Christchurch early, after the summer holidays, to find a flat for them both. She found one at 89 Shakespeare Road, behind the railway station, a considerable bicycle ride to the university but only five minutes' ride from the Metcalfs. Next she consulted the dean, Dr Derek Lawden, about her studies for the year. On 1 March 1959 she wrote a long and what she called 'gloatish' letter (gloating about her flat and degree course, and enticing new textbooks and art books):

> Prof Lawden was very interested and helpful. He said that my maths were certainly unusually high, but I told him it was all a second year of the same stuff and that I'd done nearly no work on the same. He laughed and said 'Nevertheless.' So he said was I interested in theoretical or practical physics, and the answer's theoretical, which is of course applied maths and not physics. Even the highest theoretical chemists probably haven't seen a test tube in their life. But higher maths is a really tough philosophical and logical subject (ie the pure maths necessary) with nothing to do with Stage One. He said unless I really am interested in maths I'd be a fool (my word not his) to take on the maths honours course. All the same, if I'm the type that must have proofs for formulae etc, I'd find physics quite unsatisfying. And, if it comes to doing one and picking up the other, I'd be wiser to do the maths while I can get the tuition.
>
> The trouble with me just now is I want both: the maths to reason with, and the physics to apply it to. Consequently I've left it open by enrolling for Pure and Applied Maths Two and Physics Two, without tying myself to the honours course in either. I've put physics as my major subject, but that has no bearing on anything except to make some of the Physics Dept blink. The Physics Dept has been exclusively male for years now, of course, apart from Stage One (which has labs in a different building). Dr Gregory is pleased with my course. Colin Keay, a lecturer in physics, just stared. Anyway I'm enrolled after six hours of filling in forms and standing in queues. The week is 20 hours: two or three lectures every morning, and two lots of labs from 2–6pm. Thursday will be terrific: three lectures, starting at 9am (imagine that in the winter), then labs 2–6pm, then

orchestra etc afterwards. Friday is similar but maybe (maybe) I'll be in on Friday nights.

Music. Mr Ritchie[13] has said, 'Soon, Beatrice, I want to have a little discussion with you on how much time you'll have to spend on music this year.' Answer is obvious. Little. (But I'll spend a lot.)

This was truer than she realised, because in the mail that day was an application form for an audition with the National Youth Orchestra. As Beatrice said, she could always withdraw if her conscience got the better of her and music seemed to be interfering with her science. 'Anyhow I can't imagine how I'll get past the audition which is sure to be tough sight-reading and I'm so out of practice. But no harm in trying, only die once etc.' In fact she was accepted for the orchestra, which was to meet in Wellington during the May holidays.

The Metcalfs were interested in her choice of subjects for the year. By now they knew her well, certainly better than other staff members did, and Wal Metcalf firmly assured 'the traitor' from his course that her proposed studies would not stretch her at all. She should do German as well. 'I'd rather be overstretched than flabby,' she commented to her family, and she gained a pass in Science German that year. The other Metcalf suggestion, that she should try to do a course in Russian because both German and Russian were necessary for reading scientific literature, was not taken up. The hours in her days ran out.

Nita arrived for the beginning of lectures to find Beatrice glowing with the excitement of possession: a flat of their own in Shakespeare Road, 'and the landlord's Mr McBeth!' She had warned Nita that there would be a third person in the flat, someone who had been around for a long time and who would be sleeping in Beatrice's room. Nita was curious and a little apprehensive. This person turned out to be Edward Hill's childhood teddy bear, much worn with affection, and perched on top of a cupboard. He would act as a chaperone, Beatrice said, when boyfriends came to visit. There was no man in her own life, but everyone who knew them expected that Nita would marry David Hanna eventually. Nita's happiness in his company filled Beatrice with warmth, too.

Dave Hanna was considerably older. Mature and universally liked, he had come to university with the idea of becoming an Anglican priest, but had decided against this by the end of his second year. He completed an arts degree in philosophy and psychology and, when Beatrice met him, had embarked on accountancy studies while retaining a deep interest in philosophy.

The Shakespeare Road flat was furnished, if sparsely. Beatrice had negotiated the rent to four pounds a week, including electricity. While the flat was further away from the university than they were accustomed to, they both had bicycles. 'I'll have to get a book and learn to cook,' Beatrice had told Nita. She had had little experience. Practical cookery and nutrition were not taught to girls in the top academic forms, and she had gleaned only haphazard notions of housekeeping at home. For the first time in her life she gave some thought to the business of eating. Valuable time and effort would be saved if basic supplies were kept in stock and regularly replenished, and she would follow each step in a good recipe book. Yet even a well-recommended cookbook had disconcerting gaps. 'How do you thicken with flour?' she asked Nita.

The textbooks Beatrice needed in 1959 were expensive. She told her family that she had already spent £12 and had still to buy more – this at a time when the average weekly wage in New Zealand was £14 18s. She had been given £3 by an aunt, her mother's sister, Helen Hannay, and had used it to buy a special book:

> I know I was going to spend it on a record, but then I saw in the varsity bookshop this gorgeous book for £2 18s 6d and I had to get it. It's an American book and therefore beautiful, called *Theories of the Universe*, and is a collection of writings from Plato to Einstein and Bondi, wonderfully put together and arranged, and being historical it won't date.

In her progression towards becoming a cosmologist, this book was a major factor. Beatrice kept it throughout her career, referring to it often and sometimes using excerpts to introduce her scientific papers. She was to say that it was Fred Hoyle's books which pointed her towards astronomy, together with growing dissatisfaction with the explanations of traditional religion. This new book, *Theories of the Universe*, published in 1957, builds dramatically to the most recent theories. 'I'm thrilled and dying to read it; 425 pages of small print,' she ended her long letter of 1 March.

And, in a powerfully prophetic sentence, she signed off: 'One could more or less add to it oneself.'

CHAPTER FIVE

WHAT IF

'It's hard to say what I might end up actually studying, because when things get moving in cosmology the whole aspect of the subject seems to change overnight.'
BEATRICE TO HER FAMILY.

Music remained the safe harbour for Beatrice and her parents. She could write genuinely of her ever-increasing joy and release in performance and in listening. Subjects deliberately omitted from letters home included philosophical arguments, particularly about religion. There would have been no point in trying to explain why she no longer considered herself a Christian, even if she had been prepared to risk hurting her parents. Moreover, all the time her fields of study became further removed from her family's possible understanding.

She must have been concerned at times, however, about the lopsided picture she was painting of her life. After 17 pages about her music and musical matters, she wrote on 18 March 1959:

> I suppose you get the impression I'm not doing any work at all. But actually I am very absorbed in it all and working hard when I do work, which is enough of the time, and there's no point in me telling you how thrilling electronics is or Lagrange's formula or something like that because it wouldn't get you far.
> Maybe it might get me somewhere some day.

Sharing a flat went well from the beginning, as Beatrice was sure it would. She had told her family that she and Nita were both assiduous cleaners and tidiers as well as being independent, and unlikely to be under each other's feet too often. They decided to take turns, week and week about, at getting the meals, and they shared the housework. Both had heavy study loads;

Nita was doing chemistry for a BSc.[1]

Their study habits were different. Nita would get up 'at a normal hour', about 8am. Beatrice would have been up for two hours, have breakfasted, made her bed and be deep in her books and papers which by 8am would cover the floor though in an organised fashion. Nita recalled:

> She was so considerate. She never imposed. I'm not musical so she'd never practise when I was at home.
> Often Dave would bring me home about 10pm and we'd hover outside, listening to the glorious sound of Beatrice on her violin. She'd have been embarrassed if she'd known.[2]

Nita and Dave, increasingly engrossed in each other, were careful lest Beatrice feel an outsider. Nita was happy that her man and her friend liked each other so much, and that Beatrice could shed formality, and play the fool in their company. The two young women enjoyed ganging up on Dave. He remembers an occasion when they served him a very large meal. 'My mother taught me to eat everything on my plate, but they kept piling on more and more. I managed, but at last they giggled and confessed what they were doing. I spanked them.'[3] Beatrice did not consider a playful spanking inconsistent with her views on female equality. Part of her was still a little girl playing house: she was a young 18.

Nita remembers only one occasion when Beatrice became distressed, almost tearful. It was Nita's week to be responsible for the meals. One portion of stew was left over from the day before, so Nita bought a serving of something else to eke it out. 'Beetle was upset because we were eating different meat meals. She thought we should both be eating the same thing. I was careful not to let such a situation develop again.'

Nita emphasises Beatrice's sensitivity to the feelings of others:

> Beetle was more likely to read people correctly than I was. She was very intuitive. She might say, for instance, 'You're not feeling very well, David,' and I wouldn't have noticed.
> She was light-hearted, so unselfconscious. She did things because she wanted to, using every minute. She spoke and moved so quickly and she had a bouncy, happy walk. She was happy with her whole body.
> Beetle was so intelligent that she defused any situation before it could blow up. Her rationale was that one could have more fun if everyone got on well together. She had had practice at trying to defuse situations at home.[4]

In these Canterbury years Beatrice again impressed many people as being like a Dutch doll, with her very pale skin, red cheeks, black hair and large sparkling eyes. Ruth Metcalf (later Ruth Beauchamp and a professional musician) remembers her young babysitter as full of fun and very thin, in dark clothes – black fitting slacks and a black jersey. And besides the red cheeks there were red, chilblained hands.

Fellow student Jan Heine saw her as:

> ... willowy, slim, very dark billowy hair and milk and roses skin. A small mouth that widened mightily when she smiled, and even teeth. Intense eyes that seemed to see through you. She did have a liking for pranks, little outrageous episodes.
>
> And very determined. Mostly she was diligent, a fearsome awesome diligence with her work. Lesser maths students would ask for help and she would have the problems sorted out in no time. She worked not because she had to but because she loved it, because she wanted to solve another math or physics problem. I think too that she read a good deal of philosophy because those Mobius strips and spatial maths took in the concepts of time and space, of the origins of the universe and movement of the planets and stars, ourselves.[5]

To her parents, who had taken Theodora to England, Beatrice could write freely about her feelings when she was alone in the Shakespeare Road flat, working and listening to a recording of the Emperor Concerto. 'Great joy and beauty – soaring peaks of perfection – utterly inextinguishable.' She was preparing for the following week when she was to hear Andor Foldes and the National Orchestra perform this concerto.

During the May holidays she was working in Wellington with the National Youth Orchestra, begun by conductor John Hopkins in 1954. All members had to be less than 21 years old. On this occasion he had to try to weld 78 young musicians, half of whom were still at school and mostly had never even seen one another before, into a coherent whole with purpose and flair, and do this within a very limited time. The plan was to come together in the May and August holidays for practice, then perform publicly.

Beatrice was blissful. Her friend Pip was in the orchestra, too, and she was happy being billeted with the Saunders family in Kelburn. Penny Saunders, who was also a Komlos pupil and whom she had met at the Cambridge Music School, was another who became a friend for life. At the Saunders's home there was 'no sticky conversation', she wrote. Asked what Beatrice could have meant by that, Penny Saunders (who married another Saunders, the musician Francis Saunders whom she met when studying at the

Guildhall) was bemused and said that they were just an ordinary family.[6]

The orchestra practised for three hours every morning, and for another two in the afternoon. Then came an informal concert for an hour, when friends and often those who had previously taught members of the orchestra were a welcome audience. Beatrice was at the back of the first violins. At first she found it something of a struggle, especially parts of Beethoven's First Symphony. She called conductor John Hopkins 'super', and Vincent Aspey, leader of the National Orchestra, who gave the first violins a couple of tutorials, was 'like a good-natured Hoffnung character – "take it easy, you can do it, eh?"' She found she could, and that she learned a lot.

On top of all this practice there was the daily meeting with the audiences at the informal concerts and then the evening social life. She was trying to get to bed relatively early, she wrote, so she would be fit for 'some solid mental work' when she returned to Christchurch.

The most important single source of education in Beatrice's university years – education in its true sense of leading out rather than putting in – was the club known as Soc Soc. It rarely had more than 30 members, and had been founded in 1956 as a philosophical society for non-Christians, although Christians were not barred. Ethics and aspects of morality were its staples, and the style of argument was light-hearted and witty.

Brought up on a diet of moral issues as seen through Oxford Group eyes, Rowena and Beatrice were both drawn to Soc Soc. Most of its members had viewpoints diametrically opposite from those of the Hill parents, but they took issues equally seriously. Beatrice became an active and regular member, and Soc Soc was where most of her non-musician university friends were to come from. They included Robert Ludbrook, David Lorking, Guff France, David Hanna, Nita Nitschke, Rae Julian, Brian Lilburn, Mark Sadler, Brian Tinsley, Ann Ballin, Don Locke and John Young. Rae Julian, who became commissioner of the Human Rights Commission, has noted that Soc Soc produced many who were to become prominent in New Zealand's public life. From the beginning, Soc Soc members saw that Beatrice had an acute analytical intelligence which could easily demolish any ill-considered arguments. Rae Julian has said, 'What I remember most vividly is Beetle's grasp of an issue and her persistence until she had teased out every aspect of it.'[7]

Now Beatrice sometimes asked Nita not to talk to her for a while. She needed quiet weeks. 'I've got a big think on.' The subject might be religion. She could find no intellectual basis for faith and had already decided that, logically, she could not believe in the tenets of Christianity. But she gnawed away at the big questions that Soc Soc delighted to discuss: nothing less than the meaning and purpose of life. 'You're lucky, Nita, you're a Christian,' she said. Nita, who has remained a Christian, has since said, 'She couldn't just accept the idea of faith. She thought you could work it out logically.'

A subject always safe to expand on in letters home was the natural world. The Hills, Edward in particular, had guided their daughters to an enduring delight in nature. Spring with its daffodils had reached Christchurch, and even in Shakespeare Road behind the railway station the birds sang:

> The birds have started to wake up about 6am – impossible to sleep after that, it seems, however late the night before! But I've discovered how to get unlimited (nearly) energy: even just 10 minutes flat on the back in total flop (mental and physical) has incredible effects. Last night I had 10 minutes' gap in practice before the concert, so I stretched out on a table and nearly went to sleep. Good habit to develop.[8]

That exclamation mark in 'however late the night before!' again reminded the family that she went to parties as well as making music and studying till late. Members of Soc Soc were frequent companions. She was always ready for fun, which to her meant good talk and laughter with congenial people. 'Beetle was very affectionate rather than sexy or sexual,' Nita has said. 'Sexy jokes were a bit suspect. They didn't happen in her presence.' Sex was a topic for discussion rather than practice, and conducted mainly within Soc Soc.

Some of the society's discussions as well as her own reading had convinced Beatrice of the dangers of the world's becoming over-populated. She took this threat seriously enough to become a volunteer at the Christchurch clinic of the New Zealand Family Planning Association, but not so seriously that she could not joke about it. Historian and biographer Gordon Ogilvie remembers her dancing around her flat, waving a chunk of foam rubber in the air and declaring that she was going to convert the Fijians to cheap contraception – presumably a Fijian client had been worried about the cost.[9] Beatrice could dance and fool around as if she had been drinking. In fact she drank very little at any time, but with her red cheeks and sparkling eyes and her look of being thoroughly involved she often seemed intoxicated with life itself.

With end-of-year exams looming, and her parents and Theodora due back from Britain in October, she took time to write a light-hearted account of a party she and Nita and Dave had been to.[10] Nita was lucky to have Dave to look after her, and spend an evening with: 'There are some people who really know what they want, and stick to it.' She always rejoiced in Nita's happiness. Her own deepest yearning was to be an indissoluble part of a greater whole, but this was mostly pushed to the back of her mind.

Now, to her family, she said her immediate focus was on physics and the work she might do in the coming year. She had been thinking seriously of theoretical physics, and had discussed this with the head of the department, Professor Alister McLellan. An electronics lecturer, Tom Seed, had given her some International Atomic Commission material about fellowships in various countries, so she could see the sorts of opportunities open to theoretical physicists. 'There are plenty, but one has to be really good.' First she would see how well she did in the imminent exams. She would have to consider taking on three units in 1960, and in the meantime was determined to do her 'ultra best to do some decent theory in exams'.

The Hills returned to live in Strathmore in Wellington rather than in New Plymouth. Jean Hill was a member of the New Zealand Women Writers' Society, and was gratified by her success in its competitions and similar regional writing contests. New Plymouth seemed isolated, and she wanted to be more in the centre of the society's orbit. For his part, Edward had been at a loose end since losing the mayoralty, and his researches for a proposed book on 19th-century Taranaki and the relationships between Maori and colonists could now be conducted among the contemporary records in Wellington.[11] He was also invited to chair the Wellington regional committee for state house allocation.

The house they bought was high on a hill at 31A Tio Tio Road, with a panoramic view across the airport to Cook Strait and a private 'railway' to spare everyone the steep access climb. Theodora would be studying music at Victoria University of Wellington when she finished her last year at New Plymouth Girls' High School.

This summer was the Hills' first without Rowena, who was exhilarated by her adventures on the way to Florence. The rest of the family enjoyed their usual relaxed time at the bach beside Lake Taupo. Beatrice's exam results, as ever, had been all A passes with marks in the 90s. She went for quiet walks around the lake to think about what she really wanted to do, the possibilities of the vast opportunities opening out for her. For the rest of her life, whenever she heard lake water lapping, she thought about her decision and the exhilaration of that time. She would become a cosmologist and try to find answers to the biggest questions of all.

She had just turned 19.

But whatever was a cosmologist? Her family was plaintive with questions. In a long letter at the beginning of her final BSc year, on 6 March 1960, Beatrice wrote:

> Cosmology is really theoretical astronomy; studies of theories about the universe, its origin, structure etc (that's all!) Relativity theory is absolutely basic to it, and a lot of the work that goes on in it. I gather

it is so mathematical it gets away from 'reality'; apparently why it's necessary for me to keep up with physics.

It's hard to say what I may end up actually studying, because when things get moving in cosmology, the whole aspect of the subject seems to change overnight. Prof Lawden told me that the science is due for some new advances pretty soon; I think because a lot more observational data is being analysed.

The fascinating thing about theoretical physics is that you can never learn about it fast enough because there's always more being discovered to learn! And you have to specialise as soon as possible, otherwise you'll never get to the frontier of any branch of knowledge and so be able to have some new ideas.

Now I feel everything I hear is really getting me somewhere, and all useful. It's different to have a definite aim ahead, instead of vaguely wanting to do research and not knowing what.

In the same letter she wonders what her thesis subject the next year will be; probably nothing very astronomical because the subject was not studied much at Canterbury. 'They all do either ionosphere or solid state,' as far as I can see.' Meanwhile she was working hard at two units of maths and one of physics, and already the pressure of reading was building up.

The great cosmologist-to-be also talks about the Horrocks roses dress she wore to a party: 'utterly beautiful and crisp and white, very dancey'. The multi-talented Pip had made her a sheath dress and jacket. Beatrice takes a page to draw a diagram of dress, front and rear, complete with pin-tucks and gold-edged bows. 'Pip's been arguing, of course, about not taking any money from me but I know she needs it and it was a long job. So if she won't take cash I'll buy her something.'

It is possible to detect an extra certainty in Beatrice's letters from this point, that life was now 'different' because she had the definite aim of cosmology. Certainty and a determination to crowd yet more into her days and nights, together with a new note of excitement, are revealed in an account of four days in her life, from 30 April 1960.

It was a wonder she was still awake at all, she wrote, as she had been out four nights in a row: playing in the university orchestra, at a concert, at a Soc Soc panel discussion, and playing in her chamber music group with the Metcalfs. 'Heaps of things are Happening here' – she gives it an emphatic capital – 'at such a rate that I don't know which way up I am, just about.'

For intellectual excitement it was the arguing at Soc Soc she wanted to share. In terms the Hills could relate to, she tried to convey the essence of a meeting. Arguments came from senior philosophy students who represented Soc Soc and the Student Christian Movement respectively, the university's Catholic students' priest, and the Evangelical Union. 'You can imagine they knew their stuff and how to talk about it. The Evangelical Union chap is a boy in our physics class who's really very clever but couldn't get more profound than "We believe ..." so he didn't add a terrific lot to the discussion.' The chairman was the university chaplain, the same post held in the 1940s by the Revd Merlin Davies, who with his wife had been instrumental in turning the Hills' thoughts towards New Zealand as their future home. She was 'terribly impressed by the SCM representative, who holds very unorthodox Christian views – knowing a lot about philosophy and so on – and said things fit to shock a lot of Christians (and probably did for the Evangelical Union!)'.

It was the Soc Soc representative who really impressed her, though. Don Locke,[12] son of the well-known Christchurch communist Jack Locke and peace activist and author Elsie Locke, was an atheist. This, Beatrice told her parents, was not quite typical of Soc Soc, 'but then you couldn't find any one single representative who *would* be typical of Soc Soc.' He and the SCM representative were able to argue on equal terms, Don Locke having a particularly quick mind when it came to refutation of argument.

One of the most interesting questions discussed, she thought, was what difference belief in 'a divine Being' would make in one's attitude to morality. As far as she could see, the result was much the same whether one considered oneself responsible to society or to God. 'But an awful lot was said on the subject, and then on "Absolute Moral Law".' For her own ongoing consideration, she took notes of many of the arguments.

Life was so full, with so many opportunities, that Beatrice was well aware she must limit some of the social events crowding in. She and Nita seemed scarcely to see each other in the flat except at the evening meal. This was to be the time, they had decided, rather unrealistically, when they would study. Beatrice sounded a little surprised when she said, in a letter home, that generally they could not resist talking at this meal, adding that the time everyone spent talking, especially in the Students' Association cafeteria, was quite incredible – 'but it makes life'.

It was bliss, she found, to be able to open one's mouth and let the thoughts come out, to be taken seriously, listened to and respected by way of agreement, qualification or reasoned argument – not that it was always reasoned. Foolishnesses could be trotted out, too. Wit and flippancy were often goals in themselves. Above all, for Beatrice, was the heady realisation that there was no single, absolute, revealed truth, and that she herself could be on the pathway to discovering more truths.

About this time she realised, too, that, for her, physics was more important than mathematics. 'Maths is the essential tool of the trade and necessary as such, but I haven't got sufficient interest in the analyses etc of pure maths to want to do nothing else; sufficient to know people can show the methods are justified.'[13] This is similar to the attitude she was to adopt when she became an astronomer. She was never inclined to peer through telescopes – that could be left to people who liked it. It was what could be done with these telescope discoveries that was to excite her.

In the meantime, she marvelled at the amount of physics they managed to cover each week, in terms of textbook pages and notes. 'And it is really interesting – I love it.' Her next year would be a treat with nothing but physics theory, not even any laboratory work to eat up the time.

One of the Soc Soc crowd, Brian Tinsley, belonged to the Royal Society of New Zealand and so had tickets to the Rutherford Memorial Lecture. He invited Beatrice. The lecture, on the history of X-ray spectroscopy, was given by the great experimental physicist Sir Lawrence Bragg, who had been a successor to Lord Rutherford in two posts. It was an extremely formal occasion, with nearly everyone in academic dress, and admittance by invitation only or through membership of the Royal Society. Beatrice, sitting at the back with Brian, felt very fortunate, as she wrote in a long family letter.

> Sir Lawrence and his father discovered the basic phenomenon of this science,[14] and he has led its progress all his life. The technique is used to unravel the structure of molecules in crystals, and he outlined the events from when he first discovered the structure of sodium chloride (a very simple cubic lattice with only 2 kinds of atoms on alternate corners of cubes) till its final triumph: a model of the myoglobin molecule with some 10,000 atoms in it! He had a scale model here, with every atom in its place. (Myoglobin is one of the basic molecular substances.) He also showed us slides of nearly-as-vast haemoglobin and oxy-haemoglobin, showing how the blood carries oxygen to the cells of the body.
>
> He has a most lovely personality, very simple, direct and humble – and full of humour in his talking in spite of the gravity of the occasion. No long words or complicated rhetoric, and ultra-understandable. It's exciting to hear a great man like that!
>
> We've had a feast of great scientists. The next night Sir Alexander Todd gave a public lecture. Perhaps you've seen in the papers in recent times about the great discoveries these Cambridge chemists have been making about the structure of cell nuclei, and the actual chemical processes of heredity and reproduction of life. (I admit I was terribly ignorant. Most people seemed to have a general

knowledge of the subject.) Anyway Sir Alexander Todd is the leader in this field, and one of the most famous organic chemists. As far as could be explained without getting too technical, he told us about the structure of the nucleic acids. These molecules have tens of times more atoms even than the myoglobin, and X-ray work has just about unravelled their complete structure. (What is especially marvellous is that now X-rays can reveal the structures of molecules from the very smallest up to these vast ones; and anything as big as or bigger than these can actually be seen in an electron microscope. So we have complete continuity!)

Apparently the nuclei of all living cells (ours) have a protein covering and then contain lots of kinds of nucleic acids, known as RNA and DNA. It is known that the RNA and DNA molecules occur always in pairs, and have the form of two long interwoven spirals, like a rope. These are exceptionally stable (order of millions of years) molecules, and have the miraculous property that if the right elements are present in the nearby protein, then acid molecules reproduce themselves identically, and automatically.

The proportions in which 2 molecules make different 'daughter' molecules explains the Mendelian laws of heredity precisely; and the manner in which some actual atoms from the 'parent' molecules are passed into the 'daughters' shows how radioactive elements and mutations are passed on from generation to generation. The stability of the molecules and the manner of their reproduction seems to explain completely how hereditary factors are carried on through the generations, and sometimes appear again after many; and all sorts of other things likewise! We seem to be digging at the very core of life.

Also very important is the great similarity between these structures, and cancer molecules, and viruses – which has important medical consequences in sight if the research continues at this rate. Perhaps it's also quite alarming to realise that we may soon know how to induce artificial mutations of genes.[15]

From someone who had dropped chemistry after stage I, this is a remarkably full and lucid account of two lectures.[16] Furthermore, the letter was written (without notes) for lay people. This quick perception translated into simple language was to be a strength of Beatrice's work until it moved beyond general comprehension.

She ended this letter by saying she must get back to work as it had taken her 30 minutes to write the account of the lectures, an account prefaced by a couple of pages about her delight in a concert by the cellist Rostropovich, 'the most wonderful artist I have ever seen'. If her estimate of time is anywhere near accurate, and it was usually exact, it is another indication of how quickly she could recall and write. She signed off with 'miles of kisses and hugs', and asked if it was also glorious spring in Wellington. 'You should see the Botanical Gardens here!'

To nobody's surprise, Nita and David Hanna became officially engaged. Beatrice could see that they were ideally suited, as were the Metcalfs, and she rejoiced. She was strongly idealistic about marriage, and serious about what she thought were the important things, a communion of hearts and minds. An undated letter from this time has her visiting an acquaintance who was about to be married, and who had asked for her gift to be a cake-icing set. Beatrice was astounded by the wedding preparations and the fuss over the trousseau.

Throughout the previous few months, Beatrice had been carefully dropping passing references and clues in her letters, such as lunch in the sun in the Botanical Gardens.

Now on 6 October she sent her parents a photo – 'behold two happy people' – and told them her news.

She and science student Brian Tinsley wanted to marry.

CHAPTER SIX

A DIFFERENT LANGUAGE

'They speak a different language together – on an astrophysical plane.'
Jean Hill

Beatrice knew that marriage was what her parents wanted for her; not an irregular relationship. 'We're having the most marvellous times together; getting to know each other and so much pleasure in each other's company; a Good Thing, that's what it is.' By using the tried and true family language, capitals and all, she was putting the situation into the realm of the familiar and the acceptable. Where other parents might have been concerned, or at least have pointed out that she was still a teenager with her academic work ahead of her, she knew that, to the Hills, marriage meant a safe harbour.

Beatrice's happiness shines through her news. 'Golly, life is so wonderful! I'm so happy I could fly.' At this point in her letter she swings off, still flying, to other topics. There is no description of Brian, nothing to evoke his personality. It is a strange lack, especially considering that, after hearing physicist Sir Lawrence Bragg give just one lecture, she could fly into raptures about such detail as his simplicity, directness, humility and humour, and 'no long words or complicated rhetoric'.

Sir Lawrence came to life in her letter writing. As yet, Brian did not. Although Beatrice's happiness was undoubted, it was as if the idea of marriage, not the actuality, was bringing her joy. She was certainly in love with the idea of being in love, of being a close part of someone else.

Naturally she said nothing of this to her parents, but those who knew the couple were distinctly taken aback by the news of their engagement, and said so. Beatrice was so young, only 19, and an inexperienced, unsophisticated 19 at that. The then contemporary American fashion of marrying young had not yet reached New Zealand, and had certainly made no impact on the ranks of serious students with years of study ahead. While

Beatrice's full potential was perceived only in part, and then only by her tutors and a small group of students in her field, it was obvious that she was a special person with a special future. Marriage was surely for much later.

The two people who had most often seen the couple together, Nita and Dave, were initially extremely surprised. Nita, the first to hear the news, recalls being actually amazed. She was totally unprepared for such a situation. 'Beetle came in one night and said, "Guess what? Brian's asked me to marry him!" I said, "By the look of you, I guess you're going to say yes!" '[1]

Nita and Dave were comparatively mature, and in their own happiness they could now rejoice in that of their friends. And Beatrice was radiant. There was no mistaking this. Brian, four years older, sturdy and serious, was studying for his doctorate in physics. Students saw him as an earnest seeker of truth who liked to talk within a scientific framework. He was also a well-regarded Soc Soc member, idealistic and a keen debater. 'He looked dazed by his good fortune in securing the wonderful Beatrice.'[2]

Brian Tinsley was born in Wellington on 23 April 1937. His parents, Nola and Terence, lived near Upper Hutt at Kaitoke, a farming district north of Wellington, where there was a one-teacher schoolhouse. He was the focus of his mother's energies and hopes, affectionate though she was to her daughters and to her husband, who worked as a plasterer and decorator. A strong-minded woman, Nola Tinsley ran a pious and orderly home. She attended the evangelical Church of Christ and expected her children to do likewise.

In the Tinsleys' home it was taken for granted that the males were the workers and their women were happily domesticated. Then marriage would see the pattern repeat itself. As a Canterbury student, Brian was later to write about 'strong and stable family units', obviously with his own in mind.

At primary school Brian developed asthma so badly that his mother decided to bring him home and teach him herself for a while, through the New Zealand Correspondence School. 'This gave me the solid academic grounding for my career.'[3] He had one year at Wellington Technical College, coming first in the trade engineering course. Then, in search of better work prospects, the Tinsleys moved to New Plymouth, just before the Hills arrived there from the South Island.

Brian did well at New Plymouth Boys' High School. He was promoted to one of the school's top streams for his sixth and seventh form years, to prepare him for a possible degree in civil engineering. Anyone who has changed to a different curriculum knows how hard it is to catch up on all the work which classmates take for granted. Brian worked with great determination to reach the standards expected in this top class, particularly benefiting from the physics teaching of R. C. (Horsie) Wilson: 'He ensured we weren't learning by rote.'[4]

Beatrice, in contrast, despite the deficiencies in science and maths teaching from which virtually all girls' schools suffered at that time, had been educated without interruption and mostly in very favourable circumstances, with recognition and support from all her teachers. That Brian managed to succeed, particularly in physics and mathematics, and that he came third in the province of Taranaki on the credit list for the University Scholarship examinations in 1954, was proof of his unusual persistence as well as his ability. He was awarded a Taranaki Scholarship.

Brian Tinsley was never to grow very tall – something of a handicap in a school as macho as his, where rugby footballers were the gods. He found his niche in mountaineering. Members of the Taranaki Alpine Club realised that he was an unusual boy, and encouraged him. A particular mentor was the editor of the *Taranaki Herald*, Brian Scanlan, who was also the author of a book, *Mountain of Maoriland*, about what was then called Mt Egmont.[5] Later, during his university holidays, Brian and a companion, Hans van Beers, set out to climb the mountain three times in one day, 17 January 1959, a record-breaking event which made the news pages of the *Herald*.

Brian's account of this climb, written for the Taranaki Alpine Club's journal, *With Axe and Ski*, gives insights into his character. He and Hans were lucky in having ideal weather. The climb called more for dogged persistence than for climbing skills above the ordinary, perhaps, but by mountaineering standards it was a considerable feat. The two men set off from the Old House at North Egmont and reached the summit in two and three-quarter hours, faster than either had climbed the mountain before. They kept up the pace for their next two ascents.

Characteristically, Brian appended statistics to his account.

Water drunk: estimated 2 gallons each.
Water sweated: estimated 3 gallons each.[6]
Distance walked: about 26 miles.
Vertical height ascended and descended: 15,700 feet in 16½ hours.
Number of steps taken: estimated 60,000.
Times for ascents: North Side, 2 hours 47 minutes (5000 feet);
Dawson's Side, 3 hours 15 minutes (5300 feet);
Stratford Side, 3 hours 52 minutes (5400 feet).

The Tinsleys and the Hills moved in different circles in New Plymouth, and did not meet while both families lived there.

When he went to Canterbury, Brian lived at Rolleston House, one of the recognised residences for male students. He had already switched from engineering to science, and, like Beatrice, he took time to decide what his specialist area would be. Ongoing study was revealing so many possibilities. Brian settled on upper atmosphere physics as his field of interest, using a spectroscope as the tool for his research. He began his studies of the light of the night sky, which was to be the subject of his MSc research and his PhD.

It was a field in which he had to work hard in every way, and the necessary observations at night cut across ordinary student social life. In fact, he was scarcely involved in this socialising until he became a founding member of the Socratic Society in 1956. Soc Soc answered many of his needs: 'It was such a liberating experience. Instead of talk of sin and damnation there was rational argument.' His first and only girlfriend before Beatrice, Karen Peterson (later Butterworth), was also a Soc Soc member who, like Brian, contributed to the society's magazine, *Hemlock*. When this relationship ended, he spent many weekends away mountaineering, so when Soc Soc members went to Arthur's Pass it was natural for him to take the lead in climbing activities.

Brian had known who Beatrice was from her first year at university, 1958, as he had been a part-time demonstrator in the junior physics lab. By the time he began to get to know her socially, at a Soc Soc meeting, he had left Rolleston House and was flatting. He had also done a stint in Fiji, teaching physics to Islander medical students at the Central Medical School while writing his MSc dissertation. 'By 1959 I felt I could start dating. Beatrice had gone out a few times with a physics student, Bill Moore. I checked with him that they weren't still going out before I asked her.'[7]

Harvey McQueen,[8] a student who had been at Rolleston with Brian for three years, was doing an MA in history and training as a secondary school teacher. He was flatting with Bob Patchett, another who had been to New Plymouth Boys' High School, and whose field was psychology. One day McQueen had happened to meet Brian in the street, and invited him to share the small flat. As he recalled:

Brian was the stereotyped absent-minded professor. Bob and I were good cooks and we all took it in turns to cook, week about, but Brian was appalling. The food would burn while he gazed at the stars. Greggs Instant Puddings and mince were the menu in Brian's weeks, so Bob and I lived on fish and chips. We wouldn't turn up for his big pot of mince, and it would go off.

A neighbour came in to clean us up, and brought her vacuum

cleaner. It wouldn't go. Brian said he'd fix it, and took it to bits, then re-assembled it. Still it wouldn't go. Then someone checked the fusebox – that's where the trouble was. Brian was really absent-minded. He'd forget to wash his socks.

Essentially Brian was a loner. The three of us had to sleep in the one bedroom but I never felt intimate with him. He never talked about his feelings. Bob Patchett and I talked openly. He never joined in. He didn't notice other people's emotions or needs. He wasn't interested in politics, or the world around him. And God didn't exist, so the matter was closed. Brian had solved the problem in his own mind.

He always took things back to first principles. He prided himself on his logic.[9]

Brian's unawareness of people's feelings and of local and national issues often surprised his flatmates. They thought he simply did not notice what was going on around him. One instance was Finance Minister Arnold Nordmeyer's 'Black Budget' in 1958, which led to the downfall of the Labour Government in 1960. Rumours had been circulating that the Budget would increase the tax on beer.[10] As it seemed it might be their last chance to have a drink before prices went up sharply, Harvey McQueen and Bob Patchett had bought themselves stout and oysters to enjoy while they listened to the presentation of the Budget on radio. There was indeed a very heavy tax on beer, a tax which Labour's main supporter, the New Zealand working man, took as a personal affront. As McQueen described the scene:

> Brian came in while we were listening to the radio. 'That's the end of the Labour Government,' Bob said, and of course it was. 'Why?' Brian asked. He was naive politically. He just didn't know. That was such a savage tax, then. He couldn't understand other people's feelings. He lived in his own world.

Brian Tinsley was able to help Beatrice fill in the gaps in her science education at secondary school as he, four years older, had covered these areas at university. Harvey McQueen observed how very quick Beatrice was, how she always went straight to the point, compared with Brian's laborious long sentences. Hard and diligent work was Brian's forte. The engagement between two seemingly very different people startled his flatmates.[11]

Someone else who got to know Beatrice about this time was an arts and law student and Soc Soc member, A. H. (Guff) France, who, with his future wife, Necia, became a lifelong friend. 'When they became engaged, she was blinded by romantic love. To her he was 10 feet tall, and wonderfully intelligent. She dropped off that plane but I do think she was happy at that time.'[12]

A significant part of her happiness was being able to talk within a scientific framework Both Beatrice and Brian placed huge importance on the scientific viewpoint, particularly in the questioning of religion. Their courtship was largely intellectual but included great affection, mutual gratitude and hope. There had been no hugging or affectionate touching in Beatrice's life after Nanny Gullidge had gone back to England. The women in the Tinsley household were more demonstrative, but Brian was from a milieu where males had problems displaying their feelings.

It does seem that, without consciously realising it, Beatrice and Brian each saw in the other, as people in love usually do, the perfect complement to and fulfilment of their needs. Brian saw his stature enhanced, his life turned around, by marrying what others called 'this shining creature'. Beatrice saw someone, at last, who would fill the emotional crevasse which reached to her foundations. Unknowingly, she was asking an enormous amount of Brian. Almost certainly it was too much to ask of any man, certainly someone of whom his flatmate said, 'He didn't notice other people's emotions or needs.' It did not help Brian that ultimately Beatrice was asking even more of herself.

Another key to the situation may well have been the simple fact that Brian actually asked her to marry him. Colin Broadley had asked her to become engaged, and she had said yes. She ended that engagement, but now Brian asked her, and she said yes. The yearning to belong, to be part of a greater whole, ran very deep.

John and Teddy Fardell would always be special to Beatrice. They were her link with Nanny Gullidge, and they were her 'other parents', the source of much of her childhood warmth in her first years in New Zealand. Now, with her engagement, she eagerly took Brian to meet them.

John Fardell, looking for things they had in common, as so many others were to do, said jovially, 'You're both in the same line.' 'No, Brian isn't in my line at all,' Beatrice said definitely. Teddy recalled that Beatrice's voice was quite sharp here. This added to the unease Teddy was already feeling. There was something about this relationship which disturbed her. Somehow they did not seem to be on the same wavelength. When she and Beatrice were alone together, Teddy began by remarking to Beatrice that she had had very few boy friends. None, really – Teddy scarcely counted that one-time fiancé. And she was still so very young.

Beatrice took her up on that, saying, 'You don't seem very excited about our engagement?' Teddy was not. She asked Beatrice if she really thought Brian was the man for her to marry. He had been brought up in one environment, and she in quite another. They were so different. He might bore her, and get on her nerves. 'I don't think so,' Beatrice said quietly.[13]

Brian Tinsley was similarly inexperienced. Nobody in his family circle or among his student acquaintances was in the least like Beatrice.

Teddy Fardell was the only person to try to dissuade her. Probably the adult who knew her best, she had watched Beatrice develop from childhood, and thought she could look dispassionately at this new relationship. Her unease hardened into anxiety. However could this work out? Beatrice did not appreciate Teddy's reservations, and a coolness developed between them. They did not see each other as often as before.

The Hills blithely gave the engagement their blessing. It seemed so suitable – two scientists. They planned to drive north from Wellington to New Plymouth to collect Theodora at the end of her last year at school, so Beatrice arranged by letter that she and Brian would drive up with them, where she would meet Brian's family. It would be a good way for her parents to get to know Brian, she wrote to them happily and confidently. She was particularly delighted that they had offered engagement parties even before they had been officially asked if the engagement could take place.

Edward had been appointed a part-time tutor in history at Victoria University, although a hoped-for permanent post did not eventuate. Jean was immersed in meetings of the Women Writers' Society and in her own work. Her play about the adjustment of British immigrants to the New Zealand way of life, *The Sky Is Higher*, had won a competition sponsored by the Rotorua City Council. A novel, *The Sun at Noon*, had been published by A.H. & A.W. Reed in New Zealand and by Herbert Jenkins in England. Now another novel, *A Family Affair*, had been accepted in London.

Brian had picked up enough about the Hills to have some misgivings about his reception, but they were gracious, accepting and welcoming. But what about Brian's prospects? In time-honoured English tradition, Edward Hill conducted the young man to his study so that Brian could formally ask if he might marry Edward's daughter. Edward was to say years later that he had not been expecting anyone like Brian as a future son-in-law, but he seemed 'a decent, solid chap'. He asked how Brian could support her. This was a poser. Brian had no money beyond what he earned in the holidays to support himself for the term ahead. There were hopes of a fellowship but it was far from certain. As he later told his flatmates, he blurted out that he thought he could get her a better flat to live in.

This was enough.[14] The Hills thought Brian was going somewhere, even if they could not imagine where. They laughingly protested that all this science talk was way beyond them. Beatrice was eventually to say, sadly, that not only did her father not try to understand her work, or Brian's, but also – and this she returned to more than once – he did not want to understand. Edward would get as far as the concept of black holes and then draw back. His religion did not permit him to look further.

But for the present her mind was tuned to the kind of happily-ever-after romance of her schooldays' expectations. Nobody could doubt her happiness. Friends recall seeing the usually rather staid Beatrice actually sitting on Brian's knee. What she wanted was to have a ceremony where they became engaged; that is, to have the ring put on her finger at a small family party, she wrote. This could be a day or so before Theodora's break-up at the school, which they would both attend. She and Brian would like all the Tinsleys to be taken out to dinner, and then to spend the rest of the evening at their home, getting to know one another. 'Golly, it's exciting!' she ended.

Lesley Aderman remembered this December 1960 engagement dinner well. She was waitressing during the long university vacation and was delighted to be able to serve their table. 'Beatrice and Brian were so very much in love, in a world apart. As Mrs Hill said, "They speak a different language together – on an astrophysical plane."'[15] They themselves would have described their different language as rational and logical, with a framework of proven fact instead of leaps of faith. Yet it was a language with subtexts of which the two people could scarcely be aware.

Brian joined the Hills at the Hatepe bach for a short time. Theodora showed her appreciation of the occasion by making them a 'togetherness' straw hat with two crowns and one large brim. The two of them went on what must have been some of the happiest walks of their lives, with Beatrice delighting in showing Brian her favourite bush trails and lakeside pathways. As she did not like climbing hills, the trails were mostly flat. She admired Brian's mountaineering skills but had no ambition to try to join him.

The final examination results for Beatrice's BSc were, as ever, all A-grade passes. For her MSc thesis in 1961, she was eager to work on an aspect of cosmology or astronomy, but expected to have to choose something else because those fields were not much taught at Canterbury. She went back to Christchurch well before the beginning of the university year, as consultation over her MSc thesis topic was urgent. She would normally have discussed this with Professor Alister McLellan, head of the physics department, but he was about to go on leave. First, however, he stressed that her topic should be within the department's scope.

Dr Archie Ross became her deputy supervisor. He had come to Canterbury from Oxford in 1959, with interests in plasma physics:

I have distinct recollections of how Beatrice bubbled with enthusiasm when she came to talk to me about cosmology and her MSc. I was regretful. I couldn't see how I or the department could help her do justice to this choice, though I could see she'd do well whatever she did.

Nobody at Canterbury had expertise in cosmology then. She would have been working in a vacuum. It was not thought to be wise. The strengths of the physics department in those days were solid state spectroscopy and upper atmosphere physics. Two of the staff were interested in astronomy, but really I thought cosmology was a bit risky. In hindsight, we should have let her go on without supervision.[16]

The subject eventually settled on for her thesis was 'Theory of the Crystal Field in Neodymium Magnesium Nitrate'. This is a compound of one of the rare earth elements, and she was to base her interpretation on the free ion researches of Dr Brian Wybourne. It was a subject that had no bearing on cosmology, but would be a valuable introduction to research, and to the use of that new tool, the computer. Beatrice would have to apply a new theoretical model to explain many of the observed features of the optical absorption spectrum and magnetic properties of the rare-earth compound, neodymium magnesium nitrate. This, she was told, would be invaluable preparation for her PhD dissertation, whatever its subject. They arranged that she would go to Archie Ross's office once a week to discuss progress.

Ross looked forward to those discussions. It was rare to come across such a student. Her records showed that she was outstanding, but so did her whole approach. If he asked the class a question such as 'What do you suggest?' or 'What's the justification for this?' Beatrice was almost invariably the first to answer, although never in an attention-seeking way. 'Her tone of voice was almost "Well, my best shot at the answer is ..." and of course she was always right.'

She had the extremely rare ability to take an apparently disconnected fact, something heard in a different context, and put it together with facts from a lot of different fields so that together they made a new perception, he said. She used a notebook in which she would eagerly jot down matters which struck her, even while she was taking part in a conversation.

Archie Ross, like other staff, would observe the students closely at their lab work: 'We would be leaning over their shoulders, observing them, and she was outstanding. Her general demeanour was so charming, too. She was exceptionally brilliant.'[17]

Nita and Dave Hanna had married. Beatrice had been chief bridesmaid and was temporarily in a bed-sitting-room. Brian had not got the fellowship he had hoped for – 'jolly bad luck', Beatrice told her parents – but she insisted they were not to worry about their future finances. Brian was doing seven hours of gardening a week for the family of Soc Soc member Dave Lorking, and he had heard that plenty of students were looking for coaching, profitable work with payment at 15 shillings an hour. They both felt quite safe financially, if not exactly rolling in wealth.

New apparatus to detect the radio noise connected with the auroras and airglow that Brian had been studying needed a transistor amplifier, which Brian was designing. In a heartfelt aside Beatrice said, 'Rather he than I on all that electronics!' Transistors rather than valves were needed because if there were too much local electrical interference then both Brian and the apparatus might have to go way down south to the Snares Islands.

This same letter of 25 March 1961 is largely devoted to wedding plans. Beatrice drew two sketches of her wedding dress, being made by Pip. It was of white organza over net and taffeta. Friends had offered to play chamber music in the church during the signing of the register. An early Beethoven trio for flute, violin and cello, 'Mozarty and very lovely', was their suggestion. This would be much nicer, she said, than having 'items' at the reception when people might be sentimental and polite about them. They had booked into 'a gloriously primitive and perfectly romantic and secluded spot' for the first week of their honeymoon. True to the traditions of honeymoon secrecy, she does not say where. But: 'It's only 49 days today!!! No wonder life feels so good and full of warmth. Sometimes I could burst with just simple happiness.'

Beatrice then found the right flat in which to begin married life, the whole ground floor of a house at 192 Salisbury Street. She at once impressed the owner, who lived with her husband upstairs. Beatrice's current bed-sit was in the next block, and she moved in first to get the place ready for them both. Whenever he could, Brian came around to do some painting and papering, at which he had had some experience.[18]

The landlady, Phyl Wardell, an intelligent, lively woman and a writer, took an almost motherly interest in her young tenant without imposing or becoming too close. 'Beatrice was very circumspect. No living together before marriage.'[19]

They had considered this, however, as Beatrice told Teddy Fardell, who was still politely disapproving of their marriage plans. Teddy was vehement. 'Don't do it. You'll break your parents' hearts.'[20] They did not do it. Beatrice was later to tell Theodora that she 'probably' would have lived with Brian, without marriage, if she had not believed that Teddy was right in her warning about her parents.

The wedding was set for 13 May 1961. Nine days earlier, they were both capped: Brian, MSc with second-class honours, and Beatrice, BSc. She was also one of the two Senior Scholars in her year. Plans for the wedding received Beatrice's full attention. She did not want too much to fall on her mother. The steep access to their family home made it impracticable for the wedding reception to be held there, so it was decided to use the hall next to St Aidan's Anglican Church in Miramar, where the marriage would take place. Flowers, she hoped, would transform the hall's Sunday-school look, and she planned to get up early on her wedding day and go with her father to the market to get great armfuls of chrysanthemums. Her bridesmaids were to be Theodora and Brian's two sisters, Heather and Elizabeth. Brian would be supported by university friends Harvey McQueen and David Boland. Harvey imagined he had been asked to be best man because he had already filled this role for several friends, and 'I expect Brian thought I'd know the ropes.' Harvey McQueen was later married to feminist writer Anne Else. In 1985 she was reading the chapter on Beatrice in *Springboard for Women*, and said to Harvey, 'I bet you've never even heard of Beatrice Tinsley.' 'Actually, he said, 'I was best man at her wedding.'[21]

Edward, still an ordained minister, assisted with the marriage service as well as giving away his daughter, who wore a lace wedding veil that was a family heirloom. Friends of the Hill parents were impressed by the sober aura of science which seemed to cling to Brian. This was something outside their usual ken. For the Tinsley parents, the Hills' assumptions and way of life were strange indeed. Beatrice's friends came as usual from widely different fields of interest, and included members of the National Youth Orchestra and Soc Soc, which received mention in the bridegroom's speech at the reception: 'I said we had got to know each other at the Socratic Society and that we agreed with Socrates who said, "I know nothing, but I know more than others because I know that I know nothing, and they do not." This drew some laughter.'[22]

All in all, it was the kind of white wedding of which the girls of Beatrice's class had talked and dreamed about, not so very long ago, and she, the youngest, was the first of them all to marry.

CHAPTER SEVEN

MR & MRS TINSLEY

'There is nothing so complex that it cannot be explained simply.'
ALBERT EINSTEIN

If Beatrice had given herself some years of scientific work before marriage, or had she been born even 20 years later, she would almost certainly have decided that marriage would not mean changing her name. But she was a daughter of her times and of her upbringing. The more than a hundred papers she eventually published in scientific journals are all signed Beatrice M. Tinsley. Later in her life she was to regret, bitterly, that she had lost the name of Hill. Her reputation had been won under the name of Tinsley, and her peers told her that it was too late to change. The issue continued to be so important to her that even in the very last days of her life she tried to find a way to assert her first identity.[1]

Beatrice expected a great deal from her marriage. She set out on her honeymoon at the height of happiness, believing that she too would now be part of those happily-ever-afters. If there happened to be problems, she would work on them, using all her determination and ability to reason a way through, the way she always worked.

For their honeymoon, Brian had suggested the Marlborough Sounds. They chose Crail Bay in remote Pelorus Sound, a place with gentle bush walks for Beatrice and enough climbing challenge for Brian. They came by mail boat, bringing supplies of food and bush gear to a small cabin. Beatrice was enchanted by the space and silence. In the second week of their honeymoon, when they had travelled on, she wrote to her family that Brian had encouraged her to climb with him in Crail Bay. She had amazed and delighted herself by climbing to the 2600 ft summit, the toughest climb she had ever attempted:

It was nearly vertical and with no tracks except for holes made by wild pigs. The mountain grew as we climbed it – the ridge was extremely foreshortened from below!

We set out at 10am and came back down the last 1,000 feet in sunset at 6pm. The bush along the ridge is some of the most beautiful and bird-filled I've ever seen; little undergrowth, but carpets of red-brown leaves and rich green moss and soft ferns, then countless rocks and dead logs covered thickly by lichens and moss. The beeches, mostly tall, were so covered with mosses, lichens and parasites it was almost impossible to find their true foliage.'[2]

The Tinsleys left via the fishing township of Havelock, where atomic scientist Lord Rutherford had attended the little primary school from 1883 to 1886, and where the old school had been turned into a youth hostel named in his honour. Improbably, a man known as 'New Zealand's most famous rocket scientist', William Pickering, was later a pupil at this same small country school. He directed NASA's Jet Propulsion Laboratory in Pasadena, California, from 1954 to 1976.[3]

Science writer Marilyn Head has written that Pickering attributed part of his success to the New Zealand characteristic of being an all-rounder: with so few people and resources in New Zealand, there is a compelling reason for individuals to acquire skills and expertise in a variety of fields. Pickering excelled at getting diverse personalities from different backgrounds to work together to achieve a shared vision. His understanding of both the scientific objectives and the technical requirements for achieving them was crucial to the outstanding scientific and engineering achievements of the early Deep Space missions.[4]

If Beatrice came to be described as the pre-eminent synthesiser, Pickering was also in that mould. So too among contemporary New Zealanders is Gerry Gilmore, currently professor of experimental philosophy at Cambridge. One of his strengths is seen as bridging the gap between theorists and observers by understanding how to interpret astronomical observations in terms of theoretical models.[5] But Gilmore's work, and of course Beatrice's, lay in the future.

The Tinsleys carried on with their honeymoon, exploring Nelson and the beauties of the West Coast. Beatrice's letter continues rhapsodically about the beauties of the area, with page after page of its history –'now I feel I know New Zealand miles better.'

Her effort is remarkable. Why did she bother? Beatrice's energy was prodigious, beyond dispute. Even so, it is notable that she needed to be so much the good, dependable daughter while on her honeymoon. Naturally enough, she said nothing about her feelings at this time, but it was to become more and more her habit to hide behind a screen of words, in this way building up a distorted picture of what was really happening.

In much the same way the poet Sylvia Plath wrote to her mother on average once a week for 12 years. The letters, published by her mother under the title of *Letters Home*,[6] often give an inaccurate picture. Sylvia was not close to her mother but she wrote in the persona of 'lovely little girl to a wonderful mummy'; writing, it seems, out of a need to bolster herself and her situation, and ward off self-doubt and reality. It was as if Sylvia became another person when she picked up her pen to write home.

So too it was with Beatrice, if in a distinctly different way and to a different degree. She wrote as only part of the person she really was.

Setting up house in Salisbury Street was important once they were back in Christchurch. Beatrice's first wash-day as a married woman was itself an occasion, particularly with all those 'smelly tramping socks'. By no means all New Zealand households had a washing machine in 1962. Beatrice's shared wash-house had a tub and a handwringer, and she had her own scrubbing board. Her landlady had given her a collection of washing and cleaning materials, and Pip's wedding gift, obviously cherished, was 'a lovely metal-frame ironing board'. The details in her letters suggest that at one level Beatrice was back with those satisfying games of playing at house, the games which she and Theodora had enjoyed not so very many years before. Beatrice's favourite piece of furniture was a tall chiffonier-bookcase in the sitting room. This room, with its open fireplace, was an inviting meeting place for the student gatherings soon held there quite frequently.

Beatrice was considerate – she had offered not to play her violin late at night – and was the most well-organised person imaginable, her landlady has said. Phyl Wardell used time carefully herself with a disciplined life as a writer of children's books and radio serials. She was not the kind of woman to pry, but her novelist's eye and ear for character made her particularly attuned to Beatrice:

> 'She and Brian kept their bicycles in our wash-house and ours were at the back, too. When my husband went to get his before 8am each day, Beatrice would be at the table by the window, working, with all her household chores done. She always seemed to know what she was doing. There was no time wasted in gossip or trivialities, or anything personal.
>
> One day we were all out at the clothes line and Brian, mentioning the name of one of their guests, said to Beatrice, 'I gave a false premise to him.' I didn't understand a word of this mathematical premise, but it was just so different from what other young couples talked about.
>
> Tenants usually chatter, though I myself try not to gossip and waste

time – but they come upstairs to use our telephone and usually, bit by bit, in the garden or at the clothes line, things emerge and you feel you know the other person and their background. But I don't think Beatrice ever said anything personal. I never felt I knew her. It may have been the age difference between us, of course. I'm sure it wasn't any kind of snobbery.[7]

Socratic Society members often met at the flat, sometimes 10 or more people. Calls to Beetle, that childhood name which Brian, however, did not use, and bursts of laughter would float up the stairs to the Wardells. Sometimes shouted arguments would erupt when the discussion got excited. Beatrice kept everything under control, and the Soc Soc students would do the supper dishes around 11.30pm. It was never a chore for her to prepare food for visitors; it was very simple food but she never worried about it or apologised, her landlady observed. Beatrice, too, was always matter of fact – never gushing or seeming overly impressed – and helpful if she could be. Once Phyl Wardell needed a phrase or two in a South American language, for verisimilitude in one of her books. Rowena was working in Florence and Beatrice arranged for her to ask her sculptor friend, José Fajardo, to have a couple of sentences translated.

Arts and law student Guff France decided that married life suited Beatrice:

This was the time when she used to hang on Brian's every word. Beetle was always very open, very friendly. She wasn't demonstrative – she wouldn't lay her arm on your shoulder, for instance – but she was hospitable and caring. And warm. She and Nita, and later she and Brian, often visited me. We were constantly in and out of each other's flats.

I noticed how much she was learning from Nita, who was always practical and confident. Beetle would give things a go – she was like a little girl playing house. But she was also concerned with propriety, with manners and a politeness which weren't quite New Zealand style. In her first student flat, in Shakespeare Road, she served cake on fine plates, and tea in bone china cups.

After her marriage she became much more natural, more free. Her emotional range was extended.

One time when I visited their flat I knocked and called out. Beetle didn't come to the door but called me in. She was leaning back in her chair, talking rapidly to someone, and spitting plum stones into the fire.'

This episode has remained clear in Guff France's mind. The abandonment of her position, lying back and being herself while talking and eating at the same time (that forbidden pleasure), was different enough from the earlier Beetle. But to see her vigorously and pleasurably spitting out the plum stones towards the fireplace is a memory he has treasured. She rationalised everything –'over-rationalised' – but at the same time she found a great deal of fulfilment in her early years with Brian, Guff France considers. 'And I was envious of the way she managed her life. She did her washing at 6am, put it through a hand wringer and out on the clothes line, and then began study or was off to university. With Beetle, to decide to do something was to do it.'[8]

One thing the Tinsleys had firmly decided not to do was to have children early in their careers. This decision had led first Beatrice, then Brian, to go to the local branch of the Family Planning Association (FPA) before their marriage. Very soon, in the words of their contemporary, Ann Ballin, 'They both became fanatical proponents of planned parenthood.'[9] The implication is that their zeal rather amused fellow students, but they were serious. Beatrice kept as guiding principles the tenets of family planning, and Zero Population Growth became her personal creed as well as an aim for the world.

The Tinsleys were invited to join the FPA branch committee. They realised that they could learn a lot from the FPA's resources, particularly its books. In turn they made useful contributions when the committee discussed how to educate the public in family planning principles and how to gain more acceptance among doctors and women's groups.

Brian also built a bookcase for the association's offices, and wrote to the United States to get samples of EMCO contraceptive foam. 'We rather firmly advocated that women be given a choice of several types of family planning methods, not just the one that the lady doctor preferred, the diaphragm.'[10]

Fear of pregnancy before marriage – or within marriage, for that matter – was the most potent contraceptive at this time. Naturally, nature being as she always is, there were pregnancies. Only a very exceptional young woman would decide to keep her baby. Most girls were banished to often grim 'homes' for unmarried mothers, or hidden away with a relative, and it was taken for granted that their babies would be adopted. If a woman was wealthy, knowledgeable or sophisticated, an abortion could usually be arranged. This was both illegal and expensive. The hypocrisy of all this, and the assault against social justice it represented, were not lost on Beatrice.

Soc Soc discussions of sex and sexuality, as with writings in student publications during these years, were focused on the puritan's narrow sexual codes. Even with its advanced, almost lofty, aims, Soc Soc seems never to have considered the situation of married women. Beatrice, however, was already concerned about this.

When one of Soc Soc's founders, arts student Mark Sadler, wrote an article entitled 'Ordering One's Views on Life', which appeared in *Hemlock* in the winter of 1961, Soc Soc members gathered at the Tinsleys' flat to discuss it. Mark Sadler's diary for 30 July 1961 records that the part of his article which Beatrice wanted to talk about had the cross-heading of 'Love Affairs'. It said:

> Little in life is so important as a person's success or failure with the opposite sex. Since to love and be loved is a major need, one should never reject love or spurn people simply because it might sometimes be convenient to do so. People who believe they have to fight off lovers with scorn or cruelty generally have an inflated opinion of their fatal effect on others. A human relationship, once built, should never be totally destroyed, and, while not always easy to achieve in a world of imperfect people, gentle and gradual endings to love affairs are not only morally superior, but more effective.
>
> If two people are unsuited, it should be a discussable fact, not a unilateral decision on the part of one person. I hold firmly (in theory) to an ideal of civilisation in the conduct of love affairs, even though there is in practice no sphere in which more people behave with such lawless, untutored, self-centred barbarism.

Beatrice asked: 'What if two people start going out together and then find they have nothing in common?' She and Brian had been married for less than three months.

Not surprisingly, nobody in that group of students was alerted by Beatrice's question. Having little in common was in fact the essence of the situation she now found herself in, she was eventually to say. She may not have fully admitted her feelings to herself at that early stage, although later she believed that she had. Certainly nothing had prepared her for the possibility that she and Brian might not really be suited to each other. Her pride, her fear of her parents' reaction which led her to throw them off the scent of what was really happening, her loyalty to Brian, her respect for the promises she had made when she married – all these would have inhibited her then, as they did for many years.

At the Soc Soc gathering, whatever hopes Beatrice may have had of a rational, helpful discussion of an unfulfilling marriage came to nothing. She should not have been surprised. Sadler recorded: 'I said to her in reply, "What do you mean, you find you have nothing in common? You are both human beings, aren't you?" This got a laugh from the meeting as did several of my other replies.'[11]

Brian Tinsley, although he had thrown off the religious dogma of his childhood, liked to consider issues of religion and sexual conduct. Some of his reflections appeared in student publications. The most powerful argument in favour of chastity before marriage was the risk of conception, he wrote. More worthy and positive reasons for chastity included maintaining a situation in which people could 'act from choice rather than a sense of compulsion', thus making more likely a happy home life and a rich and enduring relationship.[12]

Beatrice did not write for student publications although she increasingly took part in Soc Soc discussions. Even her rapid speaking could not keep up with her thoughts, but she was discovering the pleasure of coming into an argument under full sail and turning it in a new direction. Now, when 'Little Beetle' wanted to make a point, everyone listened. She researched her subjects. Under 'religion' in the census form, she declared herself agnostic. She was sure about that, but she and Brian attended a university course on comparative religion, deciding they should know more about other belief systems.

Three months before her marriage Beatrice had begun to follow her mother's example and make a scrapbook. The news item which prompted this had been published in the Christchurch *Press* on 11 February 1961. Under the bold headline, 'Astronomers Say Universe Had Definite Beginning', appeared this news story written by an excited journalist from Cambridge University:

> Six British scientists have announced that they have proof to explain one of man's greatest mysteries – how the universe began.
>
> They told a news conference they believed that thousands of millions of years ago all the galaxies of the universe were compressed into a very small volume, but an explosion (later called the Big Bang) sent them flying apart.
>
> Their theory of a universe still expanding contradicts the currently-held 'steady state' theory – that new stars are being created constantly in space from hydrogen atoms to replace fading stars.

Here was the stuff to grip her imagination and engage her thinking. While her attention would have to be focused on a thesis far from her heart, there was nothing to stop her reading and thinking about astronomy, and learning as much as she could about research methods.

The Cambridge team had been led by Martin Ryle, professor of radio astronomy. The team of six scientists had included a woman, Patricia Leslie, the 25-year-old wife of a graduate student. They had worked at the

Mullard Observatory using a giant radio-telescope, the most powerful yet built, and for nine years had studied the outer frontiers of the universe.

The journalist wrote that the scientists had reached these main conclusions:
– The universe is expanding.
– All matter in the universe, of which the earth is only a very small part, is rushing out into space at a fantastic speed.
– The universe had a definite beginning.
– The universe will not last for ever.

When the team was asked at a news conference how long it expected the universe to last, Martin Ryle had replied, 'I do not think it will concern us in our time. We're all right for a few thousand million years yet.'

Beatrice herself was to carry on this research, coming to have a particular interest in an ever-expanding universe. For the present, her young-woman side was uppermost. On the same page in her scrapbook as this major scientific news, the very first page, she pasted a reception and wedding breakfast card listing the attendants at the wedding of Nita and David Hanna.

Tucked away at the bottom of the main newspaper report on the Cambridge research was an interview with Fred Hoyle, whose writing had excited Beatrice's interest in astronomy when she was 14. Now, privately committed to a life in cosmology, she studied every word of the interview. It was headlined 'Fred Hoyle Unmoved – "Universe is Everlasting".' As the long-time proponent of the 'steady state' theory, that the universe is everlasting with no beginning and no end, Hoyle said that he did not accept that his theory should now be discarded. His objection, he said, was mathematical. He would be satisfied he was wrong if he could be shown that no new galaxies were forming.

To Beatrice, one of the most important aspects of this new 'expanding universe' research was the vital part played in it by that new tool, the computer. The Cambridge astronomers had used this to do more than a million and a half calculations, the report said, a phenomenal advance on the old mathematical drudgery. And almost immediately she herself was to become one of the first people in New Zealand to learn how to use a computer. With her emerging talent for synthesis, the computer now became her most exciting and essential tool.[13]

Beatrice's great good fortune was that New Zealand's second computer,[14] an IBM 1620, went to Canterbury University in May 1962.[15]

At this time Canterbury University had begun to move itself, faculty by faculty, from the desperately overcrowded central Christchurch site out to the burgeoning new campus at Ilam. The School of Engineering, one of the first schools to occupy the Ilam site, had no immediate use for a small seminar room with two even smaller rooms next to it. These became the Mobil Computer Laboratory because of a grant from Mobil Oil. The computer was big and cumbersome, with a comparatively tiny capacity, and programming relied on a card reader and punch unit. Information was punched onto cards by the punching machine, which resembled a big typewriter, and the card-processing machine was almost as big as the computer itself.

Former IBM employee Bruce Moon was responsible for the Canterbury University computer, assisted by Joan Lester, who later married Alan Williman, then a lecturer in engineering. Williman recalled the centre: 'It was very much hands on. You'd normally have to write your own programme. Bruce taught us. Then you sent in your data and your programme, Joan would punch it onto cards, and then you'd put the cards into the machine.'[16]

Everyone knew they were pioneers, in at the beginning of a mind-boggling era. The old division in the universities had tended to be the humanities versus the sciences, as dramatised in the novels of C. P. Snow. Now it was hard for the computer users not to think of themselves as the élite, the new priesthood. Competition for computer time was fierce, 24 hours of every day, as Bruce Moon remembered:

> Whether it was 2am or 3am, people queued for their turn. Joan kept the booking sheets and she was fair. When word got around of what the computer could do, every hour was used. I recall the records for one month – 744 hours. That would have been a month with 31 days as a 30-day month has only 720 hours.
>
> We had rules for how long and how often each person could use the computer. Night use was usually for people who needed a lengthy spell, and Joan issued them with a key to the building. But everyone had to take a turn. It was fair and it was fun.[17]
>
> All the bright young things rushed us but Beatrice was outstanding. You noticed her not because of her looks, though she was quite striking-looking, or because she was obtrusive, but because of what she was.[18]

Beatrice used the computer for some of Brian's calculations as well as her own, because at that time he had not learned the equipment's intricacies and was not on the licensed-user list. To her family she made light of the long bicycle rides in the middle of the night from their flat through Hagley Park to the Ilam campus.

Bruce Moon thinks that, if her family had asked her to explain what she meant by using the computer for her mathematical models, she might have put it, at its simplest, something like this. Like all models, a mathematical model incorporates only some of the characteristics of the original. A mathematical model is a series of equations representing what the researcher believes are the significant attributes of the system being studied. Usually such a set of equations is too complicated for a solution to be found by conventional mathematical means, so the researcher uses computation to find specific answers. The model and hence the theory are both developed by comparing these with observations.

Beatrice might also have said in explanation that you construct a theory that you think is right, and then write out all the maths you need in order to calculate where everything is. To establish your theory, which is speculation, you let the computer do the maths. She could have added that some theories are wrong, while others are not as much wrong as incomplete, or too simplistic. Most are merely inadequate to explain what is known in a given field, as she was to find herself when she was able to add to the models of Einstein, Hubble, Hoyle and Sandage.

But the Hills, perhaps wisely, did not ask, and Beatrice's letters home at this time skip over the explanations.

During their first year of marriage, Brian often left the city to use his spectrometer at the University of Canterbury's field station at Rolleston for his PhD studies. There he analysed the light from the aurora which could be seen quite often in the South Island skies, this being a time of maximum sunspot occurrence. It was this data which Beatrice processed for him on the computer.

Brian's motivation for this work was to increase understanding of the upper atmosphere, extending from about 70 kilometres above the earth. The work involved the use of instruments to observe the airglow and the aurora, instruments more sensitive than had previously been available.

The mechanism works as follows, he has explained. The ever-thinner air forming the outermost part of the atmosphere is exposed to ultraviolet radiation from the sun, which splits up, or dissociates, the molecules of oxygen, nitrogen, water and so on into atoms, and ionises some of them, forming the ionosphere.[19] When the ions and electrons recombine they emit a faint glow, called the airglow. Some of the airglow is due also to the fluorescence of atoms or molecules under the irradiance of solar ultraviolet light. The airglow is relatively strong during the daytime but is almost impossible to detect from the ground because of the bright blue sky, which itself is due mostly to the scattering of mainly blue wavelengths of sunlight by molecules in the relatively thick lower atmosphere.

Brian would probably have warmed to this next part of his simplified explanation, the reason for his frequent and later lengthy absences from home. At twilight and at night, when the lower atmosphere is in shadow, if the observer is away from city lights and there is no moon, the airglow can be seen as a faint glow between the stars. Similarly an aurora, that glow in the sky caused by high-energy particles interacting with the upper atmosphere, can be properly observed only well away from artificial lights.

Beatrice used Brian's absences to do her own work, although music with the Metcalfs and Pip continued to delight her. Sometimes, after their small informal chamber music group had finished playing for the evening, they would sit and smile at one another, as filled with pleasure in their own music-making as in the particular Mozart or Schubert piece they had just finished. Beatrice's landlady, calling her to the upstairs telephone, often heard her making happy arrangements to play: "Well, just play as a trio till I get there!" And she would hurry off, carrying her violin on her bicycle.

She took Brian to concerts she thought might interest him, but their real shared interests were Soc Soc and the occasional weekends at the students' hut at Arthur's Pass. She was impatient for her MSc to be over so she could return to the study of cosmology, even if this had to be through private reading.

Rowena had been writing about the Venezuelan sculptor José Fajardo whom she had met in Florence and with whom she was having a sometimes stormy relationship. She spoke of marriage as likely, although she also thought they should first have a year apart. But a rapprochement with José saw her pregnant and with him in Venezuela. Beatrice sent loving and hopeful letters, and tried to smooth the way for her with their parents. Edward began to hope that Rowena and family would come to live in New Zealand.[20]

Beatrice now began to teach science, with as much physics as she could, to senior girls at Christchurch Girls' High School. The money was useful, and contact with some bright young minds gave her pleasure. It was a traditional school with a well-defined hierarchy. As the newest and certainly the youngest of the teachers, she was made aware of her lowly place in the scheme of things. Teaching could well feature in her future, but not this kind of teaching and not this kind of institution.

A member of the staff at that time, Jean Hanlin, remembers Beatrice as quiet and self-effacing, with qualifications rare for that time:

> She was obviously producing the results. Her girls were highly successful. Often people with high qualifications can't teach.
>
> We were desperately old-fashioned, with a strict hierarchy in the staff. We didn't necessarily talk to everyone. But Beatrice made a very pleasant impression.

Girls' schools in particular were still finding teachers of maths and physics extremely hard to obtain. Beatrice's own recent school experiences made her keenly aware of the girls' needs. When on occasion she was unable to take a class, she arranged for Brian to fill in for her. 'We'd have taken anyone if he could teach some physics,' Jean Hanlin said. 'We were grateful to them both.' [21]

Not enough time was Beatrice's constant preoccupation. She tried to keep up her usual 40 hours of solid work at the department – a reasonable aim, she thought – but time had also to be taken for shopping, cooking and cleaning every week. That Brian might share these chores seems simply not to have occurred to her. It was newsworthy enough for her to make a point of telling her parents that on her 21st birthday Brian had spoilt her by doing the dishes.

For his doctoral work, Brian took his spectrometer and some radio equipment to Western Samoa in June and July 1962, to observe the spectrum of the artificial aurora produced by the United States' hydrogen bomb test 400 kilometres above Johnston Island in the Northern Pacific.[22]

Her tutors had been urging Beatrice to apply for overseas scholarships. They had no doubt about her MSc results. Dr Clif Ellyett had pointed out that she could always decline if she were offered a scholarship somewhere other than where Brian would be going. For it was plain that Brian must get overseas experience with his spectrometer work, and access to the kind of facilities New Zealand could not provide. Almost certainly this meant the United States.

Fellow students and Soc Soc members noted, occasionally with cynicism, that the Tinsleys had not joined the Labour Party or the Campaign for Nuclear Disarmament, to which most of their friends belonged almost as a matter of course.[23] But this was a careful though regretted decision. For anyone thinking about further study or job prospects in the United States, as Brian had been doing for some time, there was the fear that American immigration authorities, on guard against communist infiltration, would consider even an affiliation with the New Zealand Labour Party to be evidence of dangerous leanings towards the Red Left. (What these authorities might have made of Soc Soc seems not to have occurred to the Tinsleys.)

Beatrice kept thinking about where a life in physics might take her, and went as far as to write home about some of her doubts. It is as if she were talking to herself:

> I wonder, do I want to do postgraduate research in physics – certainly not in this [nuclear] field. How one's illusions about the nature of Research get shattered! Also, its nature changes rapidly as science progresses.

I feel that I want to understand and know as much as possible about the world, specially people; but to be engaged in some specific line of scientific research is to cut oneself off from the chances to obtain broad knowledge. Brian and I both think we're much more likely to be of use in the world if we try to understand it as a whole, and do broader things, than if we try to make a narrow discovery in a specialised field. Difficult to explain to people![24]

Beatrice used the word 'cynical' about her feelings as she worked for an exam in nuclear physics. The hydrogen bomb was an obscenity, she felt. Both Tinsleys were determined that they would never work in anything connected with armaments, although Brian argued that the medical uses of nuclear reactors meant that they did more good than the harm they had done hitherto. Beatrice had read enough and thought enough to be convinced that there could be no such thing as a limited nuclear war.

On the morning of the nuclear physics exam, the *Press* carried a report of the USSR's testing of a 50-megaton bomb. Brian tried to hide the newspaper so she would not be distracted, but she had dreamed of this the night before, so had to look.

Thinking on a much larger scale than the earth was something she seems to have returned to almost daily. It was both her passion and her solace. Whatever the disappointments and frustrations in her personal life, they were less than tiny in any consideration of the cosmos. As she wrote when thinking about nuclear testing:

> The answer to not going crazy with worry is to think on a much larger scale than the earth; the stars are comforting in the standards of relative importance they set. Yet the little, earthy, things seem so much more valuable when they are at stake![25]

Brian was spending most of his time out at the field station at Rolleston. Beatrice, still teaching and marking the senior girls' physics papers, and with her last exam looming, was elated to be offered short-term work in the university's Physics Department, helping to rearrange the stage II and III lab, for 45 shillings a day – 'pretty good for seven hours' congenial work'.

Her 15 December letter telling her parents of her success in her MSc results began in a typically discursive fashion with reactions to her parents' Christmas gift of a bathmat. No ordinary mat, this, but a joke bath-mat with huge bare footprints in bold colours, much admired. She wrote on, telling them she was having 'a relax' after doing the washing. This had included washing three double blankets, an effort which she said took toll of her nervous system as she tried to get them through the hand-wringer. There was news of Brian. He was still working long hours, and the spectrographs he had taken of the bomb test in the Pacific seemed to be yielding some very interesting information. Then at last she came to the news of her degree. She did not mention her grades, although each was an A. Instead she spoke of the possibility of publishing her thesis, if it was thought worthwhile, and said she and Brian had celebrated by having a meal out.

It seems a happy letter, and it should have been. But what it did not even hint at was her burning anger when she discovered that the university would not employ her in the coming year because it was already employing Brian. Other than for short-term little jobs, the university's anti-nepotism regulation stood firm: husband and wife must not be employed at the same time. She had had no idea that if she married Brian she would never get a job in the Physics Department or in any other department at Canterbury. Years later Beatrice was still talking to intimates about her feelings of 'betrayal' – her word. Rightly or wrongly, she thought Brian should have known, and told her. It was her life and her future, but her freedom had been lopped. True, they planned to continue studying and working overseas. But meanwhile the new year of 1963 stretched ahead while they waited to see what might be offered to Brian. And the best she could do was high school teaching ... She was strung taut with frustration. Glen Metcalf vividly remembered the anger Beatrice was still expressing even when she returned to Christchurch in 1972. A waste of so many opportunities!

Even by the time she came to the end of her MSc year, Beatrice does not seem to have given thought to the bias against women in academic employment. She was the only woman in physics at master's level. A large photograph of the Physics Department's staff and senior students at the time shows her to be the only female apart from two secretaries. She knew that she was bright (without knowing how bright), and that she worked hard and reliably. The inculcated habit of modesty probably inhibited her from standing back and trying to make an accurate assessment of her own abilities. She had made a clean sweep of every available prize in her year: the Haydon Prize for Physics as well as the Charles Cook Prize, the Warwick House Prize, the Memorial Scholarship and the Postgraduate Scholarship in her field. These triumphs she did not even mention to her family, Edward Hill making this discovery only many years later.[26]

Had the Tinsleys' roles been reversed – had she been the older scientist who was already employed in a junior capacity, although without winning prizes, and had her husband just graduated MSc with such distinction and be expecting to be employed – it is easy to imagine the confidential and ultimately self-serving discussions which would have ensued. 'Of course we must do something. Good little woman – but here's her husband coming up – brilliant chap.' And she would, surely, have been tactfully taken aside, had the situation explained to her benevolently, and then been sent home to be a good wife while teaching on the side: 'Physics teachers are always wanted in high schools.'

Bush and water in Queen Charlotte Sound were the background for a few days at the end of January 1963 when the Tinsleys attended the annual congress of the New Zealand University Students' Association at Curious Cove. This congress, like a magnified Soc Soc in its scope and intentions, had the bonus of the input of many more students from all over the country. Beatrice's scrapbook contains a pamphlet about the speakers – mostly university staff – and their topics, together with newspaper reports. These congresses had a lively tradition, and what went on was widely reported and discussed. Politics, race relations, moral issues in medical advances, and the interrelationship of science, art and philosophy were all discussed that year.

The poet Peter Bland –'New Zealand is not so much isolated as insulated' – and lecturer Margaret Dalziel on the modern novel were two people to interest Beatrice, who was conscious of the split in universities between science and the humanities. One of her main concerns was addressed by the professionally provocative Viennese doctor, Erich Geiringer, who talked about the potential menace of doctors' new powers over life and death, and thus the potentially great powers this gave governments. Infectious diseases had always naturally controlled the rate of population growth, but medicine had succeeded in preventing people from dying, and now the world was faced with a population explosion. Contraception was invented at the point when the world faced starvation, he said, and now – though once it was to be kept out of the public eye – it was a matter to shout over the roof-tops. 'Reproduction is becoming a matter of public concern.' This lecture and the subsequent discussion underlined Beatrice's lifelong commitment to the Zero Population Growth movement.

With her 22nd birthday celebrated on the first day of the congress, and her pleasure in discussion and the beauty of Queen Charlotte Sound, she returned refreshed to Christchurch with Brian and the beginning of a year of decisions. So New Zealand was not so much isolated as insulated from the rest of the world? The Tinsleys could reflect that they looked outward more than most of their contemporaries did. Beatrice had been brought up to take a world view, and both had consciously assessed their personal responsibilities from an 'I am my brother's keeper' point of view. Now their discussions became more urgent. Where could they or should they go, to continue researching and doing something of value for humankind?

An opportunity came for Beatrice to give practical help to some African students by becoming mathematics and physics coach for four men who were attempting to get science degrees. They were sponsored by the New Zealand Government as part of its programme of overseas aid, but they faced an anxious future. If they failed their exams, sponsorship ended. The Beetle who had taught her koala bear warmed to their situation. She could help them get to a position where they could make a real difference to the economies of their own emerging countries, and also – an ever-present threat to the many ill-prepared foreign students – help avert the humiliation awaiting them if they were sent home, disgraced as failures. She turned her attention to devising ways and means of discovering the gaps in their education, and building on what they already knew.

The Tinsleys' landlady watched the comings and goings of the African students with interest. Anglo-Saxon Christchurch in the early 1960s had scarcely seen non-white people, students or otherwise. Few Maori had as yet come into the city, let alone to the university, and the four Africans were highly visible:

> People didn't mean to be rude but when they walked down the street you could see heads turning in surprise. And one of them, the student from Uganda, was the blackest man you could imagine.
>
> This meant nothing to Beatrice. She worked with them perfectly naturally.[27]

Beatrice, virtually blind to matters of colour and race, was drawn to people who were trying to do some good in the world, especially if they were studying and asking questions.[28] Sometimes she coached teenage New Zealand students in physics. One, Edward Janus, now a medical director and researcher in Victoria, Australia, has always remembered her. 'She taught me physics but also conveyed tremendous enthusiasm for it. Her comment to me, "You can do anything if you put your mind to it," has inspired me ever since.'[29]

During these months some of the physics staff, Clif Ellyett in particular, had been helping the Tinsleys with approaches to possible institutions offering work and study opportunities abroad. Now, in February 1963, another possibility had arisen to which she and Brian were 'putting out feelers', she wrote. This was the Southwest Center for Advanced Studies in Dallas, Texas.

Its founder, Lloyd Berkner, was a scientist-administrator who had been a pilot for Admiral Byrd in expeditions to Antarctica, where Berkner Island is named for him. He had gone there via Christchurch in the 1950s and 60s, and had arranged for Clif Ellyett to visit Antarctica. Berkner was now recruiting scientists for the Southwest Center, and had sent some 'flashy looking literature', as Beatrice described it, about the big new laboratory for earth and planetary science.

The Tinsleys examined the proposal carefully, and Ellyett wrote off on their behalf, enquiring particularly if they could use anyone at Beatrice's level interested in relativity theory, which was part of the Southwest Center's theoretical side. 'Sounds exciting, and great for a year or two!' she wrote. 'Imagine Brian under a sombrero, talking with a Texan drawl. But I'd be surprised if we stayed there very long, though.'

By the end of March it was settled that they would go to Dallas for two years, leaving in August or September. Beatrice needed a place where she could take up her postgraduate scholarship, but her needs were subordinated to Brian's. They were optimistic. Texas was a mighty big state, and presumably had many opportunities

She was capped MSc in May. Brian, still working towards his PhD, was immersed in new research, 'some ingenious new idea which could revolutionise the efficiency of a spectrograph', she told her family. It was taking an idea, which had originated in France, for improving the sensitivity of a spectrometer and extending it much further for study of the airglow. Brian thought he could modify his existing spectrograph into this new grille spectrometer fairly easily. Although the information had come far too late for his thesis, it was a good thing to start on next.

The University Grants Committee took its time before agreeing to Beatrice's holding her scholarship at the Southwest Center for Advanced Studies. (By July she was writing about the 'Center' instead of 'Centre' in letters home, the beginning of adaptation.)

She kept up her voluntary work with the Family Planning Association. Normally she did not mention this in letters home, but one winter letter describes doing 'necessary but menial tasks' at the clinic, such as washing and sterilising rubber gloves because 'the nurse gets too rushed if she has to cope with it all on the evening clinics.'

She very much hoped that Rowena and her family would come on an experimental visit to New Zealand before she and Brian left for the US. The Hills had moved into a larger house in Moana Road, Kelburn, with sweeping views and room for all the Fajardo family, and Edward hoped that José could get a job at the art school. In the event, the family arrived just as the Tinsleys were leaving the country.

Beatrice added in her July letter that she was typing sections of Brian's PhD thesis on the spectrum of the artificial aurora seen from Samoa. The figures had to be photographed, and numerical tables typed, 'a finicky job'. (Beatrice spelt the word 'finnicky', the only spelling error detectable in many hundreds of thousands of words written to her family.) Helping Brian in this way meant she was 'learning an awful lot about upper atmosphere physics in the meantime!'

Her own paper, 'Analysis of the Optical Absorption Spectrum of Neodymium Magnesium Nitrate', had been accepted for publication by the *Journal of Chemical Physics*, so she expected to see the proofs soon, she wrote. Then, for the first time in a letter to her parents, she commented in the voice of a professional scientist: 'Such is the delay between the times when most scientific research is done and when it becomes public knowledge. The only way to get important results out quickly is to send them as a letter to the editor of the suitable journal.' [30]

She did not explain that acceptance was conditional on assessment and approval by the scientist's peers; hence the delay. Nor did she write about the excitement of first publication. Although Beatrice had more than a hundred papers published, some of them among the most eagerly received of her day, that very first publication, even though her thesis was not on a subject dear to her heart, must have said to her that here she was, at last, a scientist among scientists and on her way.

Both Tinsleys felt they were off to great adventures where almost anything was possible. When the young John F. Kennedy defeated the Republicans' Richard M Nixon in the 1960 presidential election, he had said in his inaugural address on 20 January 1961, 'Let the word go forth ... that the torch has been passed to a new generation of Americans.' What seemed to be Kennedy's call from the heart for people to commit themselves and discover a new sense of purpose had touched far more than his fellow Americans. His words had reverberated, especially among the idealistic young, and at universities everywhere.

The Tinsleys also saw President Kennedy as the first American leader with an informed interest in science. His commitment to space exploration meant they were going to a society where science, especially space technology, was given high priority in national life. Words such as cosmos, cosmonaut and cosmology were becoming common coin around the world.

Now for the Tinsleys their air tickets became the key to this new life. They could not imagine what lay ahead but they were young – Beatrice 22, Brian four years older – and, they thought, well prepared and ready to learn. It was a major step, although just how major they had no means of telling. Brian had a scientific position waiting for him in Dallas. Beatrice had been assured that there were several universities in the region where she could study cosmology. All manner of good things could happen. On that basis, on 20 October 1963 they flew off to a new world and to a new life.

CHAPTER EIGHT

TO A NEW WORLD

'We the people can make a difference.'
JOHN F. KENNEDY

It is a rite of passage for the adventurous young to explore the world, but in 1963 it was far from usual. The Tinsleys were fortunate, and they knew it. Fares represented a substantial part of the average adult annual income. Usually years of working and saving had to pass before travel was possible. A few of the academic elite had travelling scholarships, as Beatrice did. Fewer still of the newly graduated had academic appointments overseas.

Young people arriving in San Francisco five years later could take themselves off to see the flowering of the counterculture in Haight-Ashbury; 10 years later and it could have been the startling erotic movie *Deep Throat*. The adventure for the Tinsleys, tired after long hours of uncomfortable flying but too excited to think of wasting time on compensatory sleep, was to visit Finochio's, which billed itself as 'the most talked about nightclub in the world'. A waiter descended on the Tinsleys' table and swept Beatrice away to dance and be photographed expensively by the nightclub's photographer. Finochio's programme in her scrapbook is revealed as a folder for the enlarged photo. Beatrice looks wry and wary, prim in her little New Zealand dress set off by a gold locket. Typically, as in her school days, she was determined to do the right thing by joining in.

They had chosen to travel by train to Texas in order to see as much as possible of the intervening countryside. They were left with layers of impressions. As the train took them south from San Francisco, the vastness of the country predominated. After the compact and under-populated islands of New Zealand, here were untold miles of rocks and pasture dried up at the end of summer. Acres of industrial buildings around Los Angeles gave way to market gardens, nurseries and orchards being worked by

Mexicans. They exclaimed over the unending car wrecks piled up in scrap-metal yards along the line – more cars than they had seen being driven, let alone scrapped, in the entire city of Christchurch. Billboards exhorted everyone to buy. To the frugal Tinsleys, used to living careful lives without washing machine, refrigerator or television set, let alone a car, the contrast was startling. But in the whole of the United States at that time there were only about 4200 computers. The exponential explosion had yet to happen. Beatrice's experience in New Zealand put her in the forefront of the very few computer users in this new country.

On their train, night followed night in Texas, bringing home to them the vastness and emptiness of their new state. Sometimes a distant oil rig or a solitary, mean little house set down in the barren wastes, or shacks huddling in loneliness and poverty by the railway line, spoke of the contrasts which would await them in fabled Dallas with its skyscrapers and oil barons.

There they were met by Dr William B. Hanson and his wife Winona, who took them to a motel which Beatrice described as 'gorgeous – we decided to enjoy ourselves as royalty as we had no choice.' Bill Hanson had come to Dallas the previous year to the newly established Southwest Center for Advanced Studies, which was usually called SCAS or simply the Center. He was typical of the young, highly motivated physicists who were devoted to space exploration. They had become the new elite as the space race caught popular imagination. National pride was harnessed to being first in the race.

Against this backdrop the Tinsleys tried to absorb all the differences as quickly as possible: differences in light, in temperature, accents, mixes of nationalities, food and customs. Everyone spoke English, but there were so many kinds of English that it was no wonder, as Beatrice wrote to her family, that each newcomer to the Center was given a *Webster's Dictionary*. The Southwest Center was about half-way between Downtown Dallas and Richardson. It was temporarily housed in rooms belonging to the Southern Methodist University, (SMU), while its own buildings were being erected on a large campus in nearby Richardson. This land had been donated by three prominent Texas businessmen, the founders of SCAS. They were millionaires who owned a large part of Texas Instruments, itself a prominent player among contractors for the space race.

The 'gorgeous' motel where the Tinsleys were first taken, a motel routinely used by SCAS for its visitors, was their introduction to a new way of living after the simplicities of their student life. The Hansons invited them to their home that first evening, to meet a colleague, Dr Francis Johnson, and his wife. 'Very nice friendly people,' Beatrice wrote home, 'though we gained the impression they expected us to have and to use an awful lot of money; for example they said an air-conditioned car is a must.' Their own small savings already looked small indeed.¹

At first the Tinsleys told each other they need not have a car until they were well settled into their new life. A Center secretary helped them find an unpretentious, congenial apartment in University Park, only 10 minutes' walk across the campus from their work. They were taken aback to discover that they needed to use the air-conditioning unit in the evenings, even in October, and decided that in summer they would leave early for the air-conditioned offices and work on until late. But basic household items they needed at once. An iron, toaster and electric frying-pan they discovered to be so much cheaper in 'the same number of dollars here as pounds in New Zealand'. Shopping for those items, however, had also shown that life was geared to the car. Nobody appeared to ride a bicycle. The kitchen items, for instance, were bought at a vast discount store a long way from any residential area and without a footpath leading to it. The store sat alone, with its huge car park. And the apartment was not designed for tenants to do their own washing and drying of clothes. Beatrice discovered that she would have to lug everything several blocks to a washeteria, the Dallas name for a laundromat.

But she wanted to talk about science and scientists rather than domestic life. She liked the people at the Center. In Brian's upper atmosphere group – for the time being she was sharing his office – the scientists came from Australia, Canada, Ireland, Norway, India and all over the US. The mathematical-physics group to which she expected to be attached also had members from Hungary, Poland, Britain and Germany. She wrote:

> Its leader, Dr Ivor Robinson, is evidently English,² and is huge, gesticulating ... and with a plum in his mouth – extremely clever under all that, and has been very helpful to me. I'm finding Dr Wolfgang Rindler very helpful, particularly as there's an excellent textbook of his which I'm starting on. Today they had two seminars which I attended, (mainly for social reasons though I understood some), and was greatly impressed by the brain-power present and the way they operate; also they are very friendly, all of them, and most willing to tell me in simple terms about their research, and suggest a plan of action.
>
> It's great to be among people working on the things I've always so enjoyed studying – the stimulus to work hard is terrific!³

Brian submitted a plan of action that was well received. It appeared that he would be able to follow through much as he pleased. Beatrice had first to discover just what was happening in the field of cosmology she wanted to pursue, and then see how she would fit in. Cosmology, however, had occupied her almost from their arrival.

From the beginning they adopted much the same strict work-oriented regime as they had lived by in Christchurch. One change was the number of invitations to dinner they were receiving from colleagues. Beatrice felt increasingly happy, and hopeful of making good close friends as they came to know people better. Then there was music. The Dallas Symphony Orchestra, rated as one of the best dozen in the US, had its headquarters on the SMU campus. She planned to take Brian to concerts and to return to her own music as soon as her violin arrived with their other possessions from home. Brian heard that his PhD had been awarded, and they bought a car – 'I haven't dared to drive it yet,' was her aside. Brian had driven before in New Zealand, and was able to celebrate when he passed the quite stringent driving test. One way and another, as she summed up for her family, they thought things were very promising.

President Kennedy was actually coming to Dallas, she wrote home on 12 November. There was great excitement, and people from their own Southwest Center would be meeting him at lunch. She continued:

> The mathematicians around me are a very lively and interesting crowd, and in spite of not participating in their work I feel part of the group.
>
> They are having an important international symposium here on General Relativity and Cosmology (some very exciting new ideas to be discussed). The people coming to participate include Hoyle, Oppenheimer, and just about every famous living man in the field from all over the world!
>
> I feel inspired to work hard so I might follow at least a little of the proceedings.
>
> Within the USA there seems to be a lot of flying around to seminars and lectures done by the scientists; the former seem to be the focal points for discussing new ideas, and certainly show the advantages of getting together.

The day of President Kennedy's visit, 22 November 1963, began like any other day. Beatrice, up early as always, prepared cut lunches for them both – eating out was another American custom they were slow to take up because of the expense – and they carried these with their papers as they walked to SCAS. They planned to stay in the building and work throughout the day. During the morning the current of excitement which grew and ran through the streets and offices of Dallas reached the scientists. Some of them had brought radios, and one by one they switched them on, keeping everyone up with the president's progress.

The Southwest Center for Advanced Studies was proud to be one of the official hosts for the presidential luncheon, together with the Dallas Citizens' Council and the Dallas Assembly. The president's commitment to America's role in space research made SCAS an obvious focus of his interest.

Two thousand six hundred people had been invited to a luncheon in the huge hall of the Trade Mart, the biggest building around. The Tinsleys, junior scientists, of course were not there. Guests were required to arrive well in advance. Members of the secret service carefully scrutinised them and their invitations before they were permitted to enter the great hall, where they were led to their seats. The mathematician Ivor Robinson was among the guests:

> We sat at our tables and talked to and fro. I turned round to talk to people behind me, and my chair tilted. Two men too tidily dressed and healthy to be academics started up, but then relaxed when they saw I was no threat. I was pleased I had identified them.
>
> Time passed. No guests and no luncheon.
>
> Someone said to me, 'The president has been shot.'
>
> A waiter or one of the cooks had a radio. Speculation and rumour and shockwaves sped quickly but quietly – these were decorous people – from table to table.

Then the organisers of the luncheon showed what impressed Ivor Robinson as 'remarkably civilised behaviour': they ordered the meal to be served. Only when it came to an end and the unbelievable had been verified, was the announcement made.

President Kennedy had been assassinated.

Nearly all those who were adult in 1963 can remember where they were and what they were doing when they heard the news. Everyone had a story. Ivor Robinson overheard one of his colleagues. Defeated presidential candidate Richard M. Nixon was known to be in Dallas that day, too, as attorney for a soft drinks company, and the colleague said, 'First thing is to get Nixon and get him up against a wall.' 'Shush!' said his wife.[4]

In handwriting markedly cramped and different from her usual, Beatrice wrote at once to her family:

> Black date and black address. I feel ashamed to write it on the back [of her letter]. No doubt you've heard all the details. We were waiting to see him on TV at the lunch partly sponsored by the Center ... The news was incredible. We still couldn't believe it when about an hour

later his death was certain.

After he was safely through the airport all seemed well – though no-one had expected seriously that more than demonstrations would occur. The place it happened was about a mile or two from the Center, and we'd mostly decided the crowd would be too great for it to be worth going to stare.

Now it's later in the afternoon and we're at home, the Center having closed, listening to the world's reactions on the radio ... Still I can't believe that he's dead, assassinated at Dallas; the whole city seems stunned and the people in the street looked shocked.

Nearly everyone at the Center was very pro-Kennedy and is genuinely really sad for the sake of America. We're being given a lot of information on the radio about the new President[5] but in general he's an unknown quantity. The reporter broadcasting from Moscow felt that people in general there were afraid in case the new president would want war or be unable to control the extremist elements here.

Who can control lunatics? One hopes, as someone has said, that the assassin was a crazy individual and not acting on behalf of some political group ... Before America turned black we'd had a particularly good week socially and had driven around Dallas and were feeling rather good about the place. But one madman can't be held against a city (and at present someone from Fort Worth is under arrest).[6]

Have just heard some British newspaper editor's comment on the 'damnable wicked destruction'. I agree. Each new announcement on the radio sounds more incredible. Why such a good man? It is sickening to think of his wife and family, especially after her ghastly experience of catching him as he was shot.

Beatrice ended her letter, 'Forgive this shameful city.'
The Tinsleys were too junior to know about the urgent discussions going on among senior scientists about the propriety of holding the international symposium in Dallas after what had happened there. The very name of Dallas caused renewed shock throughout the world.

Ivor Robinson, one of the principal organisers, insisted that arrangements should stand. He had been invited to form a research group in mathematical physics, beginning with his own field of relativity. They would then form a group in high-energy physics. As he said: 'I got a very nice relativity group together. The symposium was because we had such a nice group here.' A relativity symposium had been held not long before, so the theme of gravitational collapse was chosen. The full title was the

International Symposium on Gravitational Collapse and Other Topics in Relativistic Astrophysics.

Gravitational collapse was a field, if a new one. Relativistic astrophysics was not. It was a small piece of scientists' mischief. Robinson explained:

> How it happened was that we were sitting round a pool, planning our symposium. A.E. Schild, very distinguished and tremendous fun, liked martinis. I left the others to go and get more gin, and while I was away they came up with the subtitle of relativistic astrophysics.
> 'What the hell does that mean?' I asked when I came back.
> 'That's the beauty of it. It covers anything we want,' they told me.'[7]

Immediately after Kennedy's assassination, however, life seemed to pause everywhere. Should scientists at such a time come together for days of talk and excitement and undoubted celebration – of ideas, certainly, but also of the pleasure of meeting one another again – and in Dallas, of all places? In the event, only one scientist who had accepted the invitation then stayed away on principle.

For Beatrice, attending the symposium was the biggest and best Christmas present ever. So many of the great names were there, some of them names from her treasured book, *Theories of the Universe*. One's first sight and sound of heroes is never forgotten. There stood the great Fred Hoyle, her first hero, and as enthusiastic and challenging as she had imagined. She could not wait to convey some of her impressions to her family, and wrote pell-mell to them on 18 December, the day after the three-day symposium ended:

> Brian and I have sat together and been inconspicuous, and understood varying amounts of the papers and the discussions. The atmosphere has been really thrilling, as the extraordinary (in every literal and extreme sense!) nature of newly discovered objects[8] was revealed by the radio astronomers from Australia and Jodrell Bank, and the optical astronomers from Palomar and Lick; and as theorists from all over the world tried to think of mechanisms to account for them.
>
> Just to hear, and see, the varied interesting personalities of so many great scientists was marvellous; they vary terrifically, from excitable and bursting with obvious enthusiasm and energy, to calm and quiet (like Maarten Schmidt, who has made more important discoveries by stellar photography and spectroscopy than everyone else put together in this field, and looks as though he can hardly see further than 10 inches).

There has been an intense atmosphere of excitement and interest ever since Hoyle's opening paper. The pooling of ideas must have been of tremendous value to the researchers, but I don't think they have any more certainty as to what the strange objects are than they did. Most of the observational evidence seems to suggest objects at distances out to about a thousand million light years, dimensions several million light years, a hundred times more luminous than the brightest of the great giant galaxies known before (though very much smaller), giving out light and radio waves of an utterly peculiar nature, about a hundred million times the mass of the sun (ie comparable to a small galaxy), extremely compressed; adding up to objects unlike any known before.

To explain the formation of such huge condensations, and the way in which they are giving out such vast amounts of light and radio waves of such peculiar nature, is straining the theorists to the utter limit. The observations may sound like science fiction, but the theories sound more so!

One thing is clear, it'll be a very long time before the least degree of certainty as to the explanation is attained – but such is and has been the case in all major new scientific advances, and anyway we can advance only from one likely (in some sense) model to another more fruitful one ...

She filled many pages of her scrapbook with press clippings and the order of proceedings – the famous names and their special subjects. To the Metcalfs in a burst of joy she wrote with more details. One 'great name' in particular had excited her:

John Wheeler has recently done some really interesting studies combining relativity with the latest ideas of behaviour of matter at heavier-than-nuclear densities, the sort of unifying, thrilling, ultra-difficult stuff that no-one knows what to expect from. He gave to the symposium here in December a talk which caused terrific excitement and controversy, and from what I could follow (small fraction) it was as fascinating as its title, 'The Superdense Star and the Issue of the Final State'.

The final state seemed to lead him to the conclusion that matter would ultimately be crushed out of existence, in a process that could conceivably be regarded as the reverse of creation of matter in a hypothetical initial superdense state of the universe. (You can imagine one source of the controversy, at that, being the presence of Fred Hoyle.) There is plenty of stimulus to dig deeper than the tiny surface scratches I've made on the theory yet![9]

John Wheeler was soon to coin the term 'black hole' for his super-dense final state, a term Beatrice adopted for her own use.

The Tinsleys did not realise that another New Zealander, the mathematician Roy Kerr, was at this conference giving a paper, a 10-minute paper presented without fanfare. It was to have a profound effect on astrophysics and to earn him astrophysicist Subrahmanyan Chandrasekhars's nomination for the Nobel Prize for Physics, although at the time few saw its significance. Chandrasekhar commented:

'In my entire life the most shattering experience has been the realisation that an exact solution of Einstein's equations of general relativity, discovered by the New Zealand mathematician Roy Kerr, provides the absolutely exact representation of untold numbers of massive black holes that populate the universe.'[10]

Kerr's paper gave the exact solution to Einstein's gravitational field equations describing the geometry of spacetime around a *rotating* star, including the extreme conditions of a rotating black hole, where gravity has overwhelmed all other forces. Until then very few scientists, including Einstein himself, had thought that black holes could actually exist and had considered this as being of only theoretical interest.[11]

Another young scientist present at this symposium, Kip Thorne,[12] never forgot the extraordinary lack of interest displayed by the audience at Kerr's presentation. The astronomical community was agog with what they were learning about quasars, and had ears for little else. As Kip Thorne has written:[13] 'The astronomers and astrophysicists had come to Dallas to discuss quasars and they were not interested in Kerr's esoteric mathematical topic. So, as Kerr got up to speak, many slipped out ... others, less polite, argued in whispers ... many of the rest catnapped. This was more than Achilles Papapetrou, one of the world's leading relativists, could stand. As Kerr finished, Papapetrou demanded the floor, stood up and with deep feeling explained the importance of Kerr's feat. He, Papapetrou, had been trying for 30 years to find such a solution to Einstein's equations and had failed, as had other relativists.'[14]

Besides trying to convey her excitement at being present among the scientific greats, Beatrice in her letter wanted to talk about her pleasure in imagining her parents and Theodora being together with Rowena and José and children in Wellington. The latter were still testing the waters, as it were, and there was a lot to consider.[15] When Beatrice was feeling far from her family, not just in distance but in the direction her work and thoughts were taking her, she often ended her letter with kisses and her

Beetle signature. This letter had seven kisses, each labelled with the initials of the three Hills and four Fajardos.[16]

She certainly did not write about everything on her mind, however, such as her growing concern about the lack of opportunity for her at SCAS. Brian was busy and stimulated. She was frustrated and alone. After the initial excitement and stimulation of the symposium, she had found herself in the – to her – quite unacceptable position of marking time. The Center's relativity research had too little application to the cosmological problems that were important to her. Perhaps she had been naive in thinking it would be otherwise. It had not been easy for anyone at Canterbury to visualise the precise scope of the Center's work and interests. People were ignorant of the scene in the US in general. Nobody warned them that they should look for a different kind of university, one that would suit them both. As one of her future collaborators, James Gunn, was to say much later, Beatrice should have been able to take up her scholarship at one of the recognised centres for astronomy. But Brian was the man, the one who had been offered a job, and Beatrice had been assured there were several universities – actually colleges – in and around Dallas. It did not take her long to discover that none reached the standard she had attained at Canterbury, let alone being able to extend her further. What could she do?

She was already certain that cosmology – the study of the universe at large, on the largest scale and over the largest periods of time – was to be her life and work. Stars were to be her tools to learn from, to answer some of the grand questions, to fill in the big, big picture. What was on offer in Dallas was so small and peripheral.

As they both wanted to see as much of the US as they could, the Tinsleys set off by car to explore eastern Texas and Louisiana during the long Easter weekend. This was a new experience altogether: alone in their own car, and being free to go where impulse suggested. They packed the car with sleeping bags, food, primus and books, 'the relaxing kind' such as Salinger's *The Catcher in the Rye*, and set off as if for a Soc Soc weekend except that they were on their own. Anywhere quiet and away from buildings and people was their aim. They found a lake that they explored the next day after sleeping in their car; agreeable and peaceful enough after Dallas, but it could not compare with Beatrice's childhood Lake Taupo.

They decided to drive on to Louisiana, but she told her family she wished that they had not picked up a Louisiana newspaper and read the two editorials, one a heavily racist criticism of the National Association for the Advancement of Coloured People for mounting a drive to end segregation, and the 'sentimental drivel' about the meaning of Easter for Christians. 'I'm sure your blood would have boiled!' Beatrice wrote, making plain that hers had.[17]

She was soon writing home again at length about the visit to Washington she had made with Brian and their friends Olav and Tordis Holt and family from Norway, Olav being Brian's room-mate from the atmosphere group. It was the first of very many journeys Brian was to make over the years for conferences or in connection with his scientific work. Like any first-time visitor to Washington, Beatrice had been impressed by its beauty. Her intense love of nature led her to write paeans about its green squares and splendid trees, and then the Washington Monument and Lincoln Memorial, and the National Gallery, where she thought she could have spent a week: 'I never realised how many of the world's great paintings are in America. The Italian Renaissance paintings and sculptures are the ones I could go back to (and did) over and over again ... I begin to see that reproductions are a tiny fraction of the artistic worth of the originals.'

She allocated special time to the Smithsonian Institution, gazing like any sightseer at the Wright Brothers' plane, but paying particular attention to John Glenn's space capsule, which he used to become the first American to orbit the earth. Outside were a Polaris missile and an Atlas rocket 'at least 60 feet high and 5 feet across!'[18]

This letter gives clues to a major preoccupation though not one that her family could be expected to notice. By now Beatrice was feeling frustrated at virtually every turn at the Center. Few things could have angered her more than her conviction that she was not being taken seriously. In the upper reaches of that male world, the attitudes she was picking up seemed clear: she was only a wife, a young one at that, and far too impatient for her own good; she should learn to wait and see what might turn up.

Many men saw marriage as having a wife to fill their needs. In the Tinsleys' case, Brian seems to have seen Beatrice as being involved in science, probably even having a fulltime job, but not having it as her main interest in life. And, without understanding what marriage would mean in practice, Beatrice almost certainly also saw herself to some extent in a conventional role. Society and her parents expected a woman to marry, to have children before long, and to live happily ever after. She knew she was going to do something special in the world of science, but as yet she did not know what.

To the senior staff at the Southwest Center the months going by may not have seemed long, but to Beatrice every incompletely fulfilled day was a waste of precious time in her attempts to find ways of advancing in cosmology and astrophysics in Dallas. Her situation seemed to her to leave her no choice: the logical thing to do now was carry out what she could of her plan to be a good wife. While Brian had to stay at the Center because of his two-year contract, she would use the time to have children, to begin the family they had taken for granted they would have one day.

Women who have made a conscious decision to have a baby, and who stop using contraception, usually expect to become pregnant right away. Educated women know that there can be a delay, but their emotions tell them otherwise. Beatrice, with her reading and her experience at the Christchurch Family Planning Centre, thought she could tell precisely which were her fertile days each month. Reason was on her side. She could be confident that she knew just when the baby would be born. But she did not become pregnant, and consulted a gynaecologist. The trouble was blocked Fallopian tubes, she was told; a minor operation should fix the problem. This was done. By the time the Tinsleys and the Holts and their two children had driven together to Washington for the men's conference, Beatrice had recovered and was confident that she would soon conceive.

As happens with women who want to become pregnant, Beatrice at this time had become keenly aware of any babies and young children she encountered. The friends had taken three days to drive home from Washington DC to Dallas, with Beatrice observing the children, and no doubt imagining her own. She mentioned a little of this to her family:

> The Holts' little girl, five and a half, has a fascinating imagination. One time in the car she was telling me the story of Thumbelina, translating from her memory of a book in Norwegian and asking her parents for words she didn't know, with marvellous embellishments of the plot and mixtures of words!
> Even the little boy showed remarkable patience for so much travelling and didn't cause much trouble except noise (mostly joyful yells) and smells when his nappy needed changing.[19]

Back in Dallas, Brian was delighted by his new spectrometer, its expensive parts paid for by SCAS. The mirror alone cost $3000. If he had stayed in New Zealand, this sort of equipment would not have been possible. For Beatrice, however, the envisioned pregnancy still did not eventuate. And time was passing, her invaluable creative time. She went again to the gynaecologist, who arranged for a second small operation. But after this she was given a verdict that was hard to believe. She was told it was virtually impossible that she could ever have a child.

Women friends and Rowena remember her state of mind. Beatrice was used to being well ahead of her contemporaries in everything she did. Not that she consciously strove to excel, except academically. She simply did. Now here was something fundamental to being female that apparently she could not do.

If she had grown up modelling herself on her mother, her shock and grief could have been huge. If she had been more fully in love with her husband, she could have felt deeply wounded, incomplete, unable to give him what her upbringing had taught her to be an integral part of a woman's life.

There seems little doubt that she did feel she had failed Brian according to her own vision of what a wife should be and do, but she was not grief-stricken. Rather she faced up to the situation. She told herself, as ever thinking rationally, that one day, in a few years, they would both be long settled in their chosen fields which would be well away from Texas. Then they would consider adopting a child, two children, because the ideal family had a boy and a girl.

That settled in her mind, she now had a clear conscience and could devote time to herself, time to build on all the reading and thinking she had been doing. Somehow she would find a way of studying for her doctorate in cosmology. The solution couldn't be in Dallas, where she was anchored to Brian. But she had made up her mind, and a whole new horizon seemed about to be opening to her. Somehow she would find a way.

CHAPTER NINE

THE DOCTORATE

'When people ask me what use something might be, I always ask them, "Of what use is a poem?" The creative scientist is as much an artist as a painter or a poet ... If it's useful, that's the icing on the cake.'
ALAN McDIARMID[1]

One incident seems to have crystallised Beatrice's feelings about Dallas and the Southwest Center. It propelled her into finding ways of studying elsewhere. This incident, involving scientists and their wives, burned in her mind and summed up this macho world for her. She was to tell the story all the more vehemently because she knew that to many people she would be seen as making a fuss over something trifling.

It happened that a phalanx of scientists' wives regularly met for coffee or tea and a range of social events for which Beatrice had no time and less interest. The protocol of such events was beyond her. Husbands and wives were both present on this occasion. Beatrice was about to join a group of the scientists deep in talk when two of the senior wives intercepted her, anchored her to the tea table and graciously invited her to preside and to pour the coffee or tea for everyone. She declined this honour instantly. No doubt she did it politely, but with finality. It would never occur to Beatrice that she could be shut out of a scientific conversation. But to many of the wives this attractive young woman was an affront. She was not a staff member. She was a wife like the rest of them. She should have known her place and stayed in it, and been suitably grateful for the honour of being invited 'to pour'.[2]

Beatrice did not think of herself as a wife, and resented being categorised as Mrs Brian Tinsley. She was a scientist who happened to be married, and a scientist determined to unravel cosmological problems, including some which might not even have been thought of yet. As for her marriage, the disappointments of Dallas had made her increasingly believe that they had both made a mistake. But, as she was to say,[3] whenever she considered her marriage and thought of leaving Brian, she stopped short because she believed her mother was 'too fragile' for Beatrice to do such a thing to her.

At no stage does it seem to have occurred to her that perhaps she herself was the fragile person in this situation, the one so vulnerable to loss of love. Her mother's torrent of reproaches if she were to leave Brian was something she could not bring herself even to contemplate. No, she would continue to try to make the best of things, and accept that she was tethered to wherever Brian chose to be. There was no reason why they could not be friends. After all, they certainly had some things in common, as she said, and the right thing to do was to focus on those.[4]

By June 1964, nine months after they had arrived in Dallas, Beatrice's investigations had shown her that the University of Texas in Austin had a fledgling astronomy department with a well-regarded staff. As yet Austin had produced no PhDs, but it had the famous McDonald Observatory with the world's fourth-largest telescope which attracted top-class observers. They would be among those producing the kind of data she would need – Beatrice already saw herself as a theorist, not an observer. Austin, too, was the nearest of all possible centres to Dallas, and there seemed no real reason why she should not commute there each week, perhaps for three or four days at a time, depending on her timetable. The return journey would be 400 miles but the travelling time could be thinking time. She wrote away for details, made an appointment, got on a bus and went to Austin.

What happened next remained vividly in the mind of the department's chairman, Professor J. Harlan Smith. He thought it was a young high-school student knocking on his office door and prepared to be graciously dismissive:

> Here was a small and shy-looking girl I took to be a high school graduate. She first had to persuade me she was old enough.
> I warned her that astronomy was a rather difficult subject. I told her we set a long and hard course for our students and that it required not only good preparation but good grades and a lot of application. She sort of smiled quietly and said how she had reasonably good grades – it wasn't till later that I found she'd apparently never had less than an A grade in her life.
> As for her idea of commuting from Dallas to Austin, it sounded absolutely ridiculous. Most graduate students were struggling hard as fulltime students to get through the programme in six years or so.
> She was persuasive, so we gave her the chance. At the end, she thanked me calmly, this little girl – which is how I saw her. I was still

doubtful, but there was a quiet resolution about her which I kept thinking about, and which had won me over.[5]

Beatrice was jubilant. To her family she wrote in triumph, giving the university its full title. In September she would be enrolling in the Astronomy Department of the University of Texas in Austin for a PhD. The American system was different from New Zealand's. She had been told she would have to attend lectures and sit exams for two years, then do a thesis. Her immediate task was to begin revising all the physics she had studied in 1960-61 for her BSc and MSc. She also had to study some introductory astronomy, as in September she would have to sit 'a sort of test' set by the Astronomy Department to pinpoint the gaps in their graduate students' knowledge. She joked that it would be a blot on the University of Canterbury, not to mention herself, if she could not at least 'choke up' everything she once knew. The test was not a pass or fail matter, but 'obviously I want to do just as well as possible.'[6]

To her family she explained the logistics she had worked out:

I expect that I will fly down every Tuesday morning, and bus back (five hours but can read etc) every Friday afternoon. I have rented a room in the home of a former professor's widow, which is nice and close. The considerable expense of all this, fares, fees and living (plus Brian eating out half the week!) can be met by my scholarship, which so far sits intact in our savings account.

It took some deciding to go back to lectures and exams, to travel 400 miles a week (though people say that isn't far in Texas), and worse to be separated from Brian for four days and three nights every week. But we really think it will be worth it in the long run. I was getting depressed at my scientific stagnation, not having a baby either, and if my life is going to be science entirely it would be a waste to continue in the relativity group at the Center.

A bit of separation for a couple of years ought to stimulate our relationship rather than spoil it! Anyway someone who feels they're achieving nothing isn't much to live with.[7]

Some throwaway phrases in this letter are worth looking at. Her 'not having a baby either' was as close as she was prepared to go in letting her family know how she felt, which was not so much depression as frustration at time passing without one of her plans being put into effect. Then there was her listing, as a drawback, that she would be away from Brian for a big part of each week. No doubt she wrote it to placate her family, who might well be distressed at this unorthodox way of being married. As for that 'bit of separation' being likely to stimulate their relationship, she may also have written it out of hope. She was to speak, later, of 'the great cloud which came over me after I married'.[8] Time apart, she was saying now, could be what they both needed. It is notable, too, that she spoke of a couple of years for her PhD, not the five or six which Harlan Smith expected students to take. Another tossed-off line, 'if my life is going to be science entirely', gives a significant clue. The possibility of adoption, even well into the future, was not in the forefront of her mind.

During September Brian drove some 5000 miles, exploring sites where his new spectrometer could be set up. He decided on a site in Soccono, New Mexico, in high mountains. The experience of driving such long distances helped them both put into perspective the mere 400 miles involved in Beatrice's commuting to Austin.

She sat her screening exam in astronomy, physics and mathematics and was accepted at once, as soon as her papers were marked. 'I passed the exam all right,'was her only comment to her family. In fact, staff and students were astounded by the quality of her work. Ron Angione, one of the students who entered the graduate school when Beatrice did, recalled:

> When they announced the results, I was thankful just to be among the few who passed. It seemed that all the faculty came up to congratulate Beatrice on her exceptional performance. Right away we all knew that she was special.[9]

One of the professors of astronomy at Austin, Gérard de Vaucouleurs, not noted for over-generous assessment of students, was to say:

> I remember clearly her preliminary examination papers; only an expert could detect minor inaccuracies in her discussion of the various subjects irrespective of whether she had taken formal classes in the subject or not.
>
> Obviously she had done extensive reading and study on her own.'[10]

As Beatrice had learned little astronomy at university before finishing her MSc, her own reading and thinking during her 'wasted' time in Dallas had been to very good effect. Most of the Austin faculty could not have done so well, Harlan Smith decided.[11]

In September 1964 Beatrice settled into the Tuesday–Friday routine of study in Austin, with three days back in Dallas each week. It was only in 1962 that Austin had enrolled its first PhD student in astronomy. Remarkably, he too was a New Zealander named Hill. Graham Hill had graduated MSc from Auckland University and had gone to the US for further academic work. He went to Austin, with Harlan Smith as his adviser, as he was rather late in applying to other, better known, universities. He was married, with little time or money for socialising, but he soon became aware of Beatrice: 'Everyone became aware of Beatrice. She was truly a phenomenon.'[12]

One of the post-graduate students in her class, John Williams, considered it a group that offered quite strong competition:

> There was the general impression that if your grade was in the mid-80s, you could get an A. Then – sensation – there was a New Zealander, Graham Hill, who got 95 in Frank Edmonds's course – gee!
>
> Then another New Zealander and another Hill – they weren't related – came along. Beatrice. She made 100 in a test, the first ever. And 100 in the next test. Then in the third test she made only 99 – some human failings after all.[13]

Some could not believe that Beatrice and her marks were real. One of the faculty, Terry Deeming, noticed a vein of mutiny among some of the male students. He went to some trouble to go over her tests again, in case he was being too generous. He was not.[14] De Vaucouleurs also saw that some students considered her unfair competition: but 'she was completely self-propelled'.

Beatrice seemed unaware of all this. Austin pleased her. The university and its staff and students, its campus, the galleries, the museum with the skeletons of the huge dinosaurs which once roamed the plains of Texas – the whole atmosphere of this city she found stimulating, even exhilarating. She also hurled herself into work. A desk and bookcase were assigned to her in a room shared with seven men; she was the only woman postgraduate student. Two of her lectures were at 8am. Each day she began work earlier than this, studying at her desk in the adjoining library or in the room she rented.

The discovery that she would have to study subjects other than astronomy surprised her. A compulsory subject was the constitution of the United States, but she enjoyed this because of the new insights it gave her. She had not realised the extent to which the states differed from one another, and how much power they had independently of Washington. The country, she found, really was a union of states. Altogether Beatrice was so enjoying this new life – being in the thick of real work and study again – that she could be amused rather than irked and indignant when she discovered that, as a woman, she needed Brian's written consent before she could open charge accounts in the town, with Brian to be responsible for seeing that they were paid.

Sometimes she realised she was 'a foreigner', as when she discovered gaps in her knowledge of US history, things which other students seemed always to have known. They enjoyed teaching her. Ron Angione had a desk near hers. He was impressed by how studious she was, but also by her friendliness and helpfulness to fellow students. He has a favourite 'Beatrice story':

> I call it the story of the prof who needed 'to look into it further'. The professor was Frank Edmonds, a former student of Nobel laureate Subrahmanyan Chandrasekhar. A monograph had just been published, 1962, by Hayashi, Hoshi and Sugimoto on stellar evolution. Edmonds offered an advanced seminar course on this monograph.
>
> Just three of us took this course: Beatrice, who hadn't done any university astronomy until then; Pete Jordal, who was already working on his dissertation; and me – I already had a master's degree in astronomy.
>
> The monograph was 183 pages long and very technical. While Pete and I struggled just to follow along, Beatrice would re-derive and check every one of the hundreds of equations in the text. I can remember a number of times when she would ask Edmonds how the authors got, say, equation 4B.35. Edmonds would stare at the equation for a while. Then he would say, 'Well, Miss Tinsley,' (we were all either Mister or Miss), 'I will have to look into this and get back to you next time.'
>
> The next class meeting he would have an answer which sometimes was that Beatrice was right and there was an error in the monograph. And then it would happen again. 'Well, Miss Tinsley, I need to look into this further.'
>
> On looking through my old copy of the monograph, I see corrections in it that are likely the result of her questions. Beatrice was not showing off. She was just trying to understand stellar evolution fully. Nobody resented the fact that she was exceptional, and this was in part because she was such a nice person.
>
> Generous, too. She would always help fellow students and colleagues. Years later, when I was on the faculty of San Diego State University, I had a grant to determine the mass of the giant elliptical galaxy known as M87. To do this I needed the latest stellar population model of giant elliptical galaxies. I wrote to Beatrice who very kindly and generously sent me her latest, as yet unpublished, data on models of galaxies.[15]

In Dallas the Southwest Center at last moved to its own buildings in Richardson. Brian, faced with a 13-mile drive to and from work each day, suggested they move to North Dallas, bordering on Richardson. They found a house at 8739 Boundbrook Avenue, backing on to a park and with its own garden. At once they wondered how they had existed for months in a small flat with no view. The young girl part of Beatrice, which used to enjoy playing house, emerged in a letter home with descriptions of virtually every detail of all the furnishings in every room.

Now she was making friends in Austin as well as Dallas, many of them 'women who were people in their own right and not just the wife of someone or other'. She worked out ways to see her friends, throwing herself with energy and care into organising her days and nights, and her lives in two different cities. Each minute was so precious, as it is when one is lucky enough to be able to work at the thing one most wants to do in life. By mid-December, for instance, during her three days in Dallas and when Brian had his spectrometer set up for observations on the Center's roof, she observed that 'life was getting all unorganised', and proceeded to organise it. One night she left Brian at the Center, then drove to their friends the Ostvaths, where she played sonatas, returned home and worked at astronomy for several hours, slept for two hours, then drove off to collect Brian when he phoned at 4am. He had intended staying until dawn but the sky became obscured by cloud. She noted with satisfaction the next day that the weather was clear so Brian would be off observing again – she would be able to do a great deal of work on her own at home.

During the 10-day break over Christmas the Tinsleys decided to visit Mexico City, travelling by bus, staying in cheap accommodation and immersing themselves in the country's history. They visited pyramids and Aztec ruins, and walked around parks and galleries and historic buildings until their feet would walk no more. With memories of their first night in the US, they made their second nightclub visit and again enjoyed themselves with determination.

Back home they found that the dog they had recently acquired, a handsome and responsive collie, had been well looked after in their absence but she was ecstatic to see them again. She had been named Rata after the glowing red flower and vine of the New Zealand bush, but the Maori pronunciation of the 'a' as 'ah' was quickly changed by their friends and neighbours into the sharp-sounding name of Rat-ter. Ratter or Rata, she brought another dimension into their lives.

Altogether Beatrice thought this new year, 1965, would be the most important of her life. It began very well. Normally, graduate students took two years of courses before a three-hour oral examination conducted by five examiners from the faculty. In January, when Beatrice turned 24, she easily passed this, having been permitted to appear before the examiners after less than five months' study. 'She was a phenomenon,' Gérard de Vaucouleurs was to repeat, almost angrily.[16] He believed that all good things take time, and that students simply should not leapfrog ahead. But there was no holding Beatrice back. The way was now clear for her to tackle the preliminary work on her thesis.

And here were the big questions with which she had begun to wrestle in her early teenage years, and which she had debated with Socratic Society friends on those heady evenings in front of the fire. What is this earth, this universe? How old is it?[17] Where did it come from? Where is it going? At last she was being encouraged to have a shot at finding some of the answers – incredible joy!

There was another reason for her happiness. She was now regularly on her own again, not constantly having to be constrained by, or adapt herself to, another person. She told her family she missed Brian when he went off, increasingly, on his spectrometer trips and to scientific conferences. Now that she was a student again, with a room of her own in her landlady's house, and precious hours for thinking with no disturbances or claims on her attention during her journeys to and from Dallas each week, she felt revived and refocused, herself again. And yes, she answered friends, one day well in the future she and Brian would probably adopt a child. Two children. This was the right thing to do. But all this would happen some time after they had found a university where they could both do the work they wanted to, she teaching as well as doing research while Brian continued with his spectrometer work.

Her adviser was the physicist Rainer Sachs, an expert in relativity theory. The two of them seemed to be on the same wavelength. Some years later Beatrice was to say, 'Rainer Sachs wanted a student who was interested in cosmology so that he could learn a bit of astrophysics himself. He more or less said to me to go ahead and find something to work on. This is exactly what I wanted to hear.'[18]

Sachs had agreed that as yet there were no adequate calculations which would begin to give an overall view of the cosmos. He was encouraging, enthusiastic even, and delighted that his astronomy department had secured such a student. This young woman with the non-American accent, who spoke so quickly that it was hard to keep pace with her, whose eyes widened and sparkled as she took in and assessed each new thread of information, might just, possibly, be the one to make a truly significant contribution to pushing out the frontiers of astronomy.

Other faculty members came to see that Beatrice's thesis was largely self-directed. Nick (Neville) Woolf, at that time a new Austin faculty member who was spending half his time in New York, recalled having only one session with Beatrice, when she presented him with the plan for her work. By then it was well along the way – too far, he believed, for him to make an effective input – and he thought it best for her just to get her thesis completed. It was plain that she was outstanding but he came to think that the faculty saw only a fraction of what she might have done had they made a better job of directing her. He said:

> It is my impression that at Austin we did very little that was useful to Beatrice. She would have benefited from much more help in the selection of her topic, and in developing it.
>
> It was somewhat like I have heard of Cecilia Payne-Gaposchkin's thesis in Cambridge – a self-starter just picking up the ball and running with it. I am just grateful that we were probably not harmful to Beatrice like Eddington was to Mrs Payne-Gaposchkin.[19]

Edward Hill gamely tried to keep up with the scientific work of his daughter and son-in-law. Brian's work seemed a little more concrete, or, if that were not exactly the word, at least a little easier to grasp. To his queries about Brian's spectrometer, Beatrice replied that he had a National Science Foundation grant for a specific observing programme, and had a number of other ideas. First he would be writing for journal publication. As he had had nothing published since coming to the US he was beginning to be affected by the 'publish or perish' complex, something she declared roundly that she disdained: 'I resent the whole system of scientific prestige-building, but rebellion doesn't help.'

Then she tried to give her family an account of what, at that stage, she thought her PhD thesis was to be about:

> Really I will be studying a whole lot of different theories of cosmology, to see which is best able to explain the observations made with optical and radio telescopes on different galaxies. The theories are based on Einstein's General Relativity (that means Steady State is excluded) and they represent different motions of the galaxies in the expansion of the universe, and different ages of the universe, different number of galaxies per unit volume of space, and so on.
>
> Using each theory one works out what to expect in observations of the faintest galaxies, quasars and radio sources. Then one tries to choose the theory that fits best. But I do not think the observations are complete enough yet for a real choice to be made.
>
> The idea is to have a systematic and extensive set of calculations to compare with the observations as they become available, as it is

a very interesting question which theory is best. An answer would have a lot of information about the past and future of the universe.

The calculations also depend on what the various galaxies etc were like at the time the light left them, the light which now gets to the telescopes, and it isn't necessarily true that they were the same as nearby objects! So I will have to include a variety of different assumptions about the galaxies themselves, which will be a matter of astrophysics.

Dr Sachs suggests that a good systematic set of calculations is really needed, though lots of people have done bits of this before of course. He is very good at seeing things straight, and seeing the wood in the trees of calculations that have been done in the past; so I think I will have very good guidance from him.

Already it looks a very interesting subject, because there are serious difficulties in trying to make any of the theories fit! (In case Father remembers how fond people were of the Steady State idea several years ago, I should mention that there have been observations of radio sources which seem to have ruled it out entirely. Even Hoyle agrees.)

Of course if it is impossible to explain the observations with any of the General Relativity theories, that will be worth finding out ... I am still up to my ears in all sorts of books on astronomy for the prelim, and wishing I could get on with my thesis because the plan of attack is clear now, and very intriguing.[20]

Edward Hill was to say to enquiring friends that it was not easy to say what Beatrice was doing. Everyone knew (or thought they did) what was meant by Rutherford's splitting the atom. Beatrice had explained that synthesising meant a gathering up and a putting together. But it was what she was putting together that was so hard to follow.

By Christmas Beatrice had a clear grasp of what needed to be done for her PhD, and was impatient to get back to it after the holiday diversions. She knew the work would be intensive; and that to her meant every waking hour. But this would be pure joy.

She worked out that it should be possible to complete her PhD within two years instead of the customary five, although some of the Austin professors continued to let her see they were not happy about this unusual and unseemly haste. The map of the road ahead seemed clear, her journey unstoppable.

Then, in the early spring of 1966, a letter from New Zealand changed her life.

CHAPTER TEN

THE BIGGEST CHANGE

'Science is the search for truth.'
Linus Pauling

The letter arrived without warning, but it had been written out of a background of fear. One of the most feared things that could happen to unmarried women in the Western world, as late as the 1960s and 70s and even beyond, was to fall pregnant. Their babies, 'born out of wedlock' and called illegitimate, brought social disgrace on themselves and their families. Nearly always they were handed over to strangers to adopt.

Brian's elders in New Zealand had sent the letter. One of the family members was pregnant, and the baby would have to be adopted. And there, over in the United States, were Brian and Beatrice, who had told their respective families that one day they would indeed adopt a child – two children, in fact. It was obvious what should be done.

To Brian it was obvious, too. He said so.

The Tinsleys had always known that their own futures would be at risk if they had a baby before they were ready. Their volunteer work with the Family Planning Association had been partly in response to this. In Dallas, after Beatrice had failed to conceive when she planned, she had accepted what she understood to be her gynaecologist's final verdict: she was unable to have children. Instead, she would put all her time and energies into astronomy until, well into the future, conditions were right for them to adopt. Thoughts of a family had then disappeared from her mind.

Beatrice was eventually to tell her father that she 'hesitated considerably' when she was suddenly asked to adopt a baby in a few months. This was an under-statement. To her friends she was more forthright – her answer was no. This was not the time. She simply was not ready. Every particle of her body was set on a different course. It was not just the external framework of her new life of study again, with its 400-mile commuting, the long hours of single-minded work to get her doctorate, and her regained freedom to concentrate on her own needs. Above all, the faculty's expectations of her and her thesis had opened the way for her own blossoming recognition of what she was doing. Apart from a handful of her professors at Austin, she was the only person who knew what she might be able to bring about: a synthesis of available knowledge about the origins of the universe itself.

To be asked suddenly to give all this up? How could she? But she knew she had to consider adoption. Brian marshalled plenty of arguments why she should.

Beatrice was not instinctively maternal. She had not been mothered herself, except by the nanny who had been sent away. In her childhood play with her koalas and teddy bears, she was more teacher than mother. Now there was nobody to talk to about the whole business of families, nor anybody to tell her how the long months of pregnancy with their slow build-up of hormonal changes help prepare women for the births of their babies and – if all goes well – release joy in them. Nobody was around to tell her how women who are about to adopt need to be whole-hearted, to compensate for this. In the 1960s there was virtually no understanding of any of these matters.[1]

Beatrice had nothing to help her sort out her conflict of feelings – nothing, that is, except the logic she lived by, and her attempts to be a good person. Logic argued both ways. She and Brian had earlier taken for granted that one day they would adopt. Here suddenly was the opportunity to do what they had planned. Logic also told her that it was too soon. This was her own year, for her own work. And so were the next year and the next.

Yet her parents' Oxford Group Absolutes had gone deep: 'Absolute Unselfishness' had been drummed in as a daily goal. Her strongly developed faculty of logic overcame her scientist self. It was logical for Brian to expect her to agree to take this child. As she could not 'give' him his own child (in the language of the day), then this would be the closest they would ever come. And this particular baby would make Brian even more pleased. It was the best possible way of compensating him for not being able to father his own child. Besides, the baby would be kept within the extended family and not lost to strangers.

Beatrice was kind. She argued with herself, challenged herself to be both a scientist and a mother, thought about what she expected of herself, said yes, and wrote to the New Zealand family. Brian was jubilant. In his recollection, 'Beatrice thought about it for a week or so but was never forthright with a no. She considered the pros and cons.'[2]

Now they had about five months to make preparations and arrange to go to New Zealand to collect the baby at the end of August. Until then Beatrice would use every possible moment to get ahead with her work. She also began to make lists to help her be a good mother. Nobody pointed out that logic is not the first thing a mother requires. The most carefully prepared lists do not give the infant what it most needs: that 'warm, intimate and continuous relationship' with its mother or mother-substitute, in the words of pioneering family psychiatrist Dr John Bowlby.[3]

This was a watershed in Beatrice's life. Her much later admission that she had 'hesitated considerably' before agreeing to the adoption – considerably longer than a week – still concealed the truth. She had consulted friends, plainly showing her ambivalence, and had waited to hear back. Everyone appears to have advised against adoption at this time, given Beatrice's circumstances, even while knowing virtually nothing of her uncertainties about her marriage. In the end, she ignored their advice.

Now that the time she had been counting on to concentrate on her doctorate had suddenly been slashed, Beatrice stuck even more rigidly to her regime, working in Austin through the summer vacation on models of the evolution of galaxies. Babies, she was told by the few young mothers she encountered, 'take up every moment of your time'. Privately she was determined this would not be the case in her household. She began buying baby clothes and a folding crib and read mothercraft books, looking particularly to the indispensable Dr Spock. These books she claimed to study as relaxation from galaxies, or so she told her parents. Other preparations included having a local welfare worker, acting on behalf of the New Zealand adoption authorities, check the Tinsleys out and approve them and their circumstances.

Midway through 1966 she was still uncertain how long it would take her to complete her research but told her parents that she should be able to do enough to finish her degree 'in a year or two'. As for her long-term plans, she hoped 'ultimately (that is, when we are at the children-grown-up stage) to look for a university plus research post – I feel very much that I would be a lot more satisfied doing teaching as well, not all research.'[4] The reference to children meant that she had also decided to have the family of her early imaginings: two children, a girl and a boy. An advantage of adoption was that this need not be left to chance: she could ensure a child of each sex.

Her mention of the satisfactions of teaching was probably prompted by the talks and tutoring sessions she had begun to give to groups of amateur astronomers, answering questions about quasars and cosmology. She particularly enjoyed talking to a group of bright and enthusiastic junior astronomers. Teaching, in some form, was stimulating, and in spite of her ingrained habit of rapid speech she gave thought to speaking at her students' level so they could follow her, and be led on. She seems to have known intuitively that specialists without experience in teaching or a natural affinity with it can have a disastrous effect on a class.[5]

By July Beatrice had come to believe that what she had been working on 'might have some usefulness – to cosmologists and towards my degree'. She had left all her results and a summary with her supervisor, Rainer Sachs, and in early August she returned to Austin to discuss them. She was still uncertain how long it would take her to finish – another year or two, maybe.

Rainer Sachs was not so moderate in his assessment. As faculty members recalled many years later, he was astonished and delighted by what she had done, and in such a remarkably short time. She was told that already she had produced more than enough to warrant a doctorate.

On 18 August she wrote to her parents:

> It seems I've done enough research to get a PhD with, and now have only to write it all up, which I can hardly believe. That's only a few months' work, even with a baby. Probably I could graduate at the end of the academic year ie. next May, though I hope the thesis will be written and oral exams over long before that. I was lucky to hit on a fruitful line of research and get some interesting results so soon.
>
> Of course it is still subject to the approval of the rest of my committee, including the Frenchman de Vaucouleurs who doesn't think anyone should get a PhD in less than five years whatever they've discovered. I'll have to sell it to him somehow – lots of references to his own work, perhaps!

This last sentence shows Beatrice being practical, as well as mischievous; she and Brian had seen something of faculty politics and manoeuverings. Meanwhile they were both jumping to attention each time their telephone rang, in case this was news from New Zealand. But their timing was good. The baby, a boy, was born on 28 August, one day before they were due to arrive in Auckland. There were more interviews with adoption authorities, and the adoption itself could not be legalised until after 10 days from the birth. They used this time to be with their respective families, but promised themselves a good long visit with their old friends in Christchurch on their next visit.

Then they were able to claim the baby. In her father's recollection, Beatrice looked 'absolutely delighted' as she carried the infant, named Alan, up the path to the Hills' home in Wellington. To be able to hold him and feel his dependence on her had swept her remaining reservations aside. She and Brian looked as happy as new parents anywhere.

The family of three flew back to Dallas. The baby slept and fed well, and by the beginning of October weighed 10 pounds, 'which is considered very good'. Beatrice read, thought and made notes while she gave Alan his bottles, and while he slept. In one paragraph in her letter home of 6 October, she moved easily from cosmology to baby-feeding:

> I've spent several hours a day on my thesis, and tomorrow will start typing the semi-final draft, to go to Sachs and Harlan Smith for approval. The pediatrician has put Alan on to 'solids' already because of his big appetite and I have a lot of fun with a few teaspoons of runny baby rice ...

Alan, a healthy, happy, easy baby, still slept well through the night, not waking until 7am. Most afternoons Beatrice brought his cot into her study for a few hours. He was happy there, 'making a jangle with his rattles' and watching her work. When that palled she played him records and clapped his hands to the music. 'This always makes him beam and gurgle. Very rewarding! I have to remind myself of that when he is screaming with tiredness at 5.30pm. He only takes about 10 minutes to get to sleep, but it seems like an hour.'

She soon received approval for the final draft of her thesis, and prepared it for typing. But she found that the public typists she approached were horrified by how it bristled with mathematical calculations and symbols. Moreover, they would charge at least a dollar a page, an alarming amount. That consideration, plus the thought of having to dress Alan up and take him with her when she visited and briefed the typist, made her decide to hire an electric typewriter and do the job herself. She could do it more quickly than anyone else, because to others it would be gibberish – virtually impossible for a non-mathematician to follow, she discovered.

Alan co-operated by sleeping well during the day, too, while Beatrice tackled this formidable typing task, and then took the pages downtown to be duplicated. On 28 November 1966, just two years and two months since she had begun travelling to Austin to work for her PhD, she wrote to her parents:

> The duplicating will cost a fortune as there are 125 pages, 20 being graphs, and the only method of reproduction acceptable to the Graduate School is photo offset, which costs $3 a page. There are 50 copies for the same price as one, so I will get 50 copies of the thing, costing 3 x 125 = $375. The copies won't be wasted as people

send out copies of their dissertations widely here, as pre-prints, to everybody in the field. It will be another year before it appears in a Journal. Hard to believe that the final stages are here. Now when Alan is asleep I am reading the literature and going through all my own notes since I started.

It's getting late and I'm getting sleepy. The days are very full but mostly with things I find worthwhile. I suppose even folding diapers is worthwhile when you think about it.

Her oral examination was scheduled for the week before Christmas but was suddenly advanced by three weeks. Such a move would have thrown most students off balance, but not Beatrice. She managed to change her baby-sitting arrangements, and got herself off to Austin in time.

In a letter of Christmas greetings to her parents, she wrote on 7 December:

Now I have my PhD. The oral was last Tuesday and I didn't feel very proud of myself after it but they passed me. So although there won't be any capping ceremony till June, I officially have the degree.

That sounds as if she may have only just scraped through the oral. In fact, the opposite was the case. One of her examiners, Bob Tull, says that Beatrice gave an exceptionally powerful presentation, so powerful that, in the discussion which followed in her absence, it was seriously suggested that they simply award her the PhD at that point, without asking her any questions.[6] The examiners did follow precedent, but for Beatrice then to tell her family that she did not feel very proud of herself is striking, all the more because she almost always knew when she was right and when she was not on firm astronomical ground.

It seems likely that she wrote rather sadly, almost dismissively, of her doctorate because she was suddenly faced with the big question: what now? What could she possibly do now, back in Dallas, with a doctorate but with no prospect of a job? The next sentence of that letter home reads: 'Afterwards I felt very anti-climactic and deflated, seeing no purpose ahead but washing and sweeping.' A few days later, however, as she continued this letter, she had talked herself into a more positive mood:

But by now I am altogether happy at the thought of doing astrophysics in the spare time left by Alan and Brian, and being Alan's mother is my no.1 priority as long as he is totally dependent all day.

I might consider applying for a post-doctoral fellowship to do research at home for the next few months, and then spend a lot of

Beatrice's paternal grandparents, Eustace Hill in Cavalry uniform and Muriel Hill when she was presented at Court. These photographs of his parents, remote from everyday life, are how Beatrice's father, Edward, remembered them.

The 'fairytale wedding': Edward and Jean Hill (née Morton) in August 1937.

Beatrice with her adored nanny, Constance Gullidge, in 1941.

Beatrice and her father, Edward Hill, 1941.

Beatrice aged three.

The three sisters in Christchurch, photographed by Carlotta Munz about 1946. From left, Beatrice, Rowena and Theodora.

Beatrice, Rowena and Theodora with their pony.

ABOVE: Jean Hill and her three daughters. From left, Beatrice with their dog Russet, Theodora and Rowena.

The Hill family on holiday in England, 1952.

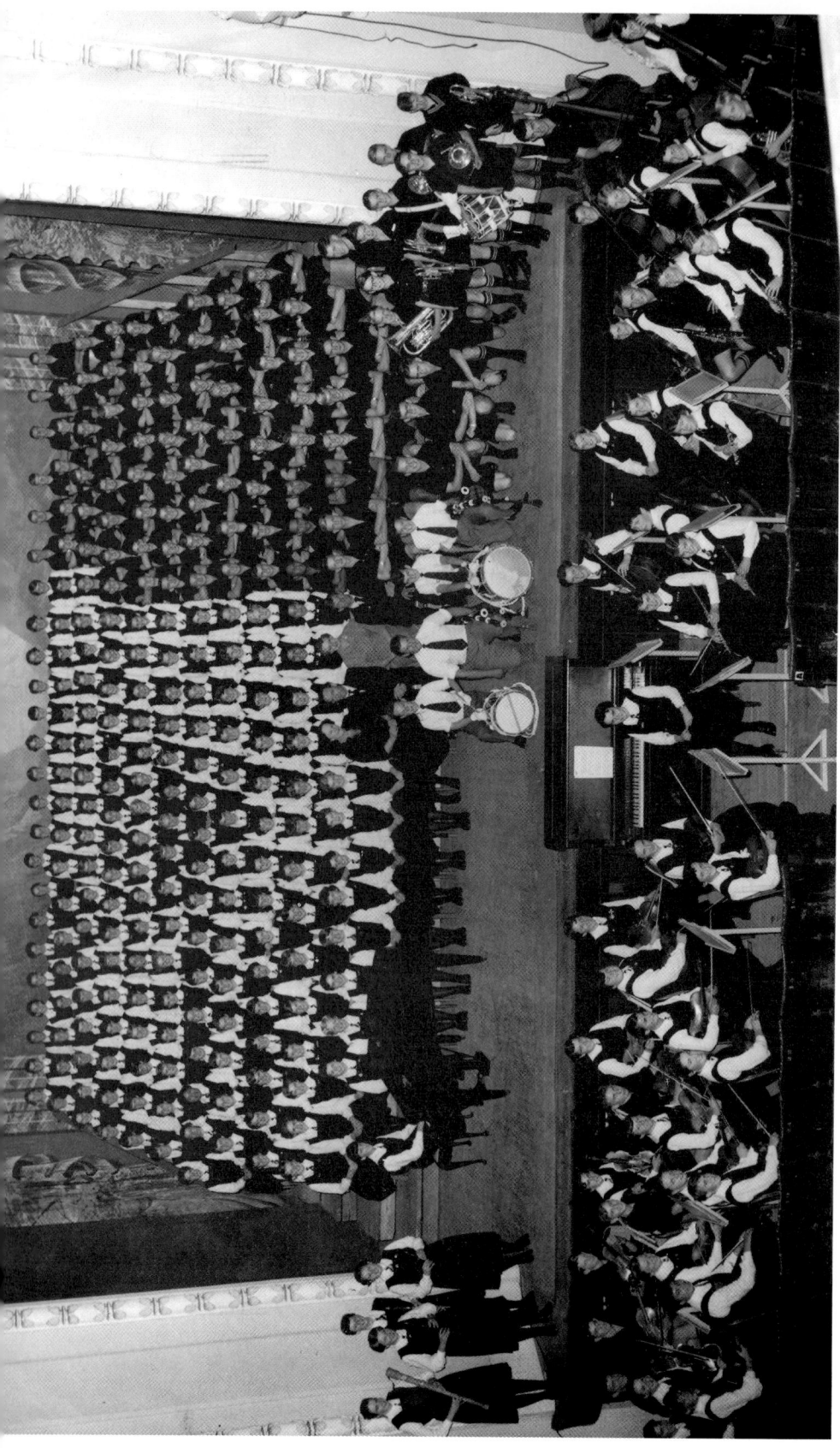

The three sisters in 1957, from left: Theodora, Rowena and Beatrice.

The first fiancé, Colin Broadley. This promotion photo was taken for John O'Shea's 1964 feature film, Runaway.

Pacific Films Collection; New Zealand Film Archive/Nga Kaitiaki O Nga Taonga Whitiahua

OPPOSITE PAGE: The combined orchestras and choirs of the two New Plymouth High Schools are led by Beatrice Hill (centre front) and Theodora Hill at the piano.

Right: Beatrice is capped BSc at Canterbury University.

Below: 'Behold two happy people.' Beatrice and Brian Tinsley announce their engagement.

it on getting daily help to do all the housework so I can study when Alan is asleep and in the evenings – most of which Brian spends at the Center anyway.

I would probably get 30–40 hours a week by swapping all the housework for astrophysics, and be much happier if no richer. Depends if the University of Texas is willing to offer the fellowship with me away from the campus.

Alan is flourishing and beautiful. He is round everywhere but not too fat. Getting very responsive to people, specially Brian and me, with a big repertoire of noises and bubbles and smiles and giggles (the last are hilarious, and we both get the giggles at each other).'

A testing time for the family came at the end of January 1967. Beatrice attended much of a 10-day international cosmology conference, another Texas Symposium although held in New York, and rejoiced in no longer being labelled in that galling way as 'a mere student'. The new Dr Tinsley had no idea that sharks swam in cosmological waters, and that her thesis, when published, would bring some swimming purposefully around her. Instead she saw blissful days of cosmology, whenever she could arrange childcare. And, for the very first time, she herself had been invited to give a 10-minute paper, on her thesis and the evolution of galaxies, because 'my work is very relevant to the problems under discussion'.

Satisfaction bubbled from her letter as she told her family that 'Brian and Alan are going to do art galleries all day'[7] while she – at long last – was the one who would be getting immersed in her own field, and meeting some of the great names in it. Brian in fact spent several afternoons at another conference, on meteorology, but he spelled her so that she was able to go twice to evening receptions at the cosmology conference, drinking in the buzz of talk and discovering who was working on what, the professional stimulation she had been yearning for. She was able to talk again with people she had already met, or had at least seen from the outskirts, at the few previous meetings she had managed to attend.

One of the first of the many people she was later to encourage or mentor made a big impression on her here. A young Japanese woman astronomer, Sachiko Tsuruta, had just got her PhD and begun working at Harvard-Smithsonian. Few women attended such meetings at that time, and the two were drawn together. Sachiko's story also spoke to Beatrice's feminism. Sachiko was the second of six daughters. Sons, not daughters, were educated in Japan at that time: 'Girls had no chance. My father decided to treat his two oldest daughters as sons, and sent us to the US for higher education. We both got a PhD. Our father wept for joy.'[8]

The outstanding young astronomer from Caltech, James Gunn, had heard of Beatrice and her work. Four years her senior, he was obviously destined for great things.[9]

Reflecting on this first meeting with her, Jim Gunn was to say, 'I found her incredibly intelligent, incredibly impressive. She had a completely fresh approach to her subject.'[10] It was a seminal meeting for them both, but as yet Beatrice had no idea of the impression she had made.

She had prepared for this New York visit by studying a guidebook for museums, art galleries, concerts and authentic low-cost eating-places for various immigrant groups. In this way they made the most of the experience. Alan had obviously been happy:

> He's travelled beautifully in the baby-sling, sleeping or gazing through galleries and museums. He's terrifically smiley and friendly now, with a gorgeous fund of gurgles and happy noises. We've had his door-swing up between our room and bathroom, and there he sits playing and singing and delighting the hotel maid with smiles. We fondly believe he's a most superior baby!

She and Alan returned to Dallas to a planned succession of 'early nights and studious and maternal days', while Brian stayed on in New York for a conference centred on Mars. Conferences in his broad field were frequent, and often in desirable exotic settings. Later that year there was to be one in Norway. The New York venture had been so successful that Beatrice wondered if it might be possible for them both to go, taking the baby.[11]

A letter addressed to 'Dr Beatrice Hill' had arrived from her father, who had long thought of attempting a doctorate himself. Now, at last, this degree had come into his family. She explained that this would have been her professional name if she had published anything before marriage, or if, as she wished, she had kept her maiden name. She was able to say that she had been granted a fellowship that enabled her to do some work at home. She was spending rather more than half of the fund employing Thelma, 'a lovely warm-hearted Negro woman who works hard and cheerfully and is great friends with Alan'. She came every weekday for four hours, primarily to care for the baby but also to do basic housework. Thelma used her earnings to pay lawyers' fees for adopting a year-old waif. The two women became friends at once.

Beatrice thought that she could make her fellowship last out until May, 'a few months' liberation from sinks and brooms ... Now I can divide the time among Alan, Brian and astronomy.' She was corresponding with some of the astronomers she had met, and was working on a paper called 'Equivalent Widths of Interest for Studies of the Composition and Evolution of Galaxies'.[12]

Now all three of the Hills' daughters were living out of New Zealand, with Theodora studying Italian and music in Florence. This was also the base for Rowena and José Fajardo, with Sito now five and Lila four. Edward and Jean Hill planned to use some Morton family trust money to rent a serviced villa on the Italian coast at Forte di Marmi, near Florence, with room for the whole family. After the reunion the Tinsleys would go to Norway for Brian's conference, then on to London.

They had busy programmes meantime. Beatrice was researching and organising material to discuss with others in her field if she could arrange to meet them. She had also agreed to talk about quasars for an educational astronomy programme, organised by the Dallas planetarium, on local television. Brian was away from home for 10 days, driving to New Mexico with extra equipment for his spectrometer.

Their time in Florence went quickly, with Beatrice saying it should have been three times as long, with more hours for her sisters. She particularly wanted to talk with Rowena, whose marriage had already gone through patches of misery and frustration. And now they were both mothers, both trying to find small chunks of time for themselves and their work. 'I do so often and so much regret that Rowena and I live so hopelessly far apart,' Beatrice wrote in a long letter on 18 September after being with her in Italy.

The Tinsleys' ambitious itinerary, which Beatrice had worked out so that they saw everything possible, took them on a three-day journey by train through Italy and Switzerland to Denmark, Sweden and on to Brian's conference in Norway. For most of the journey there was no spare seat for Alan, but he lived up to his 'superior baby' status by sleeping for long stretches in Beatrice's arms, she wrote. She, looking out blearily but longingly at the 'gorgeous countryside', could by now imagine a time when it would be her conference she was going to, her turn to decide which countries to visit, so that she could explore and immerse herself in such beautiful landscapes.

In Oslo Brian greeted scientists he had met before, and both took especial pleasure in seeing their old friends from their first years at the Southwest Center, Olav and Tordis Holt. After a week of comparative relaxation for Beatrice the family went back to Switzerland for another conference for Brian, this one concerned with the earth's magnetic field and the physics and chemistry of the upper atmosphere.

In London they stayed with Edward and Jean Hill again, in the flat in the home of Helen Hannay, Jean's sister and Beatrice's godmother. This house was just around the corner from Harrods and, as they were to arrive on a Friday night, Beatrice sent on ahead a precise list of the provisions they would need –'presuming we can buy groceries on a Saturday in London'

– for Alan's supper, and breakfasts for everyone.

Brian met members of the Hill family in England for the first time, and they went on excursions with friends from New Zealand. The latter provide a rather different picture of Beatrice as a mother from the one she was presenting in her letters. Guff France could not help observing that Beatrice was angry with Brian for not watching Alan carefully, 'whereas when they first married she used to hang on his every word'. But it was Beatrice's lack of 'ordinary motherliness' which struck her friend Penny Saunders when they met in a London restaurant. Alan began crying and struggling, and Beatrice pushed him down in his pushchair. 'I didn't have children then, but I knew you didn't do that.' They met up again in Paris after the Tinsleys had been all around the Louvre. 'Alan was exhausted. It wasn't a happy occasion,' Penny observed. 'I had the feeling she didn't know how to be a mother.'[13]

Back in Dallas, the Tinsleys made two interlinked decisions. They would find a bigger house, and they would adopt another child, a girl. Many years later, after what Edward Hill still saw only as Beatrice's 'hesitations' about the first adoption, he asked her why she had been intent on adopting a second child. She thought it had been because she had not wanted Alan to be a spoilt only child; Beatrice always wanted to do the right thing. A likely cogent reason for adopting another baby was that Alan happened to be an easy, biddable and responsive child. His obliging nights of long, unbroken sleep and his comparatively self-sufficient and cheerful days had led her to believe that this was the norm. Another baby would presumably take even less time and attention, now that she felt she knew the way. There was nobody around to tell her what most mothers know: that the work and time involved with a second baby, with its different patterns and needs, are greatly increased, and that babies come in all possible combinations of temperament.

At the beginning of December Beatrice told her parents of the decision to adopt another child. Strangely, she left this momentous news to the last few lines of her aerogramme's main page, and the turn-over. After a curious mix of topics, such as Brian's absence in Mexico, Alan's activities, her own music and work on some problems in astronomy, and comments on the books she was reading, she wrote:

> Of course you don't know our most important news yet! We're applying to adopt a little girl from a Dallas adoption agency, so will have another baby in the family in probably three to seven months. I'm terribly excited about it, although of course the prospect is for hard work and problems of sorts. But this seems to us to be the best

time to have No. 2. If I go and produce a No. 3 naturally, life will take on another aspect for some years!

We haven't actually applied to the agency yet because it's taking so long to fill in the form, including coloured photos of us and our house, and half a page each on our views on and experience with children. They say it is three-seven months usually after getting the application, before they 'approve' your request and then find a suitable baby. We could get a baby as young as five days. Strange to think she's probably on the way already, and I wish we could tell the troubled girl, who doesn't know yet what to do, that we are going to take her baby – through the agency, of course, we will never know who she is.

A brief visit to Austin involved Beatrice in a confrontation which was to reverberate through the world of astronomy. The 'Pope of Astronomy', Allan Sandage, had spoken on what was popularly called 'the fate of the universe'. Was it open or closed? Would it go on forever, or would it end? Sandage, with his virtually unquestioned pre-eminence, declared that it was closed. Beatrice, feeling certain of her facts from her recent research, stood up and challenged Sandage's premises. He had not, she indicated, carried his research through far enough to justify his claims.

Her tone had been neither hesitant nor aggressive. She spoke in the serious, reasonable tones of a fellow seeker after truth, seeming to take it for granted that Sandage would be excited by her new facts, glad to have this new knowledge. If he maintained that the universe was closed, her more extensive research now showed that it was open or ever-expanding. She sat down expectantly, waiting for one of the discussions she lived for.

It did not happen. Sandage dealt with the matter by ignoring her.

'Sandage was stunned and outraged – an outrage he was never to forget' is how Dennis Overbye describes this incident in his book, *Lonely Hearts of the Cosmos*.[14] Beatrice, Overbye says, 'resembled the cartoon Peanuts character Lucy, in temperament as well as looks. ... She and Sandage were to spend the next 10 years duelling.' It was longer than that.

Gérard de Vaucouleurs was accustomed to this duelling, and was to say that Sandage regarded him as something of an equal though not, certainly, actually an equal. It was acceptable for the two of them to argue, and they did; as the Russian theoretical physicist Lev Landau once said, 'Cosmologists are often in error but never in doubt.' But for Sandage to have a challenge from a young student, as he thought her, and a woman at that, was monstrous.[15] In later years Sandage is said not to have been able to open a paper by Beatrice without trembling.

She had submitted a paper, based on her thesis, to the *Astrophysical*

Journal in June 1967. Nothing happened. She and the astronomer friends she was making came to believe that the unusually long delay in publication was because the anonymous referee, to whom the paper had been sent for assessment, was Allan Sandage. When it did appear, in February 1968, much of the cosmology section had been cut. Who other than the Pope of Astronomy would have insisted on this, have been listened to, and been obeyed?

Nearly everyone who could follow and appreciate her line of thought, however, sat up and took notice. Who was this outsider, this woman? Ivan King and Hayron Spinrad at Berkeley were among the first to get in touch with her. Their interest and recognition came as water in the desert. Another who immediately saw the significance of what she had done was Richard Larson, a PhD student at Caltech, who was trying to finish his own thesis: 'I was working on the theory of star formation, and in a back-burner way I was interested in galaxy formation. It was clear this was an important development in the field, something nobody else had done.'[16]

So important, in fact, that Robert C. Kennicutt Jr, editor of the *Astrophysical Journal*, selected it for its 1999 centennial issue as one of the 53 seminal papers of the 20th century. Its most enduring result, he has said, was its clear demonstration that galaxy evolution was an eminently observable phenomenon. Within a decade this new subject was to grow into one of the largest sub-fields in extragalactic astronomy.[17]

On New Year's Day 1968 the Tinsleys moved into a house at 604 Laguna Drive in Richardson, virtually a suburb of Dallas, and convenient for Brian. It had more space inside and out than Beatrice had enjoyed since her childhood. Happily signing herself Beetle, she wrote home:

> I'm spoilt. The kitchen has a very fancy electric range and oven (on separate walls!) which includes a rotisserie and a meat thermometer and a mass of control dials.
>
> Also a built-in dish washer which saves the time I need to clean two bathrooms and sweep lots of floors ... I got a shiny new Teflon-lined frying pan for Christmas from Alan (because his daddy was tired of broken eggs and burned bacon).[18]

For their adoption application, the Tinsleys had one joint interview, an interview with each of them separately, and a home visit. They also had to have medical examinations. 'The interview I had was very thorough and penetrating, practically a psychoanalysis,' Beatrice told her family. As Jean Hill was not in good health, Beatrice tried to write often, with details of their lives which her mother would enjoy, such as the fact that she had joined the Richardson Symphony Orchestra. 'I was disgusted at how badly I sight-read and felt very crummy, though no worse than the other second fiddles. Frustrating to know I used to play as well as anyone there. However now I'll practise regularly.'

Her busy days, she said, included enjoying Alan – 'he's running, chattering in very-nearly-words all day and nearly always blissfully happy and keeping me likewise though lots of effort and some bad moments' – and fitting in some daily cosmology. She was also considerably involved in her old community interest of planned parenthood, doing voluntary work with the local branch, and she and Brian joined a university book discussion group. There they met congenial people with whom they could talk about social issues and aspects of American foreign policy, particularly the war in Vietnam.[19]

She was also in touch via mail and long phone calls with Terry Deeming at Austin, and had given two seminars to the small relativity group at SCAS. Beatrice thought she was quite well paid as well as stimulated because, although she had to spend several dollars on a baby-sitter, she received $25 for preparing her talk and spending one to two hours with the group of four people. All these snippets of news made happy letters home.

Around this time, however, her new friend Bea Wolf was visiting her, enjoying their conversation but finding Beatrice in fact far from happy. She seemed tense and strained, and juggling so many commitments. One day Brian came home early, to get changed before going on to some function, and could not find a clean shirt. Beatrice searched and found shirts which she had washed but had not yet ironed:

> There she was, slaving at the ironing board, while he had a tantrum and ranted at her and Alan cried in the background. I soothed the little boy, and thought that Brian should be ironing her clothes – she seemed so far above him. The incident seemed to symbolise their relationship. I soon became aware, too, of a sense of rivalry on his part, an intense desire to outshine Beatrice. This was impossible to do, of course, not just because she didn't engage in the competition but because she actually was so much brighter, brighter than all of us, as far as I could see.'[20]

Knowing that each day brought their next baby closer, and an end for the time being to her greater focus on cosmology and music-making, Beatrice increased her efforts to use every minute. This included regular attendance at their discussion group. Hardly any responsible thinkers and writers still supported the US administration over Vietnam, she wrote on 7 March. Billions of dollars were being spent on the war, whereas only a fraction of that amount could have begun to relieve the hopeless slum conditions in the cities: 'I am more convinced every day that the US should Go Home as soon as possible, loss of face and all.'

Back in New Zealand the Hills spent Easter in their lakeside cottage, with Edward fishing and Jean resting. There they read Beatrice's sad account of the assassination of Dr Martin Luther King on 9 April. The violence and lawlessness that had triggered the terrible event were heartbreaking, she wrote.

Jean's health deteriorated further. On the way to Florence the previous year she had sought cures for her migraines and spells of feeling faint by visiting 'psychic surgeons' in the Philippines. Now she suffered a severe brain haemorrhage, and died a few days later, on 7 May.

Theodora, by then studying music in London and the only one of the three daughters who was able to go home, caught the first plane she could when the news came. She broke her journey with Beatrice in Dallas, and from that grieving time she took away with her an image of a row of baby dresses lined up on coat-hangers in a wardrobe in what was to be the new nursery.

The baby girl was born on 10 May 1968, the day of Jean's funeral.

CHAPTER ELEVEN

THE FAMILY OF FOUR

'I could be bounded in a nutshell and count myself a king of infinite space.'
WILLIAM SHAKESPEARE

All three of the Hill daughters had complicated relationships with their mother, just as Jean Hill had had in her turn with her mother. Now the sisters had separately to try to deal with their tangled feelings as well as their grief. Beatrice wrote lovingly to her family, recollecting times and places and family sayings. Soon she could say:

> Here is some marvellous news: on 21 May we are bringing home our baby girl! She will be 11 days old then ... Her name is Teresa Jean. We decided on it months ago and now wish we'd let Mummy know. Teresa is to be Terry Tinsley after Brian's father.
> Somehow this is the perfect news for now. It is such a natural form of comfort that a new life begins in the family when an older person dies.

They had been given as much information about Terry as was possible without leading to identification of her birth mother, which was forbidden. The natural parents and grandparents were all short in stature, as Beatrice and Brian both were, and had the sorts of education and achievements (in the case of the grandparents, as the parents were still young) that, as Beatrice said, 'showed plenty of brains. The mother is said to be a very verbal and artistic person, in college.' As for Terry, she was 'just beautiful, and Alan is enchanted with the tiny thing.'[1]

It did not take Beatrice long to realise that having two young children meant a dramatic reduction in the amount of free time she had previously enjoyed. 'There is rather little time when both children are asleep and I'm not.' Nevertheless she contrived to find time to practise the violin, with the children appearing to enjoy it, so she could play some chamber music as well as continuing to belong to the Richardson orchestra.

News came that Rowena, who had held on to the threads of her marriage while her mother was alive, had left José, taken their two children to London, and was contemplating divorce. Beatrice was distressed for her but not surprised. The two sisters had begun to write comparatively frankly to each other since their time together in Florence. Because of her father's rigid attitude to divorce, Beatrice wrote to him even more frequently at this time, writing as Beetle and offering comfort by letting him know she was aware that he had tried to help, but emphasising that she also understood Rowena's point of view.

She did in fact feel that she understood a great deal. Her own marriage was far from that 'communion of hearts, minds and bodies' which she had dreamed about as a young girl. How could it be that her husband lived in the same house but seemed to have no idea of his wife's deepest yearnings to fulfil herself in science in her own way? Brian, on the other hand, felt he maintained a moderate interest in cosmology, whereas Beatrice appeared to give space science only slight attention. Professionally as well as personally, they were facing in different directions.

More opportunities for women scientists were opening up as a result of affirmative action programmes. Beatrice noted an increasing number of enticing positions she could fill, although none in Dallas. Realisation of what having two children meant to her career as a cosmologist, however, was beginning to hit her. Thus, after a night when Brian was home and looked after the children, she wrote to her father:

> I spent last night at the science library, reading up the latest journals in astronomy, and realising how different the current problems will all be by the time I'm active again.
>
> Astronomical problems are quite relaxing after a day of mothering and having too many serious troubles in the world of man![2]

She had to remind herself that she had made promises when she married Brian. It was a waste of time and energy to wish that her life had taken a different course. Just as it had been unthinkable to let her mother know how she had increasingly been feeling, so it was impossible to put this burden on her father. She must put negative thoughts out of her mind. Trying to look honestly at her situation, Beatrice also told friends that she missed Brian during his frequent absences. She wanted to hear an adult voice during the day. Often in the evenings, too, Brian worked until late, or he was away at conferences or on lengthy trips checking his spectrometer and its findings.

She deliberately built social occasions into her life, mostly inviting friends to visit, as baby-sitters were expensive as well as hard to find.

Summer in the scientific world is always a time of conferences. In the summer of 1968 Brian went first to New Mexico on his own work, and was then away for six weeks at conferences in Norway and Spain. Beatrice particularly missed his help with the children at that down-time for mothers, 5.30pm. The Laguna Drive house, much though its garden pleased her, was not well situated for a woman on her own with young children. Shops were some distance away and there were no footpaths. At first she found she had to get a sitter on occasions when she needed to go out and buy food. Enforced time at home led to her thinking more about international events. A number of their non-religious friends belonged to the local Unitarian Church, the 'discussion church', liberal, distinctly unorthodox and in many respects reminiscent of the Socratic Society. Both Tinsleys began attending its discussion groups. Beatrice also returned with passion to her earlier concern, the necessity for making cheap and reliable contraception widely available.

Brian was away again when Beatrice entered her third year as a mother, feeling she had little chance of doing any of her life's work. The sheer implacability of time passing – time lost for ever – increasingly set her on edge. Her world saw nothing strange in her situation. It was not, of course, a world attuned to women who wanted to work at their own careers, let alone a woman like Beatrice. She laughed to friends that, although her physical boundaries were so small and tight, her thoughts could explore the universe. What she found hard to control, however, was the toll the situation was taking on her nervous system.

Old friends Guff and Necia France noticed this when they arrived in Dallas for a short stay on their way back to New Zealand from London. In a letter on 5 October to her father, Beatrice supplies the first real written evidence, evidence which can be tested against another account, that all was not well with her. Actually, if one had only Beatrice's letter to go by, all would seem to have been very well indeed. She had keenly looked forward to the Frances' visit. Guff, after all, was one of her oldest and dearest friends. She wrote of what sounded an idyllic time:

> We had a lovely few days with Guff and Necia France ... We four and the two children all drove down to the Gulf of Mexico and rented a motel cabin at the beach at Port Arkansas for three nights. It's about 400 miles, and we did a leisurely all-day trip, returning on Monday ... The beach was beautiful, sunny and practically deserted to the horizon in both directions.

We swam and walked and talked. Alan was in the seventh heaven of delight, rushing around the sand and waves in nothing but a life jacket; Terry's at the age we had Alan in New York so I was pretty streamlined with disposable diapers and prepared food for her. The others shared the responsibility generously so I really felt free to be alone a bit. (First break since we moved house, and first swim since Italy.) In the evenings the children slept like exhausted puppies, and the four of us played cards and enjoyed the familiar company. Sad to say goodbye to them on Tuesday morning.

Guff and Necia France, however, remember this time differently. Guff's over-riding memory is of a strident and obdurate Beatrice, determined to do 'the right thing' when common sense and ordinary family experience would prompt most parents to bend the rules a bit:

We drove to the Mexican coast, to a motel, and planned a beach picnic. As we drove there we had a silly little contretemps with a traffic cop, nothing serious but he was officious. When we got to the beach, Brian brought out some sacks which he filled with sand for the kids' sandpit back home. One of them thought the cop could return and blast us for removing sand from the beach, so they decided to pack up and move to another part of the beach.

Then Alan, who was just two and delighting in all that sand and water, screamed and threw a tantrum. We'd only just got there – and he wasn't moving.

Beatrice wouldn't let Alan think he'd got his own way. The rest of us would have compromised, but we had to return to the motel. She prided herself on being cool and rational and disciplined, and she often artificially held in her stress or anger.

Later that day Brian and I had a session in the motel's woodshed where he'd been despatched to get some wood. Brian was feeling somewhat miffed. He said something in a slightly defeated way about 'people's moods'. Ordinarily I was more responsive to Beetle than to Brian who was distinctly less articulate and not the quickest in our group though he had remarkable perseverance. But here I was sympathetic.[3]

That Beatrice could see that Alan was in 'seventh heaven' as he played in all that sand and water on the sunny beach, yet resolutely remove him because he must learn to accept discipline, points up what Necia France saw as her 'carefully rational' system of child-rearing. It was as if she had learned it all from books. Missing in Beatrice was that fount of folk wisdom which is rather more than common sense, and which had not rubbed off on her in her own childhood.

Her for once unguarded writing to her father, that the others had 'generously' helped look after the children 'so I really felt free to be alone a bit', goes to the centre of her growing distress. She had to care for two young children for whom she had not been fully prepared emotionally, and care for a husband who was so often away pursuing his own scientific work. Meanwhile her own lines of thought, which she knew to be on a different plane from his, were constantly interrupted by everyday family matters.

Both the Tinsleys were serious-minded, with high ideals. Both tried hard to be good people. But it is clear that Beatrice had begun to feel she was in a trap from which only time – the children growing older, the work opportunities she yearned for coming her way – could extricate her. What Brian saw as her 'moods' came at least partly from her need for some time and space for herself.

The Frances also noticed that Beatrice talked a lot about her cleaning lady's difficulties; with no car or public tranport, she had to walk miles to get to work. She would talk readily about the problems of other people, but nothing about her own. The Frances also wondered whether Beetle's great interest in family planning came about through idealism or personal motivation. Was it possible that unconsciously she saw motherhood as something to be avoided? That at bottom she really did not see herself as a mother at all, in spite of what had been her careful plans to have two children 'one day'?

Edward Hill in *My Daughter Beatrice* observes that around this time she very often signed off her letters home as Beetle. He speculates that spending so much time with her children was taking her back to her own childhood; the death of her mother, too, might be reminding her of earlier years. Both seem likely, but so does a deep-seated wish that she could undo more recent years and be back again as the comparatively unclouded Beetle of the past, the Beetle whom Guff and Necia had known.

The children were awake from at least 7am to 7pm, 'with brief respites while they nap'. Terry was sleeping much less than Alan had at her age, sometimes seeming to be perpetually on the go. By building round their nap times with a baby-sitter or household help, Beatrice managed to give herself reading time, and a few hours at a library each month where she tried to keep up with astronomy. Every now and then she received letters about her PhD thesis and her paper published in February 1968. 'If these people could only see me now,' she mocked herself to a friend, 'the Great Astronomer ...' To her father and Theo she noted merely that such letters made her realise how long it was since she had done any astrophysics, but then pulled herself up with a brisk, 'Time enough ahead, and life is very full just now as it is.'

Brian, however, was having what she called 'a creative splurge', writing several papers. He and some men from the Center had just taken a visiting scientist out to dinner, she wrote, while she had seized the opportunity to have the other members of her quartet around to play. 'I would rather eat cold meat, then play, than spend the evening out dining – in spite of the good company!'

A meeting of astrophysicists took place in downtown Dallas in the busy week before Christmas 1968. Beatrice managed to line up enough baby-sitters, including Brian, to give her two whole days and some half days at the conference. 'So that was my year's effort at astrophysics,' she wrote home in an apparently casual aside. Actually she had managed to glean quite a few solitary hours for her work during the year, but this was very different from the thrust of talk and argument with others in her field. At this Dallas meeting she found some of the papers and some of the people highly stimulating. A few stayed on after the conference and came to the Tinsleys' first big party with some 50 friends, mostly scientists, who talked by the light of candles and the fire. This ploy was to hide the lack of careful house-cleaning as Beatrice had spent all day at the final session. As for the day at Christmas 1968 when men first went around the moon in *Apollo 8*, that was the day – 'the Great Astronomer' laughed at herself as she noted the event – when Terry cut her first tooth and learned to stand up in her cot.[4]

The Tinsleys completed legal formalities concerning both children at the beginning of 1969. They had fulfilled all the adoption agency's requirements for Terry, so the baby's appearance before a judge was a mere formality. The judge took less than 30 seconds to declare the trial period for Terry's adoption at an end, and she was a permanent member of the family. Alan, not born in the US, had to face a more time-consuming examination. He had to be medically examined and have what was called 'an interview'. This, Beatrice enjoyed telling people, was in case the US was in danger of admitting to citizenship a two-year-old communist.

Terry was still a baby when the day came that Beatrice was told that she was pregnant. She had spent some weeks of perplexity shading into anxiety before visiting her gynaecologist. Then she was outraged, incredulous and furious. How could that man have been so definite in his earlier diagnosis? This most definitely was a baby she did not want. She and Brian already had their two children, the two 'permitted' by Zero Population Growth. She fell back on her ZPG creed. She said decisively to Brian that they had been definite in their agreement that two children were the limit.

This was a knife thrust for him, a cause for real and understandable grief. He had also long since taken for granted that they could not have their own children. For this to happen now, at this stage in their relationship, was wretched timing. A child with their mixture of genes could have been someone very special, he was later to say sadly.[5]

Texas still did not permit abortions. The much-travelled route to a clinic in New York, which Beatrice had worked so hard over the years to support through Zero Population Growth and Family Planning, was still in existence. But, in Texas, women using this arrangement had to have their husbands' written consent. Now Beatrice had to ask Brian if he would formally agree. 'Of course I gave consent. What else could I do?' he was to say.[6]

Beatrice flew off on the journey she had helped make possible for so many women. She did not tell friends until afterwards, suspecting some might disapprove, but applied her rational thinking to the situation and seemed not to have let it affect her.

When news came that elated Brian, Beatrice could not share his pleasure although she could understand it. He and a group at the South-West Center had worked on a proposal to use a research satellite which NASA planned to launch. Several institutions had competed with proposals, and theirs had been chosen. It would have six experiments on board, one of which would be Brian's with his spectroscope. The satellite was to be in equatorial orbit and would be launched from a site in Kenya – but not for about two years. Accordingly, the family would almost certainly have to stay on in Dallas until then, and already Brian had had his first two-year contract renewed.

This was not what Beatrice had counted on. They had always spoken of finding somewhere that offered fulfilling work opportunities for them both, not just Brian. Their unspoken plans included a sustaining belief on Beatrice's part that her time as a fulltime housewife and mother had an end in sight. On Brian's, there was a growing determination not to go anywhere which might not offer the opportunities and increasing security and recognition that he was finding at SCAS.

Anyone who has ever worked alone without similarly involved colleagues, particularly in a field where co-operation and the exchange of ideas and theories are the norm, can begin to see what Beatrice was up against. Her isolation in Dallas was daunting. Although there was telex, this was before the advent of faxes, let alone e-mail.[7] Long-distance phone calls to likely colleagues were expensive, and Beatrice was usually without money of her own, money which she felt she could spend as of right on such things as keeping in touch. She relied on her life-long habit of writing letters, but not everyone always replied at once, as she did.

Now came a chance to test her usual careful management. Brian had just been away again on what was becoming a monthly expedition to New Mexico, to his spectroscope. When he learned he could go to a conference in Prague in May, Beatrice decided she had had enough of staying home alone. She determined to take the two children and go along too; not to Prague but to Florence, where she could see Rowena again. The problems of travelling with the children could largely be anticipated and prepared for.

Her plans for getting the whole family ready to fly to Europe – visas, passports, vaccinations, tax clearances and suitable clothes – were thrown into disarray when Brian suddenly had an important idea for interpreting his previous year's spectroscope observations. This was to be incorporated into the paper he was to present in Prague, on the hydrogen corona surrounding the earth. The idea meant 'altogether re-hashing' the paper, Beatrice said. She helped him with it.

The Tinsleys had also worked together, with Tom Patterson from the Center, on an astronomical paper, 'Distribution of Red-shifts of Quasars'.[8] They were able to send this off, in the hope of publication, just before leaving for Europe. It was not altogether an easy collaboration. There were tensions that Beatrice did not spell out. 'I don't know how some couples work jointly all their lives!' was her wry comment to friends.

The journey to Florence went smoothly enough, with Alan practising his new French phrases and some key Italian words to communicate with his Venezuelan cousins. When Brian went on to his conference, Beatrice soon understood why letters from Rowena had been sparse. She was doing all manner of translations, coaching students, teaching, doing the housework and shopping and cooking as well as looking after her children before and after school.

Beatrice did what she could to help, partly with the selfish aim, she said, of winning a little more time for Rowena so the two of them could really talk. Her efforts were of little use. Finances in the Fajardo household were at rock bottom, so Rowena was often still at her typewriter after midnight. Mood swings in the family were only too obvious. Little of Beatrice's hoped-for time with Rowena eventuated – frustrating when she had gone all that way. Neither sister, then, was wholeheartedly ready to talk.

Brian's workplace was about to be renamed the University of Texas at Dallas, (UTD), with authority to take students from their third year onwards. Beatrice thought the change would benefit the area educationally: 'I always feel SCAS has a rather sterile air, being full of research scientists and no students,' she wrote home on 5 July. Francis Johnson, head of Brian's division, was to become acting president of the new UTD. He had perceived something of Beatrice's ability and had arranged for her to be given what she called, sadly, 'a fictitious appointment as Visiting Scientist'. Fictitious or not, it brought with it a little money to pay for some computing and making pre-print copies. She tried to be optimistic; Brian's interests, she said, seemed to be turning more and more towards astronomy, so that by the time she could go back to work there might be two astronomers in the family. That time, too, was 'getting much closer now that the children are growing so rapidly' – an encouraging admonition to herself, if a case of wishful thinking. Terry was not quite 14 months old, and Alan not yet three years.

Although she had long since given up trying to explain her current work to her family she did say she was excited about 'a good idea' which resulted in a paper, 'Possibility of a Large Evolutionary Correction to the Magnitude-Redshift Relation'.[9] Old habits of modesty kept her from writing of her elation when she was working on something she knew to be good – that, and knowing her family would not comprehend. But in herself she felt almost fiercely triumphant. She would work on, she would find a way.

She was also writing sharp and well-argued letters to local newspapers about their unthinking support of American foreign policy, and to the district congressman, who had sent around 'a rather terrifyingly irrational newsletter about Vietnam'. In writing of the pervasive conservatism of Texas, Beatrice may have been transferring some of her own feelings to the political situation. But close to her heart was the saying attributed to Edmund Burke: 'For evil to prevail, it is necessary only for good men to do nothing.'[10] Beatrice was never about to do nothing if she could possibly help it.

In what may have been a deliberate attempt to shake herself out of her despondency, she began a new activity at this time, collecting clothes from the Wives' Club of the Southwest Center to distribute to impoverished Native American families in Dallas. This was something worthwhile she could do while looking after the children, who became accustomed to sharing space in the car with great bundles of clothing. The plight of the Native Americans struck Beatrice as worse, even, than that of the black people, at least in Texas, and she was particularly distressed by their vast families. Her efforts in helping spread family planning practices seemed puny indeed.

For months she had been assuring her father how much she was looking forward to his planned October visit on his way to Britain for his mother's 85th birthday. Edward Hill had been sending off a string of letters in happy anticipation. It should have been a great occasion. But it was not, not even as described in Edward's carefully moderate words in *My Daughter Beatrice*:

> I found Beatrice very strung up, as well as extremely intolerant of

diverse opinions. I was no doubt foolish to be drawn into arguments about Christianity and abortion, about which in advance I really knew we could not agree, but I was abashed to find, (as I wrote to Theodora), that Beatrice seemed to regard me as a 'nitwit' and that I found Brian much more tolerant of divergent views. I also expressed concern at the way she shouted at the children, though I had to admit that they seemed happy ...

Looking back, I believe the tremendous tension I could not but sense was due in large measure to the clash between her devotion to science and her equal or greater devotion to her children, husband and home. When she had adopted Alan and Terry she had seen clearly that some years away from astronomy would be inevitable. But simply seeing the problem was not solving it, and the clash between the two strongest forces in her personality had made her a very frustrated young woman.

In this analysis, even with the benefit of hindsight, her father was considerably astray. For 'equal or greater devotion to her children, husband and home' read 'duty'. For Brian she still had the feelings one has for those with whom one has shared years of living, with hard times and occasional triumphs, although such feelings mostly evaporated when they were arguing. Her home meant little to her. Apart from her music and joy in natural beauty, her life was mostly inward.

The children were a different matter. Alan's lively, trusting face turned up to hers, and his eager response to her favourite music and pictures, filled her with delight. His little-boy warmth when he flung himself on to her lap must have awakened memories of that same warmth and feeling of belonging which she had once had with her nanny. Terry with her big appealing eyes was mercurial, and still so young. Of course Beatrice loved them. But enjoyed them? Truly enjoyed in the sense of being able to enter into how they were feeling and thinking, and rejoice in them for their individual developing selves?

Never very self-aware in the sense of being able to analyse her own emotions or begin to trace their origins, Beatrice could see at least in part that her tensions were making her hard to live with. She had looked forward to her father's visit, his love and approval, and she could not ignore his dismay, however tactfully he tried to express it. Friends observed that her solution was to set herself to be an ever-better mother: she would read more parenting books, plan her time and the children's time more carefully, see that they had enough fresh air and sleep. In no way would she fail in her duty.

Beatrice still grasped at any work opportunity. Now that SCAS had been elevated to university status and there were students to cater for, she was able to lecture to Brian's small group who wanted some basic astronomy. Twice a week in the afternoons she taught four young people, with another three following her on closed-circuit television. This elementary teaching at least put her in touch with students' minds again, and was an impetus to arranging for Alan to attend a nursery school. By the beginning of 1970 Beatrice felt able to cast around for whatever possible part-time teaching jobs in schools Dallas offered, and applied for a part-time maths and physics post. The shock of not even being interviewed for this comparatively minor job cut deep, and pointed up her isolation.

With Brian away again and below-freezing temperatures keeping them indoors, she made a rare admission:[11] 'Alan and Terry and I are beginning to get irritable together!' She immediately followed up with bright descriptions of the children at play. Alan's great enthusiasm for nursery school, and his capacity for learning, delighted and relieved her.

In Brian's absence Beatrice was also caring for 'an extra two-year-old', a friend's child, so that this woman could get some sleep when her new baby slept. She still managed to find baby-sitters so she could attend orchestra practice, a meeting on population control and a discussion on student dissent. The Tinsleys' interest in their discussion group continued. With Brian back from New Mexico, the group met at their home to discuss problems in higher education in the US, 'carrying on at a lively pace till a record 1am'.

Brian had managed to get funds to go to a conference in Leningrad in May 1970. In spite of large cuts in government scientific funding, his job seemed secure as long as he financed his work with grants, as these resulted in a profit to the UTD. The very security of his job, and the many conferences he seemed able to attend, could not help but give Beatrice concern. The more he became part of the UTD, the less interested she thought he was in discussing a move to another university where she too could get the work she yearned for, and for which she had shown she was already qualified. Brian felt they discussed this subject a lot. An outsider might have said that Beatrice raised it rather often, but that Brian at that stage took no real steps to look for work elsewhere.

A family holiday by car to New Orleans helped raise her spirits. It came about because Brian's parents arrived for a three-week holiday. They squeezed into the car with the four younger Tinsleys and were thoughtful and helpful guests, Beatrice wrote home.[12] Alan 'adored' his grandparents, and Terry as ever charmed all adults they met. Her personal New Orleans highlight was an evening of first-rate jazz. 'I could have stayed all night, but the others' ears couldn't stand the volume.' Playing in the Richardson orchestra also gave her solace, she said. They had been playing 'very energetic things', such as the '1812 Overture' with two brass bands, 'though without the cannon'. Her own energies and aspirations, which she had bottled up for so long, continually sought an outlet. Now an opportunity came for her to help herself through helping others.

CHAPTER TWELVE

FAMILY YEARS

'A scientist has to take a mental stance that is reflective and detached, and scientific thought has to be grounded as much upon abstract information from books as on direct observation and experience.'
MICHAEL J. A. HOWE.

Most observers would have thought that an additional responsibility in Beatrice's life was the last thing she needed. She had to keep the household running and meet the daily needs of the children, and Brian when he was home. Her father and sisters were on her mind, and friends, neighbours and Nanny in London looked to her, too. She had promised herself to keep up her music. The work she most wanted to do, her work as a scientist, had to be fitted in and around any gaps in the days or evenings. But Beatrice now wholeheartedly took up voluntary work, becoming secretary of the start-up area chapter in Dallas of the movement called Zero Population Growth, or ZPG.

At this time in her life all those with whose ideas she was in sympathy, including her Catholic friends, were acutely concerned about the fate of a rapidly expanding world population, with misery or actual starvation seeming all too likely for millions in Third World countries. It seems that, consciously or not, Beatrice projected herself into the position of the world's women whose very identities were blotted out through constant childbearing and family demands. Certainly she was passionate about what would become generally known as a woman's right to choose.

Interest grew rapidly from ZPG's first public meeting on 21 May 1970. Beatrice at first wanted to keep in the background, as she thought she might be seen as an outsider trying to tell American women what to do, 'though of course anyone who understands the movement knows we are all motivated by genuine concern for both America and the rest of the world'. She saw the essential and urgent goal of ZPG as educating people to regard two children as enough, the proper family size, and to adopt if they really wanted more children. Beatrice's own contribution to this work helped counterbalance what she called the despondency which assailed her rather often.

New friends met through the Unitarian Church were helpful. Then there were congenial new neighbours, Carol Wilson, who had a PhD in biochemistry and was taking time out from her profession while her children were young, and her husband, Art Wilson, a research chemist. The children of the two families enjoyed playing with one another, and the women shared some of their care when they could. Many years later, Alan Tinsley was to look back on the early 1970s and believe they were a close-knit family then, 'even though the dinner conversation was 30 miles above my head. The parents talked about work. They'd be talking and we'd be feeding broccoli to the dog.' [1]

Whenever possible, Beatrice read. She would have liked to haunt libraries but the most she could manage was hurried visits between other commitments. If she could not take books or papers away, she made quick notes, nourishment for her thinking times. One way or another, besides her family, ZPG and music, Beatrice was able to finish a cosmology paper on the ideas that she had told her family were exciting her, 'Possibility of a Large Evolutionary Correction to the Magnitude-Redshift Relation'. Her conclusions were the opposite of those reached by most cosmologists. She was to present it at the AAS meeting in Boulder, Colorado, in the first week of June 1970. Now, immensely excited at the prospect of 'three days of being an astronomer', she told her father and Theo that Brian would look after the two children for supper, bed and breakfast. Daytime arrangements she had made with different friends.[2] This early meeting in Boulder turned out to be one of her most important ever, a turning point in all her thinking. Everyone's contributions, she told friends, left her knowing without doubt that adding to the world's stock of knowledge about cosmology was the most important thing in her life.

Edward Hill was not accustomed to men who were involved in childcare in any way. He wrote back saying he was most impressed by Brian's helping with the children so that Beatrice could have 'three whole days' away at her conference. Brian thanked him for his 'kind remarks', and added, 'I expect she will be doing this more and more now that she has more time to work on her research at home, and she has just undertaken to write another paper for a conference in Uppsala in August.'

There was no thought of Beatrice's attending this conference of the International Astronomical Union. She was collaborating on a paper with a Norwegian, Jan-Erik Solheim, who would present it, on 'Analysis of the Magnitude-Redshift Relation Including Possible Effects of Evolution'. She also prepared a talk for some junior astronomers, and 'partly edited, totally typed and partly collated, stapled and addressed' 250 copies of their ZPG newsletter in time for a meeting. In the days before photocopying or personal computers, such newsletters were produced by a Gestetner machine. The Gestetner that Beatrice was using caught fire when she went to produce a further 250 newsletters, 'but we all survived and everything got done'.[3]

Human ecology and the dire effects of over-population on water, soil and air increasingly engrossed her. 'Protecting the future', one of ZPG's slogans, was a cause she could throw herself into wholeheartedly. It also helped her, probably more than she realised, to slough off some of her personal frustrations. Friends observed her good spirits.

Then, with Brian soon to leave the country again, this time on a six- to eight-week stint in Brazil where he would look for another site for his spectrometer, Beatrice made a momentous discovery. One morning, when she took Alan to his Montessori school, she happened to overhear a teacher saying that the school actually liked taking children from two years old. Yes, they would be delighted to welcome Terry, too, before long. Beatrice had had no idea that this was possible. Now her excitement at the prospect of having whole mornings for astronomy shone through her letters home.

Brian too was elated: he was promoted to associate professor, with tenure. He was doubly delighted because a downturn in the economy, and less emphasis on space research, meant untenured scientists were being laid off in many centres. Beatrice described it as 'the current hopeless job situation in physics', and told her family the promotion was well deserved as Brian had been getting many very interesting results from his research and had been invited to do several reviews of his field. Brian's promotion, however, made it more unlikely that he would actively look elsewhere. Nevertheless she wrote firmly, 'We don't plan to stay here for ever.'[4]

Her organising ability was stretched again over the details of a ZPG conference on aspects of adoption. With a couple of helpers she spent more than one night until well past midnight, duplicating, addressing and putting stamps on hundreds of letters about the conference. One of the main topics was the push to change Texas law forbidding inter-racial adoption. A number of fertile couples, already with children of their own, were among those wanting to adopt black or Hispanic children whom they saw as languishing in often indifferent foster homes, and suffering from increasing insecurities.

In the midst of this activity came tragic news from Beatrice's dear friend from her early music-making days, Pip Harding, now Pip Jackson and living with her family in Sydney. Their older son, just three years old, had drowned. Beatrice was stricken for Pip, for all the family, and wrote to her own family that 'the tragedy is so terrible I can't imagine it.' She worked at finding at least some of the right words to send to Pip, who was always close to her even though they did not write often. Pip herself had written 'a wonderfully courageous and forward-looking letter'. The memory of this courage in a time of great trauma was to remain with Beatrice.

She still tried to keep her own troubles to herself but sometimes she unwittingly let things slip in letters home. Terry, for instance, was now talking more, and loudly, often like a little parrot. In restaurants she liked to yell out, 'Sit down!' or 'Shut the door!' Beatrice wrote, 'I have to watch my own language. It gets copied at once.'

Looking around for anything else that might interest her father and Theo, she sent them a copy of the ZPG bulletin which contained a report she had written about the conference on adoption. ZPG included among its tenets the right of women everywhere to have access to abortion if they so wished. This particular bulletin carried clarion calls for everyone to work for social justice for all women, and to change the legislation which made abortion illegal.

This was too much for Edward Hill. He immediately wrote a long reply, which has not survived. He did, however, keep Beatrice's next letter, which began 'Dear Daddy':

I've changed my attitude to abortion totally in the last four years, and feel very strongly about it – mostly because most people's attitudes are irrational and based on ignorance. Really I think it's most unnecessary to think of a less-than-12-week foetus as a 'child': at six weeks, when abortion is easy and can be done if there's no delay, the foetus is less than 1/10 inch and would pass unnoticed if the woman took chemicals to bring on a period.

The fact that it could develop into a child seems to me not so important – so could all the millions of eggs and sperms that get 'wasted'! It just depends where you draw the line. And moving the line up a bit, to 12 weeks where abortion is still possible by the safe dilation and curette operation, doesn't strike me as a sinister approach to infanticide, as the extremists would say.

I'm afraid I disagree strongly with Theo that asking for abortions is denying one's womanhood. It could mean wasting one's life in many ways to be forced to carry every pregnancy to childbirth; even if a woman uses the safest contraceptive, there is still a 30% chance of an unwanted pregnancy over 30 years!

I think the decision not to have another baby (or not at that time) is often the most responsible one possible, for the sake of the woman and the rest of the family, apart from the population which doesn't need any reluctantly added numbers. If a woman sees an abortion in this light she honestly regards it as 'preventing a foetus from becoming a baby' – just removing an unrecognisable small blob – and not as murder. Preventing something from developing isn't killing. In the long run it is frequently adding to the quality of life for others.

I think that abortions should not be the subject of law, any more than any other surgery is. The laws should be repealed and abortion treated as a medical procedure, as even the American Medical Association has said.'[5]

This letter in response to her father's stern disagreement is just that: a rapidly written reply to a man she loved and who had all the time in the world to make his case. If Beatrice had set herself to write a complete and formal rebuttal to those who would deny women the right to an abortion, she would have been impassioned but also carefully rational, logical and complete in her arguments. The playground is not level, however, if one is a loving but impatient daughter grabbing up a pen when a few minutes offered themselves, to try, tactfully, to point out the error of a father's attitudes. Did they ever speak of this exchange of views again? Edward Hill could not remember.[6]

As Terry too was now at the Montessori school in the mornings, and loving it, Beatrice was able to concentrate on astronomy from nine to noon. She had a number of research projects she had been keeping alive, just, until their time came, so suddenly her life began to change. Three whole hours a day for her life's work! She was grateful to Montessori, even more so because aspects of Terry's behaviour were increasingly troubling her. The little girl seemed unusually unsettled and volatile, with a very short attention span.

Beatrice had not taken into account the possibility of big differences in the genetic make-up and temperaments of the two children, differences which the same upbringing could not conceal. She had not thought through the nature versus nurture debate in regard to her own children. In any case, at this time the pendulum was still swinging heavily towards upbringing, rather than inheritance, as the dominating factor in children's behaviour.

She talked with the principal about 'the disturbing ways' in which Terry was behaving at home, ways which she said 'had about got me beat and worried'. The teacher was reassuring. She had obviously observed Terry closely, and had a lot of relevant advice. Altogether she impressed by giving clear reasons for the difficult behaviour at home – which, Beatrice noted wryly, was quite unlike the way Terry was at school. But in her letter home she did not spell out what the principal thought these reasons were, what her recommendations were, or whether she felt she would be able to put this advice into practice. Neighbours thought Terry was hyperactive.

Brian, meanwhile, was living in 'quite luxurious accommodation' at the space science institute in Sao Paulo, investigating several possible sites for his spectrometer. When he had made a decision he would be staying on to oversee the building of a hut to house his equipment, a six-week absence in all.

As for Edward Hill, he now had some momentous news. He wanted to marry again, although he had not yet asked the woman concerned. Edward wrote almost offhandedly, no doubt to sound out his daughter's reaction. Certainly his details were sparse. Beatrice replied immediately and warmly:

> I do hope the lady you want to marry will say yes! Not knowing anything about her except your few remarks, which leave me in the dark as to who it is, I can hardly say how much I would love her as a step-mother – but for your sake I can think of nothing better than for you to have a wife now. As you said, moving house and all would be a big upheaval, but perhaps psychologically the best way for you.
>
> It must be a tremendous effort of character to face a whole new marriage, without being overcome by comparisons and grief at the end of the first. I'm sure Mummy would have preferred you to be happy with someone else rather than be alone for the rest of your life; certainly that's the way I feel, if I were to die, that Brian should remarry. It's lonely and a strain to be alone, however many friends you have.
>
> I very much hope it all works out, and will be most excited to learn all about her if so![7]

Any parent wanting to remarry would be relieved to receive a letter like that. There is no hint of anxiety or word of caution, let alone a darker emotion. The need Beatrice had always felt to keep her parents happy most probably contributed to the warmth of her letter. Now there could be someone else to share that responsibility. In fact, as her future stepmother, Mattie, was to say, Beatrice was consistently friendly and supportive when they met, even introducing her to her friends as 'my mother', and saying to her, 'There's no need to say you're my stepmother.'[8] Mattie Fearnley was an old friend of Jean's and Edward's. A few years younger than Edward, divorced and living not far from the Hills, she had been the first among the friends who were quick to support Edward after Jean's death.

Beatrice added in her letter to her father that it was ironic that he was getting married just when divorce seemed inevitable for Rowena, but asked him to think of how Rowena would give her children a better life than if she were with 'an incompatible man'. To Rowena she sent long outpourings of sympathy and support.

Meanwhile, Beatrice was caught in the trap only too familiar to working mothers: how to ensure one's children were well looked after and happy, and the household kept on an even keel, while finding time and energy for work. Familiar – except that in her case there was no money and no employer. She arranged with friends, Robin and Ron Wheeldon, to look after their three-year-old son while the Wheeldons had their first weekend away alone together for six years. In return, Robin Wheeldon would care for Alan and Terry from after school until Brian, back from Brazil, could collect them after work. This would make it possible for her to go to a four-day relativistic astrophysics conference in Austin, the best of Christmas presents.

Her careful plans nearly fell through. Brian was back later than expected, bringing all his washing and ironing for her to do in the one day before she had to fly to Austin for the conference. She managed it, taking with her pre-print copies of her latest work. The conference was very useful. In her relative isolation from other astronomers, she was particularly grateful to be given pre-publication copies of new data.

The first letter addressed to Mr and Mrs Edward Hill – their honeymoon was a fishing holiday at Lake Taupo – included plans for 1971:

> I've spent a lot of time writing research proposals to funding agencies. The University of Texas at Dallas at present calls me a visiting scientist and provides an account to pay for my computing, publishing etc costs. But they have now formally decided the conditions under which they will pay a salary to the spouse of a faculty member. It means that to get the part-time research job I want, I have to obtain my own funds independently of any member of the institution.
>
> Elaborate way of avoiding favouritism! Rather a tough assignment, too, for someone as un-heard-of as me. But I'm trying to get the necessary funds from the National Science Foundation, which funds nearly all astronomy in the country.'[9]

Beatrice had also written, very formally considering their close association while she worked for her doctorate, to Gérard de Vaucouleurs at Austin, asking if he would write a letter of recommendation in her application for another grant, only $500 but most welcome should she receive it. His letter, she said, should be addressed to the chairman of Sigma Delta Epsilon, Merck Institute for Therapeutic Research. She enclosed a copy of her research proposal, to study theoretically the stellar and gaseous contents of galaxies, and their evolution. It was carefully referenced, and – unusual for a scientist – written in straightforward language.

In the shortest possible time she received from de Vaucouleurs a copy of his letter in support, a letter that may have clinched the award of the grant to her:

> Mrs Tinsley was the most brilliant graduate student in astronomy that I have ever met in my career in Europe, Australia and this country. She graduated from this department in 1967, within three years of entering graduate school, and her scholastic record was simply astonishing. It is a great pity that because of obsolete nepotism rules, in this and other universities, she cannot be offered the research or teaching position that her abilities deserve. I very much hope that through research grants she will be encouraged to do research work.
>
> Concerning the immediate subject of her research proposal I need only say that it is a very important and active topic. Mrs Tinsley in her dissertation work made the first quantitative prediction of evolutionary effects in galaxies by integration of stellar evolution in them. Observations in progress at McDonald Observatory and elsewhere need for their interpretation the type of computer analyses that she proposes to perform. There is no question that her research proposal is of great scientific significance, and I hope that it can be supported.

Professional meetings are always important to those working in astronomy departments and surrounded by colleagues. To Beatrice, doing her work more or less between the washing machine and the kitchen sink and totally isolated from personal contact with anyone in her field, they were life-saving. Each time she managed to juggle, scheme and pull off two or three days for herself so she could attend a meeting, of all the people there she was almost certainly the person most stimulated and most enriched.

Nor was she altogether 'un-heard-of', as she had described herself in her letter home. To hear and see Beatrice in action, speaking at a conference, meant one did not forget her. This slight young woman, her eyes sparkling and cheeks glowing, had written that remarkable PhD dissertation, people said – it was as well to listen to her. So went the word on the conference grapevine.

She was not aware of this. Instead, perhaps it was realising how much her whole future rested in the hands of those largely unknown funding agencies; or perhaps it was because she and the children all had colds and coughs, and the Texas winter was bad and she was again without support because Brian was away in New Mexico, but her birthday on 27 January she called 'miserable'. Her continuing and often exhausting work for ZPG, and an increasing awareness of threats to the environment, contributed to her depression: 'Suddenly everyone realises how terrible the air and water are, and the noise, hideous urban sprawl, acres of junk and concrete etc.' With too many people producing so much pollution, one could become very pessimistic. But perhaps New Zealand would listen to all these world voices raised in warning and do something about population and pollution before it was too late. The act of imagining the Lake Taupo of her girlhood becoming polluted and dying like Lake Erie made her realise how much she cared about New Zealand.[10]

It is a sorry outburst from Beatrice the scientist. Taupo, New Zealand's largest lake and encircled by almost uninterrupted green, does have some environmental problems but cannot possibly be compared with Lake Erie, which is targeted by complex pollution. But Beatrice had had her 31st birthday, coughing, sneezing and alone with the children. The years were passing. Was this the most she could do with her life?

With the coming of spring, however, the family drove to one of their favourite places, Port Arkansas, for four nights, and walked along the beach and dunes and breakwater with Rata the irrepressible collie bounding around them. The countryside lifted Beatrice's spirits: acres of bluebonnets, the native lupin-like flower, and red Indian paintbrushes in full bloom, although she could not help noting the 'terrible piles of litter' on the beach; this in spite of rubbish tins every few hundred yards and notices threatening $200 fines.

She was the speaker at a local high school's celebration of Earth Day, talking about the dangers of unchecked population growth and what it was doing to the world's resources. The prospect of this audience, she said, scared her 'more than talking to astronomers, except that others who had talked to high schools said they were extremely responsive and interested'.

To hear two famous astronomers, the husband and wife team of Margaret and Geoffrey Burbidge, she and a friend one day drove to Fort Worth for the dedication of the Sid Richardson Science Building at the Texas Christian University. The talks Beatrice described as 'semi-popular, interesting, and gratifying to hear my own ideas discussed in the one on cosmology – probably because the speaker knew I was there, but he did treat them with respect!'

One of her comparatively new interests was nutrition, and she was keen to pass on her discoveries. It is a little surprising that Beatrice the scientist had taken so long to realise the importance of nutrition. Now the Tinsleys cut down their intake of refined foods to near zero, adding more protein and a range of vitamin and mineral supplements, 'all enjoyed, and keeping us much fitter. I'm sure New Zealanders eat far better – less rubbish – than people in general eat here.'

A letter home on 9 June 1971 gave an idea of one day:

> My evenings are so full of things like ZPG business that it's hard to fit in any astronomy except for mornings. I was going to write to you last night, but first I did the business letters – one to an astronomer, one to the Child Welfare League of America (adoption committee work), and the minutes of a board meeting of Oasis, a trust organisation of which I'm a trustee and secretary-treasurer. It makes referrals for sterilisation operations – which are legal locally but for which it's hard to find a doctor – and abortions, which are illegal here, and for which women have to go to New York or New Mexico. By then it was 11.30pm so I gave up.
>
> Actually yesterday was a very nice day because a pianist friend was here from 9am to 4pm, and between lunch and attending to the kids we played sonatas for at least four hours. The children were quite appreciative. Terry kept getting too enthusiastic and joining in with loud singing. Alan wanted to hear every note, and hardly played outside all day – I think he's quite musical and has an excellent memory for music.

The pianist friend was Bea Wolf, now married to Walter Heikkila. Bea had come closer to Beatrice than anyone else in Dallas. Although pride and natural reticence – 'British stiff upper-lip-ism' some Americans called it – still kept Beatrice from talking completely freely about how she felt, she was able to treat her friend as the safety valve she urgently needed. Nevertheless Beatrice had a code. She would not tittle-tattle.

Perhaps detecting a faint note of scepticism in replies to the crusading zeal in her letters about ZPG, Beatrice sent off a moderate broadside to the Hills on 1 July:

> I wish New Zealanders would become widely concerned about population growth! Do you really want the number of people in New Zealand to double every 37 years, which it will at the present rate? To me it would mean destruction of the best things New Zealand has to offer the world, and a great tragedy.
>
> The American Association of University Women has just had its national convention here in Dallas, and passed by a wide majority a resolution supporting the principle of zero population growth ie. voluntary reduction of the birth rate to an average of 2.1 children per family.

Since his marriage, Edward Hill had given up his large home and moved into Mattie's house on the border of Kelburn and Northland in Wellington. It was smaller, and Mattie had her own furniture. Edward accordingly offered his daughters their choice of a number of antique pieces and soft furnishings, including exquisite rugs from Jean's family firm. Beatrice's letters show no trace of melancholy or regret at the thought of the furniture she had grown up with being dispersed around the world. Her tone is briskly practical, affectionate, and relieved that her father will be looked after.

She had organised a summer holiday for her own family. With Rata the dog they drove for hours through the 'hot, dry and dull' Texas landscape, seeing nothing but cattle, empty land and oil wells, until reaching lovely mountain country at 7000ft near Santa Fe in New Mexico. They had been there the previous year, 'and of course Alan remembered the train – ye olde narrow-gauge, sooty steam train. In fact he remembered every filling station we'd been at last year!' So while Brian carried on in the car with Terry and Rata, Beatrice took Alan for a blissfully sooty journey by train till they all met up again, picnicking in one of the flowery meadows as they drove on to their base.[11] Her letter gives no hint of any tensions. When Brian had left for a conference in Moscow, she drove the family home. She took three days so that the children, refreshed by the mountain air, did not get tired before they reached home and faced the summer heat again.

Good family news was waiting for her. Theodora had become engaged to an American, David Lee-Smith, about whom Edward – not to mention happy Theodora – had written very warmly. David came from Ojai, California, just north of Los Angeles, and had an Australian mother. He was an undergraduate at Sydney University, where Theodora was by this time teaching in the Italian Department. 'He sounds a really fine person and just the right husband for her,' Beatrice responded to her father. 'It is sad that Mummy never knew of this.' They had decided against a long engagement and Beatrice had wondered if she could possibly make it home for their wedding. That seemed most unlikely. In about a year's time, anyway, the Tinsleys planned to come to New Zealand.

She herself had good news, 'on a lesser scale':

> I got my grant of funds from the National Science Foundation, and will be a paid scientist as of next Monday – the day Brian gets back from Moscow. It's half-time work, but I'm free to arrange the hours and places as I like, so being unable to start mornings until school resumes, on 9 September, can easily be made up in the evenings – nobody will know or even care, as long as my output is satisfactory.[12]

The grant also meant that she could be designated Visiting Math Scientist and join the staff of UTD Mathematics and Physics Division, at the Center for Advanced Studies. More important to her, the grant would pay for her travel and living expenses at the forthcoming big astronomy meeting at Amherst, Massachusetts on 23–27 August 1971. Beatrice's main paper, 'Post-Main Sequence Evolution of Low Mass Old Disk Stars', gave a sample of low-mass old disk giants and subgiants obtained by combining six very similar groups.

Again she was making elaborate childcare arrangements for these four snatched days away from home: times and opportunities and reciprocal baby-sittings with other parents, preparations of whatever meals she could in advance, and general smoothing of the way for others before she could throw herself wholeheartedly into her own scientific work.

It was almost a month before she wrote home again:

> I had a great time at Amherst, enjoying the astronomy and the company of astronomers. I gave two papers, one at each of the contiguous meetings, which is always a good way of getting many new ideas, and acquaintances.
>
> A very eminent woman astronomer, Margaret Burbidge,[13] great enough to win any honours in any contest – turned down a prestigious prize offered to women astronomers only, on the grounds that special honours and discrimination for women should be abolished.
>
> It caused a gasp and then applause in the assembled throng, but the chairman cut off all comment by setting up a special committee to study the matter, and inviting 'polite letters from those with well-considered opinions'. It certainly woke up some of those men as to what problems (not all discriminatory, of course) women face, and how deeply they are felt!
>
> I've written (I hope politely) to the committee.[14]

This Amherst meeting was another landmark. Beatrice had gradually become known to astronomers and cosmologists in her field after her thesis was published. The few papers she had been able to write since then, and the few appearances she had been able to make at scientific conferences, built on this. The usual buzz of discussion saw her instigating much of it, and asking many of the questions. All the astronomy centres watched out for potential stars, and offering a short-term visiting appointment was a good way to check out a likely prospect.

This was what was soon offered to Beatrice, her first professional appointment in her own field – for she scarcely counted her part-time position with UTD. James Gunn, the young star of the California Institute of Technology in Pasadena, had sounded her out, and then got support for the idea. Caltech, with its astronomers at the Hale Observatories and the 200-inch Palomar telescope – and home territory for Allan Sandage no less – was right in the inner circle of Beatrice's work. There were always new nuggets of unrelated information emanating from Caltech. It was fertile ground for her work.

'A fantastically interesting and exciting offer' she called this January-March 1972 appointment at Pasadena. From the moment it was mooted, Beatrice must have known she would somehow manage to accept. Confirmation would have to come formally, as Caltech had to wait to finalise the contract with the National Science Foundation, but this was a virtual certainty. The children, Brian – she would find a way to arrange everything.

As it happened, Brian was due to go away again for about the period of her appointment, on another of his extended expeditions with his spectroscope. He would be mostly in Brazil, and would also visit the McDonald Observatory in West Texas. The children – well, with any luck there would be another Montessori school, or some good childcare could be arranged as well as school at Pasadena. She exuberantly signed her letter 'Beetle'.

Pasadena did indeed have another Montessori school, one with childcare arrangements which sounded ideal as she had been anxious about having to make separate plans for caring for the children when school ended each day. Their classes were from 9am to 2.30pm, but they could be looked after at the school at any time between 7am and 6pm. They would rest after school as much as necessary, and would receive Montessori's famed individual attention. Beatrice thought that on many days she would be able to stop work at 3pm and collect them then, making up time in the evenings, but to know that she would be free to work on, without nagging anxieties about them, was an immense relief. The arrangements would not be cheap, but if this was the price for being a fulltime astronomer, so be it.

Brian's plans changed. He wanted to have two dark-moon periods for observing, and made his expedition earlier, being away virtually all of November and December. This left her to cope alone with all the reorganising involved in leaving people in the Dallas house in their absence, and to deal with her ZPG responsibilities, her wide-ranging correspondence, her own work and preparations for Caltech in Pasadena, and with the children's increasingly busy social lives.

Her letters were not all to scientists or family. Descriptions of the children's activities were important parts of the letters Beatrice continued to write to Nanny Gullidge in England. She knew how much these letters meant to Nanny – flashbacks to the time when she was the centre of a small child's world – and saved up morsels of news for her. Very occasionally Edward wrote, but it was left to Beatrice to keep up the ties. Nanny Gullidge had written back at once, and sadly, after receiving the news of Theodora's engagement. Plainly she felt left out: nobody had let her know what was happening until the letter came from America. Beatrice was stricken. Of course she should write more often, gossipy, loving letters. This was Nanny. Somehow she would do this, by staying up later at night, perhaps.

Friends were becoming even more important to Beatrice as she continually arranged and rearranged her life to try to be all things to everyone yet still have some space for herself. Friends were essential in the complicated business of reciprocal childcare, and also for a guarded release of feelings: how life was, really, and how to cope. Beatrice greatly missed her friend Robin Wheeldon. In some ways Robin's situation was similar to hers, although Ron was a constant presence to give back-up with the children. As Alan and Terry had a long weekend for Thanksgiving, Beatrice decided it was worth driving them all to Austin so they could have three nights with the Wheeldons – 'a very nice break'. She returned home refreshed but also wrote an impassioned letter to Edward and Mattie Hill.[15] She was preparing a talk on ZPG, and the figures on the huge momentum of the world's population growth were frightening:

> It seems impossible to stop before even more damage, vastly more, is done, and countless millions more people have died of malnutrition. The world population will almost certainly double between 1970 and the year 2000:[16] another three and a half billion people (1,000 million), as many as it's taken all of history to get to date. I seem to have letters about it in the local newspapers monthly. So do other ZPG people – it's a campaign strategy, which wins some friends and makes some enemies!

Her good neighbours Carol and Art Wilson looked after the children for two Sunday mornings while Beatrice addressed Unitarian Church congregations on ZPG and world population pressures. She was determined not to let her organisation down when she went away for three months, so took on extra assignments to give others a break before she left. One way she could gain more time, when Brian was away from home, was to have an early meal with the children, and then, after Alan's bedtime at 7pm, work on into the night.

Brian was back again for Christmas. Two days later they all set off on the long drive to Pasadena and what Beatrice said the children were already speaking and thinking of as a fairy tale place, a 'mythical' California. For her it was to mark the beginning of a new direction.

CHAPTER THIRTEEN

ROCKY YEARS

'In the simplest sense, one becomes what one has experienced.'
EDGAR LEVENSON

Beatrice's priority before she began her three months of astronomy at Caltech in Pasadena was to find a house or apartment where the children would not be able to keep each other awake. She wanted to do everything possible to get them to go to sleep early, and stay asleep. Then she could be sure of the evening hours for her work. The house she found was shabby, but this was entirely secondary to being in Pasadena, her mecca. Caltech's honey-coloured stone buildings and wide avenues lined with trees pleased her, too.

Brian stayed with them in Pasadena for most of the first week of 1972 and then left to do some work at the McDonald Observatory. He planned to return briefly in February, before going off for a major expedition in Brazil. The two parents had been able to oversee the children's settling into their new Montessori School, and to solve the mystery of Terry's strange and sudden illness. Terry had become ill, vomiting, as soon as they arrived in Los Angeles. A bug, they thought – or maybe 'the vile air pollution'. These two scientists finally realised that the small girl might be scared, and reacting to all these changes in her life. The day when school began, Beatrice told her family, 'Terry ate no breakfast and shook like a leaf (and must have lost many pounds in the previous five days), but we decided to believe it was nerves, and took her to school. Result: she came home very happy and well, and has been eating three times more than usual ever since, to make up. She was disappointed at no school on Saturday morning.'

James Gunn, Caltech's star all-rounder, had been quick to see that Beatrice herself was a star: her calculations of galactic evolution were of the greatest importance for cosmology, and he wanted to work with her. He had sung her praises to anyone he encountered in the same field:

> Bev Oke[1] and I had a grant for some observational astronomy at Palomar. We had observations of distant galaxies and we needed to interpret them. Some people had produced complete nonsense about galaxies. Sandage, for instance, spoke as if he had had a divine revelation and he was dismissive of others. We saw that Beatrice would be ideal to work with us.[2]

The two men got the data, and Jim Gunn and Beatrice worked together to interpret it.

As Jim has said, the main thrust of what they were trying to do was to take the observations of galaxies and of stars in our own galaxy, then find out what mix of stars made the spectrum of the distant galaxy.

It was while Beatrice was at Caltech that Jim began the long programme to get observations of stars:

> This was done at tag-ends of nights on the 200-inch when the sky was too bright to do faint objects, and I accumulated a very large amount of data. I also got observations in more 'prime' time of several nearby elliptical galaxies taken with the same instrument, Bev Oke's multichannel spectrometer.[3]

For Beatrice, to be in Jim's orbit and to be accepted by the inner group of like-minded astronomers meant that she began living in a whirl of intellectual stimulation such as she had never known. Her work and thinking can be summarised like this:

> It's been known, since people began using general relativity to describe the universe, that if you look at a distant object you are looking back in time, that is, into the history of the universe. So looking at a distant object is better than a history book – you are seeing what the universe was like. Galaxies were different from the way they are today because they are made up of stars, and stars change. That is, they live, and die. Furthermore, the universe itself is different today. Because it is expanding, it was smaller then.
>
> From the perspective of general relativity, however, an evolving universe is much more natural than one that looks the same at all epochs.

Her family could have followed this. What she would not have said, and indeed it was still not completely recognised at that time, was that she herself had begun to make this whole idea into a science. Beatrice was the first to attack the whole problem in a systematic and quantitative way and make detailed predictions.

She was able to use spectroscopic observations to calculate the rate of stellar birth, hence the rate of stellar death, and therefore the rate of galactic evolution to give some basis for modelling 'the story of the universe'.

Caught up in a new life, her life as a scientist at last, she wrote home on 9 January 1972:

> I'm working very hard, enormously enjoying life being surrounded by eminent astronomers and astrophysicists. This is a very prestigious place, of course, owning the 100-inch and 200-inch telescopes.
> Lunch times are very fruitful. The astronomy department has a big table at the faculty club, where the food is excellent and inexpensive, but overlooked in the conversations. I'll be going back to Dallas with enough ideas to last for years, not to mention many good friends.

That sounds happy, very. In fact, nobody at that time would have been able to understand how utterly happy and fulfilled she felt. Her code forbade her to talk to others about the deprivation she had been feeling for years. Among her Dallas friends there was nobody who could fully comprehend her passion for astronomy, let alone the contribution she now knew she could make. Being able at last to go out each day to connect with other scientists in work she believed was the most important in the world brought her to the highest pitch of happiness she had ever experienced.

Beatrice was honoured to be working with Jim Gunn. He and Jim Westphal at Caltech had won a contest to design an electronic camera, to be a prototype for the first camera on the Hubble Space Telescope. With money from NASA, construction to its final form took years. Larger than a refrigerator, it was called the 4-Shooter.[4] It was useful in the work on very faint and distant galaxies that Beatrice and Jim began to do together. Beatrice was not interested in the camera itself, but in what Jim was discovering with it.

In fact, Beatrice had practically no interest in tools of any kind, nor ability with them. As Jim once remarked:

> Astronomers carry Swiss Army knives. Not Beatrice. She had never been used to tools. At one point she let it be known to an incredulous bunch of us that when she was growing up her family didn't even own a screwdriver. In this respect she was like some of the Asian

students we knew. They came from wealthy families, with servants, and they'd never done anything practical for themselves.[5]

Jim Gunn may be the only astronomer who has distinguished himself beyond argument in three fields: as a theorist, as an observer and as a builder of instruments. In his days at Caltech he was also a well-known charmer – 'a magnet,' a colleague has said, 'to whom women were drawn like iron filings'. But it was his mind that charmed Beatrice, as hers did him. From early on she thought him a genius, even though scientists are wary of saying that about anyone: what does it mean? There are so many areas, some vital but so small ... Friends could joke, however, that Jim was the only astronomer to be 'certified as a genius' as he had appeared, so designated, on the cover of *Time* magazine.

His Texas father, James Edward Gunn Senior, was an itinerant oil prospector who taught Jim from early childhood the joys of using tools 'to build ... well, almost anything we thought of '.[6] Many parents make things with their children. This father and son did not so much make as create, going to the boundaries of practical and fantastic creation. Together they savoured the joys of making something that had not existed before. The father gave his son a book on astronomy when Jim was only seven. The next step was building a small refracting telescope together, searching the skies and identifying what they saw.

The sudden death of this exceptional man, when Jim was 11 years old, appears to have plunged him into much the same dark hole of grief as Beatrice had experienced with the loss of her nanny. Beatrice knew nothing of this then, but it is perhaps not too fanciful to suggest that this shared experience of loss might have been an element in their future relationship, and eventual firm friendship. Jim has said, 'We both had fantasies. Our relationship wasn't meant to be.'[7]

Another of her new colleagues was the French astronomer Jean Audouze, a post-doctoral research fellow in Caltech's Kellogg Radiation Laboratory, headed by Willie Fowler.[8] Beatrice worked with them both. One day she gave a colloquium on the chemical evolution of galaxies, a topic that greatly interested Audouze, 'but her strong New Zealand accent made comprehension hard for French ears'.[9] At the end of her talk Audouze introduced himself. As she was often to do, Beatrice took the initiative and said she would enjoy collaborating with him. She had in mind applying her models of chemical evolution of galaxies to the light elements – lithium, beryllium and boron – on which Audouze had published a few papers concerning their nucleosynthesis. Both scientists were to be at Caltech only until the spring. Immediately they began collaborating.

Beatrice had been struck by a 1969 paper by Richard Larson of Yale, on the formation of galaxies. Now she wrote from Caltech, asking him for a reprint. When he sent it to her, he said a little about the extension of his work. She responded at once – 'but of course. She never let the dust settle.'[10] Richard Larson had read her thesis paper in the *Astrophysical Journal* when he was at Caltech as a graduate student and just finishing his own thesis on star formation, 'Dynamics of a Collapsing Protostar'. He was immediately interested in Beatrice's work. 'As I read I thought, "Hey, this is something! It's really novel. Gee – this hasn't been done before." This was clear even to me at my stage.'

He and Beatrice soon collaborated. The first Larson-Tinsley paper, 'Photometric Properties of Model Spherical Galaxies', took a theoretical model of the dynamics of these galaxies that Larson had created in an earlier paper, and used this to work out the ways the mixture of light, emitted by these galaxies, changes as they grow older.

This paper was preceded by 'a blizzard of correspondence and phone calls' from Beatrice at Caltech. As Richard Larson recalled:

On the phone I used to hang on for dear life to pick up her words as they rushed past me. She had so many ideas and was so impatient to get them out. Those who didn't know English properly had a tough time following her. And her speed and her accent – an upper class English version of the New Zealand accent – just made it harder. She couldn't go slowly. She didn't have a slow gear.[11]

Richard had made new models for galaxy formation, and Beatrice predicted what they would look like when seen through a telescope in different wavelengths of light. (As to her family's wondering what a model is, Beatrice had once explained to her father, 'A model is a recipe, a table of numbers to, for example, describe what a galaxy is made of. Graphs are results of models.') Their joint paper, which combined her stellar population models with Larson's hydrodynamic galaxy models, was eventually published in the *Astrophysical Journal* in September 1974.

Another interested astronomer was Virginia Trimble. Now working at Maryland, she had been a graduate student at Caltech. A bubbling, confident young woman, Virginia Trimble liked to 'drop by' Caltech to catch up by talking with the astronomers and attending whatever was going on. Now she met Beatrice for the first time. The two women responded to each other at once, and did some work together.

Beatrice and the children soon settled into their new existence. Her feelings of fulfilment spilled over into better relationships with the children, and her colleagues made it plain that they greatly valued her and enjoyed her company as well as her contribution. The children also quickly made new friends. Even the winter in Pasadena seemed warm. This was indeed 'mythic California'. Jim Gunn has said:

> If I had done nothing else, bringing Beatrice to Caltech would have made my scientific life worthwhile. She had had no contacts and the reverse of stimulation in Dallas. Nobody knew about her. Unless someone notices a student, say at Oklahoma State or wherever, this student could even be 'the second coming' but his or her paper heads to oblivion if it isn't noticed. These days there are far, far more people in the field and it's hard to keep up with everyone but it is so important for a student to be noticed. After Caltech, Beatrice was most certainly noticed – everywhere.[12]

Soon came visits to the University of California at Berkeley, and to Santa Cruz where she was able to take the children with her. 'Millions of new ideas' came from these visits, she wrote exuberantly on 27 February. The Santa Cruz campus in particular, with its scattered buildings hidden among redwood trees in what seemed a vast area of farmland, reminded her of Taranaki, an idyllic place. It was even more special because of the people she was able to work and talk with, including the astronomer who was to become one of her closest friends, Sandra Faber, who was professor of astronomy at Santa Cruz's Lick Observatory. In her introduction to Edward Hill's *My Daughter Beatrice*, Sandra Faber wrote:

> One might say that I first met Beatrice Tinsley by reading her PhD thesis. I was struggling with my own thesis at the time and was consciously seeking a good model to emulate. I read the thesis diligently every night between 1 and 2am before bed, and in a week I had finished it, deeply impressed.
>
> Beatrice's distinctive scientific style was already remarkably developed in this, her first major publication. The arguments were lucid, the presentation meticulous, the writing itself graceful yet economical – the net impact could best be described as 'lean'.

Not only was there its economy of thought and presentation, she said, but scientifically the thesis was so outstanding that virtually every astronomer who read it could see at once that here was a shining new talent in the field of research on galaxies and galaxy evolution. Sandra Faber added: 'It was also uncomfortably clear that the rest of the world, including myself, would have to run very fast to keep up.'[13]

At Santa Cruz Beatrice gave two colloquia on her work, and talked to as many people in her field as she could possibly manage in her short visit. She decided that California, either Santa Cruz or Pasadena, was where she would like to stay as long as possible.

Jim Gunn and his wife Rosemary had been allocated a three-bedroomed guest cottage on Mt Palomar, and they invited Beatrice and the children to join them for Jim's last weekend of observations. He gave them a guided tour of the observatory by daylight, impressing them all greatly by the vast scale of the 200-inch telescope and its dome. The children, she thought, would now have very inflated ideas about what telescopes could offer.

For her last day at Caltech Beatrice organised a seminar, taking her usual pains to make it a memorable event. She wanted to give as well as to receive. The whole Californian experience, she felt, had immeasurably enriched her. But Dallas was where she must return, and the children were now counting the days until they could be reunited with their own school and friends, their dog and all the familiar things of home.

Beatrice had long hoped to attend the three-day meeting of the American Astronomical Society in Seattle, but it was about to begin. Her concerns about managing family schedules disappeared when she discovered the children were so excited to be back home that they even insisted on going to their Montessori school the next day, and they were happy to find that Brian was home, too.

In her relief at being able to get away to Seattle without anxieties shadowing her, Beatrice filled her aerogramme[14] home with descriptions of the children being so ready and happy to be at school all day that she now felt able to try 'rake up' more research funds, so she could work fulltime. At both Montessori schools the children had flourished, 'not surprising when I talked to their teachers and found what dedicated loving people they are'.

The meeting in Seattle must have consolidated her feeling that she was now moving into fulltime astronomy. It made possible an almost immediate resumption of the exchanges of ideas and assessments that had so delighted her in California. Jim Gunn was there, and at last she was able to meet Richard Larson face to face after all their letters and phone calls. Beatrice sought out Sandy Faber for a breakfast talk, having by now read Sandy's own thesis. 'Beatrice was older and further along than I was, but she welcomed me, the newcomer. I was charmed, flattered, honoured.'[15] Out of these meetings came lifelong friendships.

When Beatrice was among her peers and tried out ideas, the more the ideas came tumbling from her. Words shot out, propelled by excitement and the sheer joy of being in her own rightful world. She seemed unstoppable. As Sandy Faber put it:

Many times Beatrice would turn up bright and early in the morning, trailing graphs and computer output that fully answered the questions you and she had barely managed to formulate the afternoon before.

On one occasion at a scientific meeting Beatrice presented me at breakfast with a draft of a paper she had finished the previous night. She requested my comments, and since she was invariably the most careful reader of my own papers, I wanted strongly to return the favour. I therefore stayed up late that night reading the rather complicated and lengthy analysis. Coming down the next morning, I was armed with cogent observations and ready for a good discussion. To my surprise, however, Beatrice presented me with yet another paper on a totally different subject but, like its predecessor, completed just the night before. My cherished chance to interact with scientific colleagues at the meeting was fast evaporating, what with reading Beatrice's papers! The forebodings of earlier years were indeed coming true – it was proving hard work to keep up.[16]

The more Beatrice met other astronomers, the more she yearned for this stimulation to be a permanent state – not of course the intense stimulation of formal meetings with time limits, whose dates now shone on her calendar, but the lower-key, work-absorbed daily talk of her peers. Then came a real opportunity. The University of Maryland offered her an 18-month appointment as visiting lecturer in astronomy. She was elated.

Brian was not. Her base should be in Dallas with him. She was going away too much. They argued.

Beatrice said she would find a house big enough for all of them. Brian could come and stay whenever he wanted to, just as he had when she and the children were in Pasadena. In any case, he was away so very often on his own work ... At night their angry words pinged back and forth when they thought the children were both asleep. But six months was the limit Brian would agree to, not 18. He was adamant.

Beatrice looked again at her situation at UTD. For years her title had been Visiting Scientist; by this time it was Visiting Math Scientist, a category which still excluded her from so much that Brian had access to as a matter of course. Now she made another attempt to be promoted to an assistant professorship at UTD, although there she would be alone, isolated in a working life to which interaction and group dynamic were integral.

Nepotism regulations had meant that nearly always, when couples applied, the woman was refused a position so that her husband could be employed. Now the women's movement of the 1960s and 70s and some celebrated cases had made it harder to defend the indefensible. A factor that gave her hope was, paradoxically, the overwhelming maleness, still, of this Texan enclave: 'currently one faculty member of 52 is a woman,' she observed. Federal authorities would surely step in if the numbers and status of women on the staff did not improve. Try as she might, however, nowhere could she find a job at UTD.

Brian was satisfied with his own position and prospects there. His was not so robust and unique a talent that other universities would automatically open their doors to him. He was secure where he was, valued, and could not see where else he could find an equivalent position. Beatrice did concede there had been a turn for the worse in the scientific job situation. Their original assumption that they would soon be moving on from Dallas was silently passed over, on Brian's part if not always on Beatrice's.

In spite of her arguing, she had a distaste for the role of the nagging, wronged wife. That was not the way to behave. Perhaps if she did more community work she would feel more satisfied. Zero Population Growth remained her main cause, and she was co-chairman of the Dallas chapter for the year. She was also arranging a programme for the Richardson Unitarian Church on the problems of local population growth. ZPG had its office at this church, conveniently almost next to the childcare centre where Alan and Terry were looked after out of school hours.

In her discussion group at the Unitarian Church, which valued challenges to orthodoxy, she learned of a new book, *Open Marriage, a New Life Style for Couples*, by Nena and George O'Neill. She bought a copy. It pushed the line that 'love and sex should never be seen in terms of duty or obligation, as they are in closed marriage.' Rather, love and sex could expand to include other people. These new relationships in turn could 'enhance and augment' the marital relationship of the couple in an open marriage. A rigidly adhered to, traditional marriage, in short, was obsolete.

Unsurprisingly, Brian could see no merit in these arguments. Beatrice said she could – apparently not taking into account the well-nigh universal reaction of jealousy in such cases. Could a marriage be open, and remain happily so, for each of its partners? English novelist Aldous Huxley of *Brave New World* and his wife were one of the rare couples who were to declare they had succeeded. At all events Beatrice was to revert to the topic of open marriage from time to time, even drawing Brian into discussion, in letters to him, as late as 1974.

But in her letters home Beatrice's language, content and tone remained, as ever, that of a cheerful and well-adjusted young mother with some unusual scientific interests, which was how her father perceived her – 'Beatrice is so absorbed in the stars.'[17] By now, however, each new meeting of astronomers added to her resolve to join them on equal terms, fulltime, instead of living through the weeks or months before she could snatch another few days of life-sustaining discussion and exchange of ideas face to face.

She looked forward to first-hand news of Rowena from Edward and Mattie as they were due in Dallas some time in the autumn on their way home. But she urged them not to come towards the end of October when she would be away in Pasadena at a Caltech meeting for three days, a specialist meeting in her field that she could not miss. Fred Hoyle – the one astronomer whose name her family would know – would be present.

The Tinsleys were having a busy summer, visiting the Wheeldons in Austin, celebrating Alan's sixth birthday with a big children's party and having a bigger party of their own to farewell their English scientist friends, Michael and Valerie Shaw. This went on until 2.30am, with Beatrice having to get up the next morning and go to the Unitarian Church, where she was in charge of her programme on population control.

Brian was once again away for some weeks, this time to Japan. On his way home he took another two weeks in Alaska to do some observing. By this time it was almost a novelty for the children to have two parents in residence simultaneously. But he was back for the visit of Edward and Mattie Hill, helping Beatrice make it memorable for them. Edward and Brian got on well together. The younger man was respectful towards Edward, not inclined to argue over aspects of religious belief, as Beatrice now seemed constrained to do. Edward after all was still an ordained clergyman. Many of his beliefs were diametrically opposed to those of both Tinsleys. Beatrice, who was nearly always restrained, almost docile, in her letters home, found the actuality of her father's presence at times an irritant she could not altogether suppress. She liked Mattie from the beginning, however, and made a point of introducing her to her friends and people at the children's school. Mattie, who had been apprehensive about meeting her, was charmed.[18]

Edward Hill has said of this visit, 'We found Beatrice so totally organised that it was virtually impossible to help with the household chores, to which we inevitably added.'[19] The Tinsleys took the Hills on a round of sightseeing, with Alan and Terry enjoying having grandparents. The oldest and youngest enjoyed one another's accents. American visitors particularly liked listening to Edward's public school English, and comparing it with their own, as in Edward's pronunciation of the Tinsleys' dog's name, Rata. Rah-tuh, or Rat-ter? Visitors encouraged Edward to talk about his high-born English friends, such as the Marquis of Anglesea. 'Quaint' was their word for these reminiscences. Beatrice was irked by this. But the Hills took back with them memories of laughter and what seemed a busy, secure family life. It was a brief visit, but they would all meet again in a few months, as the Tinsleys were coming out to New Zealand.

Beatrice immediately flew off to Pasadena for her specialist conference. Her account of this meeting was included in a letter on 9 November:

> I had a great time at Pasadena, scientifically and socially – more invitations from old friends than I had time to accept. My papers were both well received.
>
> The paper about Hoyle's theory was also accepted by all but Fred, though nobody took his lines of defence seriously. I had a very good conversation with him about it, and we basically agreed that Jeno Barnothy[20] and I showed very serious difficulties for his theory,[21] though not insuperable ones.

Her position as visiting lecturer in astronomy at the University of Maryland was confirmed that same day, and Beatrice ended her letter very happily, saying that an astronomer friend in Maryland, with young children of her own, was hunting up a good school for the children. Her thoughts running on women astronomers coping with children, she added that she had just heard from a friend at the Lick Observatory – Sandy Faber – that she had taken her baby, Robin, born in September, on a three-night observing run, so that the breast-feeding routine would not be interrupted. 'Must be about the first time in history that's been done!'

She said nothing to her family about having to tell Maryland that she could come for only six months, not the 18 they were expecting. In her keen disappointment and frustration, she felt herself inching closer to a decision. Leaving Brian sooner or later was becoming inevitable if he would not leave Dallas. But the children were still so young. She should struggle on.

Again she tried to create work opportunities in Dallas. No door opened. Then came a ray of hope. UTD's executive dean of graduate studies, Dr Tom O'Dell, suggested she work on a formal proposal for an astronomy department there. For years Beatrice had been badgering the authorities subtly and not so subtly about this need. Not only UTD but also the State Co-ordinating Board had to be convinced. She and Brian had been teaching a junior astronomy class, and this might have helped to change official opinion.

Now she swung into action. Her researches led her to recommend a programme leading to a master's degree, MSc, with astronomy as a major, and a programme for one of the master of arts degrees in teaching, MAT, offered with astronomy as the major field. To her family she described her proposal as 'a complicated business, and foreign to me'.

Among those to whom she wrote for suggestions as to the kinds of programmes and courses she should propose was Gérard de Vaucouleurs. After outlining what she had been asked to do, she stressed that she was warmly inclined towards the possibility, particularly as 'O'Dell is interested in somewhat innovative ideas. In particular, each of these master's degrees is to be regarded as a complete course of study, not just a step towards a PhD, or a booby prize for failure in a doctorate!' [22]

She had been told that, if the programmes were approved, they would expect to hire another astronomer as well, who would –'necessarily' – be someone of a more practical bent than she was. All those from whom she sought guidance appear to have agreed with her ideas.

The lead-up to the Tinsleys' departure for New Zealand on 22 December 1972 was extra busy with pre-Christmas social activities and shopping for gifts for their respective families and friends. Beatrice warned the Hills that they might receive some calls and letters before she arrived, as she had given their address as a contact point for Family Planning and Zero Population Growth in New Zealand. She had agreed – 'I've stuck my neck out' – to help them publicise the cause by giving interviews to newspapers, radio and TV.

Almost nothing about Beatrice's scientific work, let alone her eminent position in extra-galactic astronomy, had appeared in the New Zealand media before she was interviewed during this visit. If Edward had remained in New Plymouth, where he and his family were so well known, almost certainly he would have contacted the local newspapers with excerpts from Beatrice's letters. They would have run stories that in turn would have been picked up by larger papers, following one of journalism's most basic tenets, 'local man or woman makes good'. That Beatrice was already 'making good' was an accepted fact in world astronomy.

After the move to Wellington, however, Edward lost his previously central place in society, and the attention of the media. He was not successful in interesting journalists in his daughter, and nobody knew enough to seek him out to learn what Beatrice was doing, although he would not, of course, have been able to tell them much. His lack of knowledge of scientific matters, and – always much worse in Beatrice's eyes, as she told Rowena, his 'not wanting to know' – had meant she had long ago given up on any real explanation of what she was doing and what she was most interested in. For intelligent people, nature lovers too, to put science and the arts into two separate categories was something she found hard to understand and to accept.

The newspaper interviews Beatrice gave in New Zealand about the ZPG movement – interviews illustrated with photos of the handsome, confident-looking family of four – did contain a little information, sometimes garbled, about what she was doing in astronomy. She wanted the focus to remain on ZPG, however, so she deliberately did not talk much about her professional work.

The Tinsleys were welcomed warmly, with everyone exclaiming over the children and marvelling at their accents. In Wellington they were concerned by Mattie's ill health. To everyone's sorrow, she had been diagnosed as suffering from multiple sclerosis. Beatrice did not want to add to the Hills' burden by staying with them with two lively children for more than a few days. In any case the senior Tinsleys were to join them for a tour of the South Island's West Coast, where Brian took some 200 slides of the scenery.

The American family then stayed in Christchurch, for a special reunion with Theodora and to meet her husband, David Lee-Smith, for the first time. Like everyone else, they immediately warmed to David. He and Theodora had moved from Australia to New Zealand in December 1971, and attended Christchurch Teachers' College the following year. They were about to move to Invercargill where David was to teach at Southland Boys' High School, and Theodora at Southland Girls' High School. Beatrice was immensely relieved that Theodora, at least, seemed to have made the right marriage. She herself was so enjoying being back in New Zealand that she was even able to feel a little more optimistic about her own future.

The first of the friends they visited were Guff and Necia France. When the Frances were in Dallas, Necia had admired a nylon fake-fur coat. Now Beatrice had brought her one, and was gleeful because she had naughtily got it into the country without having to pay customs duty. She had put a crumpled tissue in one of the pockets so that the coat appeared to have been worn.

To her old friends Glen and Wal Metcalf she seemed the same bubbling Beatrice, as full of joy in life as ever. Alan and Terry romped over their lawn, dousing each other with a hose, while the adults talked and laughed their way over the gap of nearly 10 years. One thing, however, surprised and concerned Glen Metcalf: Beatrice's continuing anger (Glen used the word 'bitterness') that she had not been employed by Canterbury University because her husband was on their staff already.[23] Now she went over the story again to Glen: the Physics Department – and Brian – had betrayed her (Beatrice's expression) by not telling her that if she married him she would never get a job in this department or anywhere at Canterbury, 'not even as a lab demonstrator'. She felt the information had been kept from her. She had not been consulted, and her freedom had been curtailed. If the two women had had more time together, Glen might have been able to get Beatrice to talk more openly and intimately, but circumstances were not right for that kind of discussion. In any case, Beatrice had never been confiding on that level. Pride, reserve, and an almost desperate resolve to hide behind the façade she had constructed, all stood between her and even trusted friends.

At the back of her mind in New Zealand was the thought that this might be their last visit as a family. She said nothing to anyone – she had promised herself she would not. She very much enjoyed meeting up with family and friends, and the beauty of the countryside worked its old magic on her. She was soothed enough to feel that, maybe, she and Brian could work something out for a while yet, at least.

To Teddy Fardell, however, she was not the same Beatrice who for years had been part of the Fardell family. While Brian took Alan off for time with his family, Beatrice brought Terry, now aged four and a half, to stay with the Fardells. Terry refused her evening meal. 'Right. Then you'll have it for breakfast,' Beatrice said sharply. To Teddy, intervening out of Terry's earshot, this was something carried over automatically from Beatrice's own childhood. Teddy had been concerned when Jean Hill had done this to her young daughters; probably, she mused, because it had been done to her. To see this behaviour being repeated was distressing. She was further distressed when Beatrice put Terry to bed, 'with no bedtime story, no hug, nothing'. What had happened to their little Beetle, their merry, affectionate and open little girl? Clearly she was very stressed, 'running on her nerves'. What was this marriage really like?

Mulling over these incidents again, years later, in the light of what happened not so very long after this visit to New Zealand, Teddy Fardell[24] thought perhaps that Beatrice, however wrong-headedly, had been trying to make Terry more amenable, more independent of her. At the time, however, it was plain that all was not well.

Nobody else seems to have queried the Tinsley family's stability. Probably only Teddy, of all the New Zealand family and friends, had any idea of what was so soon to happen. But then Teddy had always been intuitive where Beatrice was concerned.

CHAPTER FOURTEEN

THE UNRAVELLING FAMILY

'Our way of looking was changed forever.'
JILL KNAPP

The time spent in New Zealand, crowded though it was with people and sights, had given Beatrice the chance to distance herself from her day-to-day problems and concerns, and see them a little more clearly, even optimistically. She arrived back in Dallas with determination and hope – enough to join Brian in a 3am champagne toast to 1973 and their future. They realised it was only 9pm by their internal clocks after they had put the children to bed and discovered the chilled bottle their house-sitter had left for them.

The year as a whole looked exceptionally promising. Beatrice felt that she could postpone any significant family decisions now that she was poised to take advantage of contacts with some of the best people in her field in the world. She did not realise that the more she was acknowledged and sought after elsewhere, the more her frustration with her Dallas imprisonment would grow. Her greatest continuing disappointment, and it was bitter, was that Brian still refused to let her take the children and accept the full 18-month appointment she had been offered at Maryland. Six months was the maximum Brian would agree to, even though his own plans now included being away from the US for quite a lot of that time. He had also set up work possibilities to enable him to be based for periods with the rest of the family in Maryland. Beatrice had stressed that he could come as often as he liked while she was there, but it was as if he could not let Beatrice out of his sight for long; she had to be anchored, preferably in Dallas. Earlier she had consulted a lawyer. Could Brian really prevent her from taking the children to Maryland for the full period on offer? The answer was that she would first have to have legal custody arrangements in place.¹ She still hoped that she would not have to take such a step.

True to her decision not to criticise Brian to her family, Beatrice wrote to them only that he would help her drive the children and their special possessions to Maryland on 24 February, then leave for his spectrometer work in Peru. But before that she was to visit three universities, giving lectures on galaxies at each: five days at Chicago and Pittsburgh, with a week at Caltech.²

Although various astronomy centres were now vying to have her come and lecture, in Dallas it was uncertain whether her plans for an astronomy department at UTD would be accepted. She had to finish the work on her formal proposal before leaving to lecture, however, so worked extra hard to get it done to her satisfaction – which she translated for her family as the satisfaction of the authorities who would consider it. The good straightforward English she had been taught, and which she used for her scientific papers (however incomprehensible much of them remained to lay people), was not acceptable for a formal presentation to the university or the state co-ordinating board. She wrote in disgust:

> The proposal has to be written in the most nauseating jargon. I read three other proposals through, drank a glass of wine, then translated my straight English into a fairly good imitation. (It's getting sober appraisal too.)³

That aside was Beatrice laughing at herself. She drank very little, a habit which made her stand out during the social side of astronomy meetings but which helped her wake early and clear-headed each day. Brian similarly steered away from much alcohol; in this they continued their habits of student days. Their idea of a party also remained much the spartan same: lots of good talk, with the simplest and most easily assembled food and a little wine or a not very strong punch. One sub-freezing night at the end of January they held a party for more than 50 people – 'the Christmas party we didn't give, and for my birthday, though we didn't advertise the fact' – where Brian showed many of the spectacular slides he had taken of New Zealand scenery. 'Everyone squashed in, enjoying warmth from fire and candles and hot punch'.⁴

Brian would be out of the US for quite some time but first he would be in different parts of the country. Beatrice prepared for her own time away, leaving Dallas as soon as he returned and could be with the children. Alan and Terry by now had become well used to having only one parent at home at any one time.

Beatrice's week at Caltech, she told her family, was most enjoyable and fruitful. She stayed at the faculty club, the Athenaeum on the campus, which was convenient for work in the morning and for computing late at night – 'that's when the queue goes home, and jobs get in and out quickly.' With her French collaborator, Jean Audouze, she worked on a small research project and found their favourite theories were compatible. She also made progress with her 'very lengthy and meaty work' on distant galaxies with Jim Gunn.[5]

What she did not say, could not say – and could scarcely admit even to herself – was that her great intellectual attraction to Jim was growing into something more, and that this had probably begun before she consciously realised it. The attraction was serious and bound to be long-lasting, based as it was on deep respect and admiration as well as keen enjoyment of their working together. Some of her friends thought Beatrice fell in love with minds, not bodies.[6] Certainly she made no bones about being attracted to the very best minds, what was inside people's heads, how they looked at things. So for every reason she was drawn to Jim. He was the leader in the work most dear to her. Now, in exchanging bursts of ideas, one would pick up the other half-way through a sentence in the most satisfying duet of mutual understanding and recognition.

For some time the small, ingrown world of Caltech had had something of a reputation as a marital merry-go-round. Beatrice had never been part of such an unconventional scene. She was aware, though, that Rosemary Gunn, whom she liked, and Jim were among those couples said to have an open marriage, that relationship she had tried to discuss with Brian. Meantime almost nobody knew about the strains in her own life.

Beatrice was glad of Brian's help on their cross-country drive to Maryland. Their car broke down twice. Terry was sick, partly a recurrence of her usual fears when travelling into the unknown, and the dog Rata got into all the food. Alan and Terry were resilient, however, and when they visited their new school the next day they begged to join their classes. Brian soon left for Peru.

Beatrice shared an office in a basement with her acquaintance from astronomy meetings, the effervescent Virginia Trimble. The two got on well, although Virginia has said she was 'somewhat frightened of Beatrice – she was very bright, and very quick to see when some idea might not be quite right.'[7] Virginia had worked with her on a project at Caltech, when Jim Gunn had helped them both.

That 'very quick' also applied to her use of time. More than ever Beatrice now planned to extract the most from every minute. She could scarcely have imagined the volume of work and the number of work-related discussions that began to come her way. Maryland was her base but she also visited other astronomy departments, including meetings at Austin and Yale. Each time she tried to pass on to her new grad students what she had learned.

One of these was the young Scottish astronomer, Gillian Knapp, most often called Jill. Her husband, Stephen Knapp, built astronomical instruments and was frequently away. Jill was surprised and felt honoured when Beatrice spent time with her – teasing out strings of ideas and getting her to join in assessing them. Her attention was focused on Beatrice:

> I think she was the first great person we students had ever seen in action. She was so busy always, but we hung around her when we could. At that time it was a strange thing for a female to be an astronomer. It was a source of enormous strength to know that Beatrice approved of you.
>
> She gave us the big picture, the whole picture – she'd paint in big brush strokes, make us look at things differently. She changed forever our way of looking.[8]

The two women responded quickly to each other, in an appreciative relationship that soon had Jill becoming one of Beatrice's closest friends. The chance to make new friends among astronomers was one of the great benefits of this Maryland experience.

Beatrice in turn felt honoured when the astronomer Vera Rubin,[9] whom she had long admired especially for her work leading to the discovery of dark matter, invited her to visit her lab in Washington DC. She went several times, and sometimes Vera came to Maryland, including attending a meeting in honour of the famed astronomer Henry Norris Russell, where Beatrice talked about the evolution of stars and hence galaxies. Apart from giving her talk, Beatrice was intent on listening and learning, almost visibly soaking up new information and ideas. But this was just one side of the coin. She also took deep pleasure in spinning webs of possibilities, linking material and trailing hypotheses for others to consider.

Vera Rubin was one of those who listened to her:

> Her work was remarkable, and she was a born teacher, at least in talking to other astronomers.
>
> I learned important things every time I spoke with her.[10]

Brian came back again, staying with the family in Maryland for almost three weeks. He had discovered there was no point in returning to Peru immediately, as he had planned. His equipment had still not arrived because somehow it had been sent off on a circuitous route around Cape Horn. Then he received the news that he would have two months back with the family in Maryland as well, as he had been given a grant enabling him to work at two nearby government research institutes.

Many of Brian's colleagues from Dallas came to an American Geophysical Union meeting in Washington DC, so the Tinsleys had an unexpectedly social few days. Their friends Walter and Bea Heikkila were house guests, and Beatrice squeezed another two young scientists 'into spare corners', to help them save accommodation costs. One evening they gave a party for Brian's associates from around the country, people he had met from earlier meetings, so Beatrice at last could put faces to names. She had one friend there of her own, Sachiko Tsuruta, who was working in the NASA/Goddard Space Flight Center in Maryland at the time.

Brian shared Beatrice's office for a week, and worked on a lengthy review paper that she called his magnum opus, 'Hydrogen in the Upper Atmosphere'. He in turn could meet the people she worked with and ate lunch with every day. Eating often seemed a secondary matter at the university. Lunch times were when people gathered and ideas could be tossed about. Much of science is talk. At Maryland eight or ten people could be seated around the table, some merely chatting to their neighbours about what they had done the previous night, or the baseball scores. Then Beatrice would take over. In Jill Knapp's recollection:

> She would dive right in about some wonderful paper she'd just read. She busted to talk science. Her voice would get higher as she talked about it, overlaying the baseball, and higher still until everyone was involved and talking too. She was like the sun coming up, unstoppable, and drawing everyone into the discussion. 'Have you seen this? What about that?' And she would get you to talk. Every day was potentially exciting.
>
> One particular day it was about a paper published in the *Astrophysical Journal* by Jay Frogel and Eric Persson. It was on the first spectroscopic measurements of stars at infrared wavelengths. They had built a new instrument. Well, so what? All of us had seen this paper and had thought nothing of it.
>
> But Beatrice arrived for lunch in high excitement. Dwarf and giant red stars of the same colour – stars at different evolutionary stages, dwarfs being long-lived, giants short – show quite different spectroscopic behaviour in the infrared, as those authors showed. Thus, with a single measurement of a galaxy you could say which kind of stars dominated its light, and thus get a good handle on its evolutionary stage.
>
> Next day she had the draft of a paper written. Typical – she'd spotted the importance of this result immediately. No one else had.

Bea was lightning fast. She knew everything, had tremendous clarity, spotted connections instantly. She was just plain dazzling – no wonder she inspired such devotion.

It's easy to get into a comfortable rut as a scientist – we know so little! You can master a technique and spend your life observing one thing after another, or study one kind of object to death as new techniques come along. Not Beatrice! Why she was such a huge and exhilarating inspiration is that when you told her what you were working on she immediately threw out a dozen reasons why this was interesting in other areas, or gave you a result from elsewhere that really tied in, that showed you how your work connected to everything else – this was just wonderful.

I don't think anyone who encountered her was ever able again to get back into the rut, think inside the box, and so on. Did the people who studied white dwarfs realise that their results were crucial to understanding galaxy formation? Not till Beatrice came along, I bet! It made you feel so good about your work, and feel inspired to do better.

And another thing. You always knew where you were with her. If she was cross with you, she was cross. But she was always your friend.[11]

Although her working life was centred on Maryland for the present, Beatrice continued to accept invitations to other astronomy departments. In a quick rundown of her working weeks she ended a letter to her family with an understated 'No chance of stagnation.' It was not long since she had realistically faced up to the problem of getting some research funding by saying matter of factly that she was 'so unheard of'. That situation had changed dramatically, and it was not easy for Brian to come to terms with it. A combination of significant published papers and her being able to go, at last, to key conferences meant that so many people now knew or knew of her. Professionally she had arrived.

In spite of her scientific work and the new stimulation, Beatrice was still missing music. At lunch times she would bring her violin and join Chris Simonson who played the recorder. The students gathered outside the room to listen, then asked for the door to be left open. Beatrice promised herself to practise more, and also to get to every possible concert – 'essential nourishment'.

With childcare organised, she was able to fly to Austin to a meeting organised by Dave Schramm and Dave Arnett. She and Jean Audouze were able to present a paper containing the first results of the work they had begun together at Caltech. They were both excited by it. As Audouze saw it,[12] to Beatrice it was an argument supporting one of her favourite hypotheses, that the galaxies were continuously receiving some still-primitive gas which was not evolved. She called it 'the infall of primitive or unevolved interstellar gas'. For Audouze it was another convincing argument that lithium 7, lithium 6, beryllium and boron were formed through the interaction between the interstellar medium and cosmic rays. They had to finish this collaboration by mail.[13]

It was a stimulating meeting. Moreover, Beatrice had been offered, and had accepted, a half-time position as an assistant professor of astronomy at Austin, beginning in September. She was not wholeheartedly happy about this, although pleased when Gérard de Vaucouleurs told her it was a unanimous departmental recommendation. Her heart and mind were set on permanent work at a top cosmology centre where she herself would be constantly learning as well as teaching.

She very much wanted to attend the International Astronomy Union conference in Poland in early September, visit Jean Audouze in Paris, and, particularly, stay with Rowena in Florence en route. But first so much had to be organised in Maryland before she could consider going. She had to find a room for her nights in Austin, and work out a new course, introductory astronomy for non-scientists. The main problem, as ever, was how to care for the children once she was based in Dallas again. She had arranged for her Austin schedule to be squeezed into the first part of each week. By flying down to Austin on Monday mornings and back on Wednesday afternoons, she would be away for only two nights. 'The wonderful Bets Espeset', whom she had met through the Unitarian Church, would be looking after the children again, an arrangement which made everyone happy. 'Bets is truly marvellous with them,' she wrote home, 'one of the very few people who has no tendency to spoil Terry.'[14]

Terry's big appealing eyes and her habit of slipping on to women's laps and saying 'Will you be my mommy?' marked her as a very needy child, with an almost bottomless need for affection, observers said. She was manipulative, too, with her high intelligence and quick ability to sum up a situation. To Beatrice this was being spoilt. She noted that the children had always seemed calm and organised when they returned from being at Bets's place, and full of creative ideas about what to do. 'Everyone should have a babysitter or grandmother like that!'

Would it be too much for the children if she were away in Europe for two solid weeks? Brian would not be back from Japan and Alaska until the end of September. On the other hand, her colleague in Paris was expecting her, and she would be letting him down if she did not go.

The problems were resolved before she left Maryland after a most satisfying time, the longest she had ever been able to spend in the work situation she craved. She tried not to think of how much more she would have done, and learned, had she been able to accept Maryland's invitation in full.

Back in Dallas at their Richardson home, the children were settled in different schools as Alan had moved on from kindergarten. Their reunion with Bets Espeset, with whom they were to stay, was obviously happy all round. Everything else fell into place. Rowena cabled to say she would meet Beatrice at Milan airport, so she flew off on 4 September, with Brian leaving for Japan the next day.

This time she found Rowena not quite as harried, and alone with her two children in a big old stone house, beautiful if primitive in terms of basic amenities, and with a long view over the Tuscan countryside towards Florence. Sito and Lila she found wonderfully warm-hearted children, a delight to be with. Mostly the sisters used the time together to talk over their lives, talking on into the night and getting little sleep before Beatrice's IAU conference in Poland.

Then a cholera scare hit Italy when a woman who had been eating seafood in Naples became ill. Poland, where the IAU conference was to be held, required all travellers from Italy to have a vaccination. Beatrice promptly had this done but nobody told her it would not be valid until six days had passed. This meant missing the conference. She was wretched with disappointment. There was nothing for it but to go straight to the Audouzes in Paris. Jean had to leave for the IAU the next morning, promising to bring her back the main strands of the presentations and discussions. His wife, Françoise, a marvellous person, Beatrice thought, was a distinguished archaeologist who was currently supervising a dig for bones and tools near Paris. They enjoyed each other's company whenever Françoise was free.

When Jean returned he introduced her to his extended family – 'charming people' – and helped her make up for her lost opportunity in Poland.

> I met many French astronomers who had just been names before – with some surprises. There are a lot of women and they sign their papers by initials only, so people I'd imagined to be men had the wrong sex. On the last day I gave a formal colloquium which generated a lot of interesting questions (the most satisfying response) and an invitation to dinner with an eminent couple of astronomers.[15]

This is Beatrice's plain, unvarnished account of her talk, but what she said is still remembered vividly. The eminent astronomers were Roger and Giusa Cayrel.[16] The dinner was held after she had given her evening colloquium at the Institut d'Astrophysique de Paris. Her subject was the chemical evolution of galaxies. Everyone agreed it was a triumph of a presentation.

She was particularly happy to meet women colleagues of Jean Audouze and learn about their work. They included Suzy Collin-Souffrin, whose work was in the evolution of galaxies and quasars; Jacqueline Bergeron, X-ray astronomy and interstellar medium; Monique Spite, stellar abundances; Sylvie Vauclair, nucleosynthesis and stellar evolution; Monique Joly, galactic evolution; Lucienne Goughenheim and Lucette Bottinelli, galaxies, radioastronomy and cosmology.

Back in Dallas again, and fortified by her new contacts and the ideas generated, Beatrice thought the children were happy and calm after their time staying with Bets Espeset – a great relief, this, as Brian was still out of the country. With Bets continuing her part-time role of child-minder, Beatrice felt free to begin her weekly flights to Austin in her role as assistant professor.

She soon began to feel happier there. Being able to fly, instead of enduring that long twice-weekly bus journey, made a difference, as did having colleagues on the spot. Some of Beatrice's peers have since said how unfortunate it was that de Vaucouleurs[17] was temperamentally so pedantic, so set in his ways, that he and Beatrice could never really get on the same wavelength and enjoy a spontaneous exchange of ideas. If they had been able to work together, there could have been exciting developments.

To her family she wrote that she was getting experience in teaching. She was doing far more than getting experience, according to Greg Shields,[18] who was then in his first semester at Austin, and also an assistant professor in the Astronomy Department. Beatrice took the initiative, asked about his work – the hydrogen regions in galaxies – said she was interested in it, and generally encouraged him. Out of this eventually came a joint paper.

Shields's idea was that the mass and temperature of the hot stars that produce ultra-violet light, which heats and ionises the gas, may be determined in part or influenced by the chemical composition of the gas. He said:

> My expertise was in studying the gas. I knew nothing about stars, which Beatrice did. So we combined. It made sense. We began a paper and had a few discussions but did most by letter later when she was at Santa Cruz.
>
> Some of the most detail-filled letters I've ever gotten were from her. They came fast, signed 'Cheers, B.' She considered everything, exhaustively. I have more pages of her letters in my files than from anyone else. She made me feel I was appreciated – a good feeling.[19]

Their joint paper, 'Composition Gradients Across Spiral Galaxies: Part 2, the Stellar Mass Limit', was published in 1976 in the *Astrophysical Journal*, with Beatrice telling Shields that it seemed to her it was more his work than hers. 'But she was just being polite. She was always so generous.'

While Beatrice commuted to Austin each week, Brian had gone to Japan but now had his spectroscope in Alaska where he was doing some auroral observing in Fairbanks. Although Beatrice had to take over his usual yard work – the lawns, garden and odd jobs – she had found a maid who housecleaned once a week. This gave her more time for the children when she was home, and each week she tried to give them at least one excursion or special treat. One weekend they all went camping at Tyler State Park with a group from the Unitarian Church. 'We sat around a vast bonfire – the stars were very bright – and Terry asked me if there were an infinity of them. Maybe that's something I'll help find out, in the next few years,' she wrote home.

In the part of the week she was in Dallas she was trying to negotiate a gift of two telescopes to UTD, a 10-inch and a 16-inch reflector. A local physician wanted to donate them but the authorities were worried about the admittedly considerable cost of building domes to house them. Beatrice remained hopeful. Having two telescopes would support her proposed astronomy programme at UTD. It would also give her a better chance of a permanent job where she could steer some of the activities towards her own field. It would also help – surely, she thought – if she had been instrumental in getting the telescopes accepted. These could be used only for star-gazing, however, not research, as the city was growing around the campus, and the skies would soon be too full of light to see anything but a few bright stars and planets.

What she did not tell her New Zealand family was that she felt this might be her last chance to make her life in Dallas bearable:

... And that one talent which is death to hide
Lodged in me useless ...

To have one talent, a still-expanding talent which she knew could help extend the world's knowledge of its origins, of the origins of the universe, and yet to be anchored to a place which was indifferent, if not hostile, to what she had to offer, was eating away at her. The teaching at Austin was only a stop-gap.

When Brian was home, she led the talk back in the evenings to what they had originally planned and what they had told each other they would do – have no more than two years or so in Dallas. Now it had become 10 years. And they had agreed to find a university or a city which could offer them both – not just Brian – employment in their separate careers. She wrote away, and persuaded Brian to do the same. At astronomy meetings she made soundings: was this a centre that could accommodate them both?

Brian was not whole-hearted in his enquiries – understandably. He had a safe and solid niche at UTD. He was among people with whom he fitted in. His work offered constant opportunities for exotic travel, although the sites he selected for his spectrometer observations were often bleak, even for an experienced mountaineer. To give all this up for an uncertain future somewhere else? Long since forgotten was his declaration of 'no more than two years'.

The Tinsleys' situation was not unique among academics, but certainly nobody then seemed to have thought of a situation where a woman's intellectual abilities and creative talents were far ahead of the man's. As a colleague was later to say dryly, comparing Beatrice with all the rest of them, 'There was Brian, fiddling away perfectly adequately in the back row of the second violins, and there was Beatrice out in front of the orchestra, the Yehudi Menuhin of us all.'

Again Beatrice rehearsed, by no means always only in her mind, how it would be if she took Alan and Terry away from Brian in Dallas, away with her to one of the astronomy centres where she already knew she would be welcomed. She and the children had already managed very well at Pasadena and Maryland. Brian really was away so very often ... covering the world, it sometimes seemed. How much would he actually miss them? Was it not more that he would miss the *idea* of having them waiting for him at home in Dallas? And of course she would give him every possible access, have him come and stay – as he had so very often done during the short-term appointments she had already had. She would share the holidays, keep his memory always fresh with the children.

But he was possessive. He even had some special family right to be, she acknowledged to Rowena. For her part, Rowena remembered how her younger sister as a child had been capable of causing the most almighty commotion if she were frustrated in something she had set her heart on. If formidable then, she would surely be unstoppable now, when she felt her very life was at stake.

Alan at seven years old, highly intelligent, precociously attuned to the angry current connecting his parents, listened, kept quiet and, as he was to say much later, kept his head down.[20] Terry, younger in most ways at five, was attuned differently. She needed to be at the centre of everything. Sometimes, or so it seemed to people like Bets Espeset, Terry grew more needy all the time.

None of this crept into Beatrice's letters to her father and Mattie. Some of her anger, however, she unleashed on 'the crooks in government', mainly President Richard Nixon and Vice-President Spiro Agnew, as the scandals of Watergate were being revealed.

Jean Audouze arrived from Paris to see Beatrice, stayed with the family, and discussed their current collaboration. She drove him down to Austin. After another astronomy meeting in Tucson, Arizona, in the first week of December, she told her family:

> It was full of scientific and human enrichment. By now I have great friends from all over the US and Canada, whom I meet once or twice a year, and exchange papers and odd phone calls with, in between.
>
> It's a dispersed community with much in common that therefore keeps strong bonds over time and distance – a diaspora, as Toynbee would describe it.'

Then almost immediately Beatrice returned to Austin for the last week of classes, to be followed by the final exams. It was a crazy schedule, a word she used herself. 'I had to get the exams graded for a crazy deadline, so spent Saturday night plus Sunday plus Sunday night going over 130 three-hour papers!'[21] And that was it until after Christmas when she flew off to Caltech before New Year for a week, 'squeezing in some hard work on my continuing research with Jim Gunn out there before classes start again'.

Her first letter home for 1974 was largely about this 'marvellous week' and the work they had done. The Gunns had taken her to two parties where 'the eminent company had its hair down.'

This time what Beatrice omitted was considerable and significant. She liked Rosemary Gunn and had no wish to hurt her. The Caltech crowd, living what often struck Beatrice as a life of extremes, had made it plain that Jim could be attracted to more than one woman at a time, and that this state of affairs had been going on for many years. Jim, so brilliant and so intense, seemed at times to be driven by private demons and depressions. When he and Beatrice were deep in work together, they were in the same electrical field, different, apart from the world, unique – just as lovers feel they are. In this case, though, they actually were unique, the only man and woman in the world who were breaking barriers and pushing out the frontiers of galaxy evolution together. The remaining barrier in their personal relationship was getting closer to being broken, too.

But then it was time for Dallas again, and the same, seemingly intractable, problems at home. Immediately after she returned, Brian left for a two-week meeting in his field and then flew on to New Zealand to see his family. The children again stayed with Bets for the times Beatrice had to be in Austin. 'What complicated lives we lead,' she wrote.

Rather more than complicated. Each of the two Tinsleys had a schedule which would put a strain on almost any marriage, let alone a marriage containing two such schedules. It was likely to work only in a relationship where the two people actively enjoyed each other, felt close, and took it for granted that they could count on each other for full understanding and mutual help. It was small wonder that Beatrice, who had carried the major weight of family and household responsibility for so long, and often alone, was reaching the point where she could no longer carry on.

Almost certainly it was about this time, as she looked at the new year of 1974 unfolding, that she began moving, at last, to a definite resolution.

CHAPTER FIFTEEN

DECISION

The scientist's research is akin to that of the religious worshipper or the lover; the daily effort comes from no deliberate intention or programme, but straight from the heart.
ALBERT EINSTEIN

Beatrice continued making professional plans for the summer, as if all in her personal life were well. The invitation she particularly prized for 1974 had come from England, from the Nato Advanced Study Institute in Cambridge. Her fares would be paid (an important consideration when she was still on a part-timer's salary, much of which went on childcare), and she was to spend three weeks there during July–August. 'I'll probably be giving a lecture or two, and learning a lot from others, and getting a lot of research done with people from all over the world,' she wrote home.[1]

She also hoped to take Alan and Terry with her to Aspen, Colorado, in June for a similar workshop, provided she could find a good holiday programme for them there, such as a day camp which they had enjoyed in Maryland. But this would depend on Brian's plans.

Her work at Austin was passing through a disagreeable phase. Perhaps triggered by the taking stock of attitudes and activities most people go in for, if briefly, at the beginning of each new year, academic and personality arguments had broken out in the Astronomy Department:

> It has been a rather harrowing place the last two weeks, with bitter disputes between the young turks and the old guard on the faculty, as to what is relevant and useful astronomy to teach our students.
>
> I'm lined up against one of my old profs, who always was lamentably conservative and encyclopaedic in his approach. I suppose we (the turks) will subvert the old-fashioned system somehow, and save the kids' minds from atrophy, and the old fogies from total loss of face.

> The Boss, Harlan Smith, is on the Right Side, and has a difficult diplomatic job. [2]

Lapsing into the family's old in-joke of using capital letters for part-mocking emphasis, Beatrice may have unconsciously been trying to strengthen her bond with her father, in the light of what she knew she would have to tell him before very long.

There was another reason for her to tell her father it had been a harrowing time at Austin, although it was not something she would write to him about. One morning as she had walked, as usual very early, to the department to begin work, taking a short cut through the car park, a man had thrown himself at her and tried to rape her. She fought and screamed, and the man ran off. Craig Wheeler, also on the faculty, was the first person she met after this incident, and she told him briefly what had just happened: 'She talked to me for a few minutes. I thought at the time what strength of character she had. She was an incredibly special person.' [3]

Back in Dallas, having arranged Bets Espeset's help with the children, Beatrice accepted an invitation to go to a summer school at Erice in Sicily for the last two weeks in May. She thought there could be good astronomy talk, judging by the list of other people invited, and she would be lecturing to an international group of selected students. She would have to put in considerable work, however, on the 'massive lecture notes' needed for such a group.

No hint of a rift in the marriage appeared in the flow of letters home. Instead she talked about the problem of finding enough money to enable her to attend meetings. The Annie J. Cannon Prize for proposed research offered worthwhile funding which would be of the greatest help. In between joining in her children's activities she worked on her proposal.

For her family she put this research into easily understandable language. The point of it all was that if the universe were 'open', it would go on expanding forever, but if it were 'closed', it would eventually stop expanding, and contract again (in about 20 billion years). The work she was doing with Jim Gunn, she said, pointed heavily in the direction of openness, but they still had a lot of work to do for that particular section of the problem to be properly solved. Her theoretical contribution was one of the necessary parts for answering the question: is the universe open or closed? They felt this question to be the most interesting in present cosmology. It rested on interpreting the observations of distant galaxies in terms of Einstein's general relativity, 'which may or may not be the correct theory of gravitation – but there is no better theory yet. Anyway, it is a lot of fun and very fascinating.' [4]

She also wrote tenderly to her father when she had news of the death of his mother. Edward had not at that time shown Beatrice his memoir, 'Half a Life', nor was she as sensitive as, say, Rowena, to his obscurely expressed regrets, but she knew he would be deeply affected by his mother's death, and all the more – as Rowena saw – because old conflicts could never be resolved now. At this time, too, she had news of the death of her uncle, Edward's younger brother, in an air crash near Paris. Francis Hill had been on a trade mission to the USSR. Beatrice said later that his death seemed doubly tragic when Edward told her it was just chance that he happened to be on that plane. 'So much to bear,' she wrote to her father. 'How frail life seems, when we all travel so much and expose ourselves to death.' Mattie's multiple sclerosis also hung over them all. The most they could hope for were long periods of remission in this crisis. Beatrice's letters home continued to contain references to Mattie's health, and how Edward was managing to look after her. Still there was no reference to any crisis of her own.

Jill Knapp and her husband Steve now came through Texas and had a night with the Tinsleys on their way to Caltech. At Maryland Jill had felt almost hero-worship for the older woman, and was deeply honoured to be counted her friend. She was in no way prepared for what she found in Dallas, the jagged rift that was dividing Beatrice and Brian:

> The atmosphere was awful. At Maryland when I'd met Beatrice and the children I had thought of her as the perfect woman – she had everything. So it was such a shock to see her in such an atmosphere in her own home. By then, I realised, she was beyond trying to cover up.[5]

Then came a decisive show-down. In March Jim Gunn came to Austin while Beatrice was there. They worked together for a few days. In spite of their great intellectual and emotional attraction, they had said nothing explicit to each other. She spoke only of not being able to bear her marriage for much longer. As Jim saw the situation, her thoughts of leaving Brian had little to do with him. 'Leaving was inevitable.'[6]

They had earlier arranged to drive together from Austin to Dallas in time to meet Rosemary Gunn at the airport, Rosemary having flown in from visiting her parents. The Tinsleys and the Gunns spent the evening together, carefully not saying anything that was in the forefront of their minds. Then Jim and Rosemary flew back to California, and Beatrice wrote to her family that the Gunns were 'exceptionally good friends' whom she was glad to see.

Three fraught days followed before Brian left for Peru on 20 March. Beatrice had finally found she could no longer hang on. She had come to a decision after what she was to call 'a landslide crisis'. No matter how she had tried to hide her feelings, her absorption in Jim and their shared work was evident. Brian erupted. Beatrice reacted and lashed back.

After so many years of indecision, of telling herself she could and should keep on trying to make the marriage work, she had reached the end of the road. Suddenly she crossed the line she had long since drawn. She consulted a lawyer. Everything then happened very quickly.

She wanted a divorce. She would of course take both children, and Brian would have the fullest possible access to them, always. The two of them must remain friends, as much as possible – this was the logical thing to aim for, for the children's sake, too. Being friends should not be difficult. After all, they had been able to remain on reasonably amicable terms for much of the time in their years together, and had shared so much of their lives.

The lawyer said they should sign divorce papers as they would both be away and apart so often during the summer. To Beatrice's surprise, Texas law did not prevent their continuing to live in the same house – even after they had signed these papers. To Brian, even though he had signed the documents, this may have seemed that it could not be a real divorce. In any event he appeared not to take it as final.

Beatrice's letter home on 30 March 1974 began, 'Things have been hectic here,' and said later, 'Brian went to Peru about 10 days ago and I'm not sure if he's coming back this weekend or not.' This from someone who was always so sure of dates might have been a give-away if her father had had any suspicion that all was not well. But he had no such idea. She had taken good care of that.

Writing next, Beatrice noted that, with less than a month before she left for Sicily, she had to put in a whole week in Austin, taking oral examinations, then fly to give talks in both Illinois and Indiana. She had also been asked to lecture in Brazil at the end of June, but that sounded as if she could not take Alan and Terry as she wanted and needed to do, and Brian had a meeting of his own. 'It's fun being in demand at exciting places.'

Brian – 'poor Brian', she said – had to go into hospital for minor surgery. She would be home to look after him for a few days before going to Sicily in May, the last time they would be at home together until July, just before she left for her Cambridge visit. She would, however, be home all of June with the children, and all four Tinsleys would meet up again in Peru during the two weeks she planned to be there with the children when she joined Rowena. This was Rowena's new plan: Peru could be the country for her children. She wanted to give them more space than she could afford in Europe, and did not feel ready to go back to her husband's

Venezuela. The idea was for the sisters to be there together while Brian did his spectroscopic observations. Edward, trying to jot all this down on his calendar so he could follow her travels, was bemused.

By this time the impact of Beatrice's move to end their marriage was striking home to Brian, although he did not fully believe she meant what she said. There was one point, however, where he would stand firm. Never would he let Beatrice take the children away from him. Night after night they argued, when they thought both children were asleep.[7] Perhaps, Brian said, Beatrice could take Terry, a mother taking a daughter. But Alan – never.

Beatrice was adamant that the children must not be split up. Everything she had read was against this. It also seemed to lead to the danger of making each of them an 'only' child, the reason she had wanted to adopt a second baby. She conceded that Brian had a special link with Alan, but of course she would do everything possible so that Brian could come and go and be with them, just as had happened so often in the past. Nothing would change except that the marriage would have ended. In any case Brian could not be with the children very much. He was so very often away, out of the country. Surely he realised this? The right and best thing for everyone would be if the children were with her. The new regime would not be so very different from the old. She promised he would always be welcome.

Nothing made Brian give way. It seemed to some that his pride was outraged by the very notion of Beatrice's leaving him – he blamed Jim Gunn – and that if he were to let Beatrice have the children, his concept of himself as the male authority figure would be eroded. Unless he held on to them, his stature would be diminished. It seemed that he stopped listening to her.

One night he roundly declared something which brought her up short: 'I will fight you in every court in America!'

This changed everything.

No mother wants to see her children fought over in court, or asked by frightening figures of authority what they feel, or which parent they want to stay with. Beatrice had at last got around to admitting to herself that all the arguing was distressing the children, and that the parents' raised voices were affecting even the dog. Brian's ultimatum was something else again. She could not fight him, she thought. He was the one with the salary, the resources for legal fees. If this was what he was going to do, she should not oppose him and upset the children still further. She would have to give way.

In fact, she could have fought him, and almost certainly she would have won. The children could have been told that their parents were arguing because each loved them, and both wanted to have them. Many people would have helped Beatrice, and lent her money for court costs. And, whether she was aware of this or not, in the 1970s it was rare for a judge to award custody of children to their father. Even a judge in macho Texas was most unlikely to do this. She had not left the children without care. She had tried hard to create circumstances in which she felt she could keep the marriage going. She had a strong legal case, including the facts of Brian's weeks away from home, absences which would continue.

It does seem that once again Beatrice did not look deeply into her own feelings, and face them. Part of her, deep down, must have suddenly envisioned what it would be like to be free, living alone, her own unfettered person again. There is no doubt that she did feel deeply for the children, as later events were to prove. She was sure it would be bad for them to be separated. When Brian suggested he keep Alan, and Beatrice take Terry, she quoted her authorities about the bad effects on children if this happened. Certainly, whenever over the years she had had thoughts of ending her marriage, it had always been a scenario where she had both children with her, giving Brian every opportunity to visit them, have holidays with them and remain their father.

Now he seemed to be standing over her, making the ultimate threat That part, that essential part, of Beatrice which put the pursuit of science really above everything, whether she realised it or not, now began to invade her consciousness. If Brian really, finally, would not let the children go – whatever his own reasons – then she would become the parent who visited, who had the children for holidays, who wrote and phoned and sent money for their upkeep.

As for the children, like most in a situation where a divorce is pending, they leaned towards going with their mother. What they really wanted, naturally, was to stay on with both parents in their own house and go to their own school with their own friends. This she could not give them. She would do everything possible to make the children feel she truly loved them – and Brian might still change his mind and see the sense in her proposal. But nothing, now, could stand in the way of her leaving him.

Still no hint of her distress, her anger or concern about what best to do appeared in any of her letters home. Her letter home of 13 June 1974 flowed on calmly:

> I still seem to have a lot of work to do before the Cambridge Institute, as I've now been asked to present an introductory survey lecture as well as the talk on my own work; and review talks tend to tread on sensitive toes in this field if you don't think as much like a diplomat as a scientist while writing them.
>
> I'm also spending a couple of days at the University of Chicago

next week, which is one of the places that I might move to next year.

UTD seems to be an utterly hopeless prospect. For reasons best known to them (though guessed by me) to a formal request last March, they have not even had the courtesy to reply to me. Now I find that a number of first-rate places in the country would like to have me. No plans yet but I will definitely be on the Dallas-Austin trek September-December.

It was against this background, having made her usual careful childcare arrangements, that Beatrice flew to Sicily for her summer lecturing. Alvio Renzini, the organiser of the summer school, had not met Beatrice before but had been impressed by the freshness of her papers. He had accordingly invited her to give a series of lectures on 'Synthesis of Stellar Populations'. He was further impressed by the way she arrived with a thick manuscript containing all her lectures beautifully prepared – the only one of the lecturers to have done this, he noted.[8]

Beatrice found Erice a respite from her family trauma and her turmoil over the children. It was a charming small town containing a medieval village on a steep mountain with views over Sicily and the Mediterranean, 'a delightful little dream-town'. Then she and Rowena met up again in Rome, to talk about their lives. That is to say, Beatrice talked about her decision to leave Brian and their battle over custody of the children, but she also wanted to talk about her science, with Rowena doing her best to understand as they sat at a table at an outdoor café. It was a windy afternoon, not the best of circumstances in which to concentrate on Beatrice's evidence for an open universe. Altogether they talked so much during their brief time together that they had little sleep, discussing what would be the effect on their children, with or without men, and whether Brian would make good his threat to take a stand in law. Rowena talked about Peru's advantages. Brian, by now reasonably familiar with conditions there, had been helpful and had given Beatrice information to pass on to her sister.

When she returned to Dallas, Beatrice received news she knew would please her father. In a quick letter home she wrote that she had been awarded the Annie J. Cannon Prize in Astronomy. In his pride in his daughter Edward at once wrote back, asking for details. Beatrice knew she could delay her main news no longer, but she filled up most of her next letter, on 26 June 1974, with the information he wanted.

The Annie J. Cannon Prize used to be for senior women astronomers, she explained, but changes in the terms of the award meant that for the first time it was an open competition among all women astronomers in the United States. Winning it was a great surprise, and the $1000 that went with it was 'not to be sneezed at' although the prize was mainly an honour:

> Really one can't do much research on $1000: to publish a long paper in a classy journal costs that much. I recently ran up a bill for one paper of $922, and the computing costs alone for what was in it must have been several hundreds more. But I suggested I could use the money for travelling to my far-flung collaborators, which I will do.
>
> It was specifically for one project that has occupied a lot of my energies since we started it at Caltech in 1972, and which should after another year or two be a substantial contribution to cosmology.

Then Beatrice steeled herself to tell her father. She was 33 years old, she had just been recognised as one of the leading women astronomers in the US, she had a fair appreciation of her own formidable abilities – and she then filled up the rest of the main page of her aerogramme with odds and ends of family news before coming, at last, to what she knew she must say.

In the remaining space, on the very last line of this page, she wrote 'Maybe you ...' And on the 12-line space available on the flap of the air letter she continued:

> ... have been too insulated from us to see it, but combined with the hopeless job situation here has been a steady erosion of my relationship with Brian over the last several years, and the erosion came to a landslide crisis earlier this year. We are really quite good friends; but not good as people to live together. (I doubt he has mentioned this yet to his family so please keep it to yourself and Mattie.) You can imagine I've been through a lot of agonising and soul-searching lately. Having taken years and years to make the decision to go, it feels much better to have decided. I will try to explain more later. I am staying here until December. Love. B.

The effect on Edward Hill was as devastating as Beatrice had expected. In Edward's world people simply did not divorce. True, his oldest daughter had divorced her Venezuelan artist husband, but there had been so many dramatic leavings and reconciliations and altogether so many other matters to disturb Edward that he had been well prepared, even to some extent relieved. It was also true that Mattie had been divorced, but Edward and Jean had accepted this, and Mattie had been a friend of them both.

Before he could receive this thunderbolt from Beatrice she had flown off with Alan and Terry to Peru. There they would meet up with Brian who was doing his observations, and together they would help settle Rowena and her children in a small village on the outskirts of the city of Huancayo, high up in a mountain valley. At least, that was the plan.

The coming together of the two families had been put forward as a holiday, with Beatrice in a supporting back-up role, but Rowena found her sister almost unbearably stressed, unable to relax and enjoy her surroundings. The high altitude affected both women and Terry severely; Rowena suffered so much that eventually she was to leave the area with her children. On this occasion the two sisters talked about the directions their lives were now to take, with Beatrice saying her marriage was totally on the rocks; it was over. As they talked, they walked through what Beatrice called, in a family letter, 'patchwork fields farmed by ancient picturesque peasant methods. I could hardly believe that such things still exist outside story books – peasant women taking a little herd of animals (a mixture of cattle, pigs and goats) out into the unfenced fields, and tending them all day ... Rowena is in love with the place.' That all sounded charmingly romantic, as it was meant to do, to allay Edward's misgivings. In fact it is a startling example of Beatrice's omissions and ameliorations in her letters home. By now both women felt desperate. Beatrice was later to admit to Rowena that it was one of the worst times of her life, while Rowena was already wondering what she was letting herself in for with this change of country. Brian arrived at their small house and was with them for far longer, and more obtrusively, than the sisters had optimistically envisaged. Beatrice's exchanges with Brian were nowhere near friendly. In response to his invasive presence she became hysterically aggressive and defensive, Rowena found.[9]

With better weather for his observations, Brian stayed on in Peru when Beatrice took the children back to Dallas, to settle them with Bets Espeset who would look after them while she went to England to work in Cambridge. First she wrote a reassuring letter to her father, from whom she had still not heard in reply to her letter about divorce. Rowena – she said hopefully – did seem to be finding in beautiful Huancayo the relaxation of soul she needed after Italy's complications and corruption. As for herself and Brian, she was firm about not wanting to get acrimonious by trying to describe all that had gone wrong:

> We hope to part and remain friends ... I don't think it will be any harder for the children having only one of us at a time than it is having two of us together creating a bad atmosphere, to which they are as sensitive as the dog is.
>
> I would as soon teach them that a marriage relationship can honestly come to an end, than that parents are bound to each other no matter what the cost.[10]

Whether she had consciously contrived this or not, the timing of her first letter with the news of their divorce, and her subsequent far-flung travels, meant that she was gaining a respite from what she dreaded and what she knew would come: the outpouring from her father. He had not yet caught up with her.

When she reached England, she spoke candidly for the first time to her musician friend, Penny Saunders, about the mistake she had made in marrying. Penny was later to say:

> Women didn't usually talk intimately to each other in those days. 'What would people say?' was something that hung over you. Most women kept quiet. Parents were judged by their children's behaviour, and Beatrice's parents knew nothing of life, and what to expect.
>
> One small incident stuck in my mind. Beatrice happened to tell me that she hadn't wanted to attend some recent function in the States with Brian. He had said to her, 'Well, you have to come because you are my wife!' She didn't go – but she told me that earlier she would have gone along with Brian's assumptions.[11]

They never talked astronomy together, Penny reflected. Beatrice had many different circles of friends, and she was different but true with them all, without making them feel they were getting only a part of her.

Beatrice did not stay long in London. She was particularly intent on getting the most from her immediate Cambridge experience. This was the first time she had been her old, single self, a scientist among her peers. An invitation to work on in Cambridge for up to three years gave her enormous pleasure. It was a real honour. But, whatever their final legal agreement turned out to be, she knew Brian would never permit her to bring the children with her, and she could not bear to be so far away from them. She would have to say no to Cambridge.

Word that she might be available was spreading quickly, and invitations to be a guest lecturer were coming in from all around the US. One was from Cornell in its beautiful setting, and a faculty where she would enjoy working if they could offer a job. A Chicago offer was good, but the city did not appeal. She was immensely pleased to be able to accept a six-months' job at Lick Observatory at Santa Cruz, with Sandy Faber and her team. She wished that this short-term work, scheduled for after Christmas, would turn into a permanent job, but there was no vacancy. So many enticing invitations in so short a time, with so many opportunities to learn more and contribute more – she would have been less than human if she had not felt justified in her decision to leave Brian. At the same time she told herself she must try to see things from his point of view, and remain friends – 'and most of the time we are,' she told Rowena.

The Tinsleys' divorce became final at the end of August. Brian had not given way about the children.

Her father had been firing off a succession of shocked, distressed and disbelieving letters. On 7 September she felt impelled to try to answer them, point by point:

> Perhaps I should have warned you years ago that I knew we would break up eventually, as it would have been less of a shock now.
>
> When we were in New Zealand two years ago, it was a rather happier time than normal (no wonder), and at that time I thought I might 'stick it out' for a good many more years, but things fell to bits rather rapidly after that.
>
> I see no point in trying to attach blame. It is a failure of a relationship, that is, each of us relative to the other, rather than faults in either one.
>
> I had thought to save you the knowledge that we are now divorced, since I know it is bad news to you, but there might be legal problems so there you are. We set it in motion last March, and it came through at the end of August. Brian is the legal parent, called the 'managing conservator' under Texan law, and I am known as the 'possessory conservator' which means just what is written into the agreement (sending money, having visits etcetera).
>
> I went through all the agonies ages ago when I knew it would happen, so now it is a relief to have all the decisions made, explanations over, and to be free from trying to live in a very suffocating relationship.
>
> It is thoroughly symptomatic that Brian has felt much less than me the tensions and gulf in communication between us. I have been very unhappy, and am no longer. It is far less lonely living in solitude than with someone who is psychologically far away.

After three months of trying not to say unpleasant things about Brian, Beatrice probably thought, and certainly hoped, that this candid letter was enough by way of explanation and reflection.

She had returned to Dallas with almost more ideas than she could cope with. But it was time to tell the children, in terms they could understand, that their parents were going to live apart and that they were still trying to work out what was best for them. It was a gut-wrenching time, with Terry tearful and Alan quietly stoical. Even at his age he had seen a number of family break-ups among his classmates, and he had heard enough at home to be unsurprised. Thinking back, he said:

The divorce didn't surprise me one bit. There was only one thin wall between my room and the living room. For hours, for ages, every night they argued. They tried not to fight in front of us ... They had different ideas of their roles, where they were going. I think Dad expected all of Mum's attention.[12]

Brian was again away a lot. Beatrice found a room in a house nearby, which she moved to during his periods at home. It was so much easier, she found, to do concentrated work away from the tensions and uncertainties of domestic life. In particular she needed time for the work she had been doing with Jim Gunn. It would have a public airing at the big conference coming up in Dallas in December, and she had to prepare. Being alone was balm to her. Regardless of her feelings for Jim, solitary living was what she now craved.

Writing to Brian about family matters, Beatrice said:

Living alone seems to be very good for me. I enjoy solitude and the inner relaxation one can find. I always said that if I left you it would be to be alone, and to go to some place to work, not to be with another person. This is still true, tho' it is also true now that I wish I lived somewhere near Jim.

They sat down together and sorted out family photographs, making sure they both had one of each. Brian retained all the furniture, including family heirlooms as well as the house.

It had taken time for Beatrice's long letter of explanation to her father to reach him, and for his reply to come back to her. She hoped it would have reconciled him to events. It had not. Brian, hurt, bitter and baffled when the divorce finally came through in August – he had seemed to believe that Beatrice would change her mind – had written to his father-in-law with his own version of events. She knew nothing of this until yet another stream of reproaches arrived from New Zealand.

This time she would tell her father at least something of what she was really feeling, a cry from her heart. In five closely hand-written pages on 21 September 1974, she began by saying she had thought she was the first to write to him about the divorce's coming through:

I certainly didn't expect Brian to do so. It wasn't out of any wish to keep secrets from you that I didn't tell you, back in March. I have your letter in which you say you don't understand how everything has developed so fast. I knew you'd be upset, and I wanted to wait until I had a secure job set up, to relieve you of what has always been a major worry with Rowena.

The lawyer advised us to get the original request for divorce in soon especially since we would all be travelling so much in the summer. I expected back then to be able to live with Brian up till Christmas, but it became quite impossibly contentious. I can understand his feeling bitter, but it only exacerbated enormously the problems that had been poisoning our relationship for so long.

As I probably told you, I knew years ago that I'd have to leave him sooner or later – it happened faster than I might have hoped because (mostly) of the worsening job situation in Texas. The six months in Maryland were something Brian resented greatly. (I could have gone for 18 months but he absolutely refused to let me take the children. I called a lawyer then, and learned I couldn't help myself short of legal custody arrangements.)

Trying to work here [at UTD] had reduced me to a state of mental anguish. Hard to explain! I *am* a good scientist, and among my peers I'm treated like a full and respectable person and full of *worth*.

UTD has kept me at the nearest possible level to nothing, and there is *no one* who knows enough about astronomy to care in the least for my work. Austin has helped but it is a second-rate job (underpaid, half-time) at a department much worse than I'm worth.

This isn't supposed to be boasting. To be rejected and undervalued intellectually is a gut problem to me, and I've lived with it most of the time we've been here, apart from extended visits to Caltech and Maryland and shorter trips and meetings and so on.

Then, on the other level, I've been continuously malnourished (and felt rejected) at a spiritual level by trying to maintain a close relationship with Brian. Also a really gut problem! Now, looking back, in spite of all the happy times, I wonder we survived together so long. If he can't understand this, that shows the depth of the problem.

From your letter it almost sounds as though Brian gives the impression that I loaded the responsibility for the children on to him. I have *always* said and very deeply meant that I would gladly take the children – it would have been difficult, but of course I would have managed them somehow, out of love and responsibility. Brian became completely adamant about taking them, and finally I realised it was the rational choice to make for their sakes, in terms of continuity and security in their lives. (The one compromise I could *not* make – it almost seems for my own sanity – was to stay any longer with him!) It has been for a long time, and still is, an agonising thing to do.

Nothing hurts me more than to have Brian accusing me of burdening him. He *chose* that.

I told nobody, except one friend in Dallas, where my thoughts had been for so long; and when things came bursting out in bad moments with Brian, he never took them seriously. He realised only when the decision was final and I was doing concrete things like applying for jobs.

I was responsible for deciding to end the marriage (though I refuse to *blame* myself or him for the breakdown), and I would have taken, and originally expected to take, the responsibility for the children. With pleasure!

This is all very difficult to write. One of the things that kept me from making a break sooner was knowing how sad it would make you, and other people in both the families.

Thank you so much for all your efforts to understand, and the things you have written.

This raw *cri de coeur* with all its emphases and exclamation marks seemed to get through to Edward. He still could not fully understand, but he settled for accepting the situation as best he could, at least for the time being. Beatrice was his much-prized daughter but he remained on affectionate terms with Brian, and of course the children.

Not until later[13] did Beatrice tell her father more, the ultimate truth as she saw it. She had tried for so many years, she then said, both before and after adopting the children, to do anything she could to make the marriage workable, and to keep her wedding vows and not complain. She had tried at first to bury her own thoughts and interests, then to find all possible ways, no matter how makeshift, of fulfilling them. She had tried not to get into the swamp of recriminations, or to spell out those aspects of living with Brian which had at last proved insurmountable.

She reserved her final words to her father for the real reason for their break-up: 'I knew, two weeks after our marriage, that I had made a mistake.'[14]

CHAPTER SIXTEEN

TO A NEW LIFE

'Beyond the ramparts of this world …'
Lucretius

Alone as she liked to be, in her new small room in Dallas, Beatrice had time to think about her new situation and the institution of marriage, as she now saw it. She felt she had now become a different person, although in one way she had always been that person. She had to become used to living alone, but, again, in one way she had always lived alone. Now she felt that she was Beatrice Hill again, and rejoiced. Perhaps from now on that could be her name. Friends she talked to about this change of name advised against it. Her reputation, her papers and lectures, had been built under the name of Tinsley; a change would lead to confusion. She reluctantly dropped the idea but told these same friends she had promised herself to live alone from now on.

Inexperienced, dizzy with freedom and possibility, she could not look clearly at Jim Gunn and their situation. Sparking ideas and tracking them down with him was the purest joy. Neither then nor later could she have found a collaborator on his level. And he had sweetness, humour and a lightness of touch long missing from her life. But under the weight of her expectations and intensity, most men would have begun to feel uneasy. To step back, to look around for a way out of a life lived at such a pitch – this would almost inevitably happen sooner or later, but it was not something Beatrice could be expected to see.

Her father could not stop writing to her about her divorce. She tried to reassure him by describing how she talked with the children, who were remarkably philosophical, she felt. Many of their school friends were from single-parent homes. Their friends Walter and Bea Heikkila had recently divorced, with Bea reverting to her maiden name of Bea Wolf. She thought that Terry, and the more reticent Alan, seemed able to talk openly about how they felt.

By October both Yale and Chicago had offered her a job as assistant professor. 'Very tough choice!' She had to make a decision soon. Big-city Chicago, or Yale in small, highly cultured New Haven? Chicago was possibly the better scientific institution but Yale had a select group of very congenial astronomers, including her valued collaborator Richard Larson. 'I'll probably choose Yale for its civilisation!'

Pierre Demarque, then the chairman of Yale's Astronomy Department, had not met her at that point. But he had heard her speak and had never forgotten her. It was at an American Astronomical Society meeting in Austin in December 1968: 'The lady with black hair and eager manner – very original and interesting. Of all the papers there, over three to four days, hers is the only one I remember'. Her paper was not close to his particular field, stellar evolution, but it excited him.[1]

Beatrice's collaborations with Richard Larson, who with Pierre Demarque and Bob McClure made up the department's 'key triumvirate' – the only three astronomers to have tenure – had made her a bright star in Yale eyes. These men knew a little of her reputation for being fearless in the pursuit of scientific truth, a passion which put her offside with some of the loftiest names in the field when she outspokenly questioned some of their conclusions. When Yale first looked at making an appointment, they had not thought that Beatrice might be available, and were surprised when Ivan King, of the University of California at Berkeley, told them that she was looking for a job wherever she could get the right one. As Richard Larson said, this was a revelation to his department. 'People were very surprised when she left her husband. There had been no hints or clues. Ivan told us he had put Beatrice on the top of his list for Berkeley, but there were no jobs then open.'[2]

There nearly had not been a job at Yale. The position had been offered to, and accepted by, John Arons of Berkeley. He and his fiancée had been looking for a centre where they could each find the right job, and they had found them at Yale. Then they had received more attractive offers from Berkeley so had stayed there. Yale had then thought that they would have to find a job for Brian Tinsley, too. But when Pierre learned that Beatrice was alone, he and Bob descended on Richard in his office. 'Tell us everything you know about Beatrice Tinsley!' He did. Their joint paper, 'Photometric Properties of Model Spherical Galaxies', had been published earlier that year in the *Astrophysical Journal* so the others also had recent knowledge of her work.

Yale's offer of an assistant professorship was effective from September the following year, 1975. Beatrice mulled it over, talked to friends, and wrote back with what she called a slightly amended proposal. As she put it to her father and Mattie, she was really more drawn to Yale than to Chicago, 'for the place and the people', but various friends insisted that she was selling herself short at Yale as Chicago had offered a much better salary. But she would still be an assistant professor at Chicago, whereas Yale might agree to the next step up for her, appointment as an associate professor. After all these years, she was not about to sell herself short.

She was due at Yale in mid-November, anyway, for a long-planned lecture and consultation visit. When she arrived in New Haven she quizzed Richard Larson about Yale's advantages. 'I mentioned music,' he has said, 'and she perked up.' Chicago was probably stronger scientifically, still offering more money, and still a strong contender for her, he realised, but Yale and New Haven seemed to her more appealing, more on a New Zealand scale: 'she idealised New Zealand.'[3] For her part, as she told friends, Richard's calm manner, his aura of quiet dependability, and what she had soon recognised as his exciting and first-class work, struck chords. By the time she had met the other people in the department, looked around and been shown something of New Haven, she was ready to say yes if Yale offered her an associate professorship.

Yale did. The offer and her acceptance had a few committees to go through before everything became fully official, but she was soon able to tell her family: 'With a bit of politicking, Yale has now offered me an associate professorship, initially for five years, but I expect I would get tenure in the meantime if I want to stay there.' The department, she explained, was small and very good, with only nine faculty and about the same number of research staff and as many graduate students. They seemed extremely nice people and united in research interests and spirit – unlike some places she knew, she said darkly, where rivalry dominated relations between factions of the department. The chairman, Pierre Demarque, a Canadian of extraordinary personality, was quiet, calm, gentlemanly, 'like a friendly father', and his peaceful atmosphere pervaded the department. After her experiences elsewhere, however, Beatrice assumed that there would be at least some difficulties, for she ended, 'No doubt the subterranean problems will sooner or later become apparent to me, but now they seem like a delightful and welcoming group.'

She was not the only one thinking of subterranean problems. At a meeting not long after the job had been offered and accepted, Pierre Demarque ran into Allan Sandage. 'Oh,' said Sandage, 'I hear Beatrice Tinsley is coming to Yale. Well, I'll have to visit you guys in a year or so and see whether she's wrecked your department.' The implication was that Beatrice was so dominating and headstrong that she would over-ride 'lesser males'.[4]

A busy and stressful December faced her before she could begin her new life. She had work to finish in Austin and had all her gear there to pack up. There were the children's pre-Christmas activities to attend, and preparations for taking up her six-month appointment at Lick Observatory in Santa Cruz after Christmas. First, though, would come the big international conference in the biennial series of the Texas International Symposium. This time it was being held in downtown Dallas, and she and Jim Gunn would be speaking. She had worked independently on her topic, the life history of stars that form galaxies, as well as conferring with Jim. Many friends would be there from all over, and she planned a big party for them.

Meanwhile, as Brian was away at a conference of his own she moved back to the family home to care for the children. One night, when the children were asleep, she wrote home again. 'They are so full of energy and love it hurts. But I can tell they are OK, and lots of people love them.' And she had some gentle advice for Mattie, who had written of her exhaustion when preparing a dinner party. Friends would be coming to see her, Mattie, and not for fancy food, she emphasised. What with her own packing and the conference, she herself had decided against any fancy entertaining. The 40 or so people at her party 'would feed themselves with large quantities of make-your-own sandwiches out of sliced stuff from the delicatessen' which she would pick up on the way home. The thought of the party really excited her: 'Toynbee is right about groups of scientific specialists making a diaspora. We meet a few at a time, and enjoy it!'

Then she wrote on, more as if she were making an entry in her diary:

There are some strange astronomers in the world. Allan Sandage has just got enormous newspaper publicity, based on some really fine research he's done, but set up to imply that he's done far more than he has. His last published estimate for the age of the universe was 10 billion (10 thousand million) years, and now he has a new one of 16 billion years. Nowhere does he mention that people like me and my colleagues have been suggesting values like 16 for several years! He is a strangely jealous person, hates anyone else to make discoveries in cosmology (as he more or less told me once). Not only doesn't mention anyone else in his interview (which the journalists wouldn't ask for, so we couldn't blame him for that) but not in his scientific publication, either. If you read about all this in the New Zealand papers, add a good many grains of salt.[5]

Sandage, the would-be inheritor of Hubble's mantle, had never forgotten being challenged by 'that young woman', Beatrice, when he was giving a lecture at Austin. Now, just before the conference in Dallas, and sure to have been read by everyone attending, the *Astrophysical Journal*[6] had finally published the paper 'An Unbound Universe?' Beatrice and Jim had conceived it first, then been joined in the writing by Richard Gott and Dave Schramm.[7] With its several lines of evidence that the universe would expand for ever, the paper was another challenge to Sandage's theories. It caused a sensation among cosmologists, and, as all four authors were under 40, it also heralded a generational shift in cosmology.

'An Unbound Universe?' asks what New Zealand physicist Richard Easther[8] of Yale calls a 'taxi driver question'. He says:

> If you are travelling and the cabbie asks why you are in town and it comes out that you are a cosmologist – and if the conversation continues past that point – one of the first questions you will hear is 'Will the universe expand forever?' It is a wonderfully simple question, and can be posed by anyone who has heard of the Big Bang. The two remarkable things about it are that we can tackle it as a technical, rather than a philosophical, question, and that we have a definitive answer – namely that the universe will expand for ever.
>
> The achievement of this paper was not that it definitively answered the question – the last piece in the puzzle arrived in the late 1990s with the discovery of 'dark energy', or a cosmological constant. Nor did Beatrice and her collaborators pose the question for the first time, as it is as old as the idea of the Big Bang itself and had been discussed in the 1920s and 30s. Rather the achievement was to show that the then conventional understanding – that the universe would eventually reach a maximum size and start to collapse – was flawed. They put the question firmly on the field's agenda.[9]

In a note to *Scientific American*, Beatrice wrote that it was not until Gott came from Caltech to Austin to give a talk that they realised that 'pieces of evidence from our various research specialties could be fitted together into a surprisingly clear case for an open universe. There were several subsequent visits between Texas and California and many long telephone conversations while the work on our joint paper for the *Astrophysical Journal* was in progress.' As the authors were listed alphabetically, Beatrice's name was last although she did most of the writing – 'Beatrice was the glue,' Jim has said. She was the one who chose to challenge Sandage pointedly, by selecting from her copy of *Theories of the Universe* the quotation from the Latin poet Lucretius with which the paper opened:

> Desist from thrusting out reasoning from your mind because of its disconcerting novelty. Weigh it, rather, with a discerning judgment,

then, if it seems to you true, give in. If it is false, gird yourself to oppose it. For the mind wants to discover by reasoning what exists in the infinity of space that lies out there, beyond the ramparts of this world ... Here, then, is my first point. In all dimensions alike, on this side or that, upward or downward through the universe, there is no end.

In a three-column report in the *New York Times*, science writer Walter Sullivan wrote a lengthy account of the Dallas conference.[10] Jim spoke before Beatrice, maintaining that many lines of thought pointed to the universe being open, and outlining much of the work he and Beatrice had done together. She presented the Gunn-Tinsley analysis of the life histories of stars forming a galaxy. From this they had concluded that the galaxies seen farthest away were shining more brightly, when their light began its long journey to earth, than galaxies in the vicinity of the earth. A point Beatrice emphasised, Sullivan reported, was that the implied correction to deceleration estimates indicated an open universe 'by an embarrassingly large margin'.

Sullivan also seized on what he called 'a critical problem' that Beatrice had pointed out. This was the likelihood that galaxies became dimmer with age. Had they always shone as brightly as they now did, the derived expansion rates would suggest enough slowing to allow for a closed universe. His *New York Times* report, trying to sum up, ended: 'The cosmologists and astrophysicists who had come from as far as South Africa, Britain and Poland then headed home to perform more observations, more experiments and calculations in the hope that, when the next of these meetings is held, the answers may be more definitive.'

Before heading home, special friends and colleagues stayed on for Beatrice's party to celebrate the conference and her own farewell to Dallas. It was an emotional occasion. As she described it to Rowena, she managed to pack in 'nearly all the people I love most', including old Dallas friends as well as many of the visiting scientists. 'It was a wonderful end to 11 years of Texas.'

So much was happening in her life that, instead of trying to record it all in her diary, on 23 December 1974 she sat down to write at length to Rowena:

This last week has been even more of an emotional drain than I expected. Yesterday I spent the afternoon and supper with Alan and Terry and was fighting tears all the time. Terry keeps on kissing me and saying frantically emotional things – like giving me messages on pieces of paper to remind me that I'm Terry's mummy when I go away. They are lovely people. Brian is happier and more relaxed. He sincerely invited me to go and live in the house after my things are packed – though I won't need to.

The astrophysics symposium was of mixed value scientifically, and an incredible experience personally. Maybe I'll spill out a lot to you. You are interested in people and their relationships! Jim has become an even closer friend and we are even more loving of each other, but there is no hope of simplicity. (There never would have been with a man of such emotional complexity as Jim.) Since I stirred things up last spring, he and Rosemary have been seeing a therapist and improving their life together enormously. I was rather surprised to find this has made me very happy. I have theorised for so long about being unpossessive, but now the feeling is there at a very gut level, partly because I love Jim's honesty that he can tell me how he feels about other people …

Rosemary said she wanted to talk to me, and we spent two or three hours in a secluded ladies' lounge learning things we didn't want to (didn't want to be true, but glad to know what is true) about each other's feelings. Thank goodness she is open and intelligent and we don't have to play games with each other. I had lunch with both Rosemary and Jim the next day, then I had to give my big talk. This talk was generally interpreted as meaning that Jim and I are making very important discoveries in cosmology, which we love doing.

Also had lunch alone with Jim, in a state somewhere between laughing and crying. Really very joyful, because we aren't afraid of real feelings and because our love for each other is so good.

If ever you thought scientists were unemotional impersonal eggheads, change your mind! Jim and Jerry Ostriker are by far the best astrophysicists of our 'generation', i.e. 25–45ish, and have gone through a complex history of loves and hates and jealousies for each other, which now seems to be on a very good swing, maybe because they are far apart geographically. I suspect Jim is such an intense person that that is almost necessary. Anyway I am awfully happy at the way things are, as long as I can avoid fighting Rosemary. She wants us all to go to their therapist together when I pass through Pasadena next week but I am not sure how much good it would do. Or how much it ever does. If he can't be open with her, something's going to happen to change things while I'm only 300 miles away at Santa Cruz.

Funny to think back to a year ago. It's wonderful to think how life changes – I wouldn't change all of this for the past!'

There is no good day on which to leave one's children. Beatrice with inexorable logic had worked out what would be the best time for her to leave. She decided it would be on Christmas Day. The children would be so busy with their presents that they would scarcely notice her departure.

She bought rather more gifts than usual, wrapped them well and added them to the accumulating pile. Then, on the day, while Alan and Terry and Brian opened their parcels, she left the three of them sitting in a sea of ribbons and cards and Christmas paper. It was a scene that became engraved on all their minds.

Beatrice began her 2000-mile drive, first to Pasadena and then on to Santa Cruz, the first step of her new life. She was filled with projects, ideas and hopes of collaboration. For the first time she felt free to let these take over her mind, and she exulted. Yet, even as joy began to course through her, she was tugged by an undertow of grief at the loss of the children, and by her own guilt. To the end of her life she was to suffer replays of this. Now, as she drove, the scene that she had just left in Dallas constantly recurred to her.

No doubt she reminded herself of how she had set in place a network of childcare arrangements because Brian's frequent absences from home would be continuing. There was nothing new for the children in all this, except that it was the first time she would not be coming back. She would have made herself review the situation factually. Alan and Terry would have the security of their own home and school. She would keep a stream of loving letters coming. Every holiday they would come to her, and she would make these times extra special. And, of course, if she were a man leaving a family, few people would give the situation more than a few minutes of regret and disapproval. Men did this all the time. Because she was a woman and people would not know the circumstances of her leaving, judgement on her would be harsh. (In fact, judgement tended to divide along gender lines. Many men condemned her for 'deserting the family'. Women wondered about the reasons.)

Now she also had another world to think about. There was the reception given the papers she and Jim had presented at Dallas, together with her enormous pleasure in working with him. Colouring everything was the fact that they had become lovers, but for the intense Beatrice to have written to Rowena that Jim was 'such an intense person' that space might be necessary in any relationship with him shows that already she was aware of possible rough waters ahead.

In Pasadena came the heady excitement of meeting up with him again. Beatrice had never seen herself as the kind of woman who could be part of a triangle. She did not want to give pain, she wanted both Rosemary and Jim to be happy, and to be happy herself, but neither did she find it possible to deny her own feelings. At Rosemary's instigation the Gunns had been seeing a psychologist for some time to try to sort out long-standing problems in their marriage, problems which pre-dated Beatrice's meeting Jim. On this post-Christmas visit she went along with Rosemary's request that she go with them to their psychologist. As she told Rowena, she was probably guarded in what she contributed, but she could tell herself she had done what Rosemary wanted, and maybe they had all gained something. And now that all three of them had talked openly, it was much easier to see Jim, to see them both.

On New Year's Eve, as she left Pasadena and drove on up the Pacific Coast Highway to stay with her friend and colleague Sandy Faber and her family in Santa Cruz, the coming year seemed filled with promise. She observed Sandy with her lawyer husband, Andy, and their young child, Robin. Their remarkable rapport had them seeming to be intuitively in tune. This marriage reminded her of her friends and mentors in Christchurch, Glen and Wal Metcalf.

The campus of the University of California in Santa Cruz, one of the loveliest in the United States, was 'lost in a beautiful redwood forest'. Her appointment there was as assistant research astronomer and lecturer, and her office in Lick Observatory looked out on to this forest. Her apartment had an even more spectacular view of a tree-lined landscape down to the sea. 'I'm paying a lot of rent for the view, but after all these years in the middle of Texas it seemed worth having the sea to gaze at,' she wrote home at length on 4 January. The apartment had to be big enough for the children's visits. She did not mention that she had to watch her spending because she was paying child-support to Brian, the amount based on her taxable income. Nor of course did she say that here was an apartment where Jim could be with her.

By now copies had come of the extensive press coverage of their papers at Dallas. Caltech had put out an official news release on their joint paper. This appeared in *Time* magazine and many newspapers throughout the United States and even Mexico, she wrote. She was able to track the progress of the news release because of 'the torrent of crank mail' she and Jim had been getting. The science reporter from the *Los Angeles Times* did not rely on the news release. He had done an excellent job, Beatrice considered, talking to Jim for several hours, and, for another viewpoint, even taking the trouble to phone the Princeton astronomer he called 'their friend and theoretical antagonist' – a realistic description of Jerry Ostriker, Beatrice noted. Some of the 'crank mail' also appeared as letters to the editors of many papers. Their speculations had touched a nerve, lots of nerves. It was fun being spectacular sometimes, she continued, but the exposure had shown what could happen 'when you start blowing up the universe'. Her own daily life became 'filled with peace, about and within'.

Unknown to her, some opposition had cropped up earlier to her invitation to work at Lick Observatory for six months. This was because Beatrice was a theoretical astronomer and not also an observer. The problem, raised by one of the staff, was bureaucratic. The money available for her salary was 'observational money' – in other words, it was intended for a visiting observational astronomer. Sandy Faber, with faculty member George Blumenthal and others, argued that Beatrice was by far the best person to appoint. They considered her a unique resource, and a real facilitator of discussion and interaction.

George Blumenthal had known her only by reputation before she arrived at Lick Observatory. Like most other astronomers, he described himself as having been bowled over by her PhD thesis:

> Scientifically she was unique. She spanned the gulf between theoreticians and observational astronomers. She really paid attention to observations. She really made us think hard about galaxies.
>
> Before Beatrice there had been a gulf between the two. Sandage had the seeds of galaxy evolution in his paper from the 1950s – he used the 200-inch telescope at Mt Palomar. Beatrice took those seeds and made them grow into a reality.
>
> Non-scientists may not know the difference between the two kinds of astronomer. Theorists consider, for instance, how babies are made. Observers see the mature people who have grown from those babies. Beatrice was unique in trying to see the birth through to adulthood of galaxies. This was crucially important to astronomy.

Although Blumenthal was new to the staff when Beatrice was at Lick, her stay there left a strong impression – and not only of Beatrice the astronomer. The young Blumenthal looked up to Beatrice the hostess with awe. The parties she gave were unforgettable. She would welcome crowds of scientists and encourage them all to talk. 'She was trying to re-establish her life. We all knew that she had left her husband, but she never referred to this. She just got on with what was obviously a new life for her.'[11]

Lick Observatory professor Bill Matthews remembered her brief stay there with equal clarity:

> She was an inspiration, and not just because astronomy needs women scientists, but because she herself was such a strong force. She moved in an influential direction and influenced so many people. Echoes of her work are found in so many places today.

Beatrice was charismatic with a strong sense of her own worth. She never saw herself as a victim, but projected confidence, speaking with assurance. It was plain she was high-strung, but was always friendly and with enormous energy.

It was a marvellous time when she was with us. There was a great feeling of excitement, of being born again. They were great days.

Then Bill Matthews, still thinking of Beatrice, paid two women the greatest compliment. Speaking of his colleague Sandy Faber, he said, 'Sandy has a very strong intellect. Sandy is our Beatrice.' [12]

Beatrice kept up a flow of letters to her children and to a wide list of people. She put as much energy into letter-writing as many people do into their regular jobs. The letters seemed spontaneously tailored to each recipient. It was as if, at the back of her mind, there were instantly available files of the arc of their interests, their abilities to comprehend and the extent to which they were prepared to read, absorb and understand.

Rowena did not write often. She wrote poetry, not letters. She had decided, after all, that Venezuela, not Peru, was the place for her and her children, and her house and work in Merida seemed right for her. Now Rowena, too, poured out her feelings in unguarded pages that affected Beatrice deeply. But it was only six weeks since Beatrice had left Brian and the children. Much had happened, and she needed to talk about it. From Lick Observatory on 3 February 1975 she wrote to her sister:

> Dearest R, ... My own life continues to be full of the profoundest emotions, from despair (in a week of total gloom, since lifted with the help of time, Jim and Rosemary) to the kind of happiness that's close to tears. Though mostly it's just working here and enjoying the lovely place; no particular friends yet, tho' pleasant people at work. One really good thing is that Rosemary and I have got past being icy (casting each other as The Wife and The Other Woman), and past treading on eggs, and can really communicate; so we cope with our own feelings towards each other, and Jim doesn't get loaded with shit from both sides; not that there is much as a result.
>
> Perhaps I told you I visited a clinical psychologist, whom they've been seeing for months, when I was in Pasadena after Christmas. Seemed rather unhelpful then, as I was sure in advance he saw me as in the way of his task to help their marriage. Last time I was in Pasadena we all went back again, and he was enormously helpful. Really takes the 3-fold relationship seriously; and helped us towards where we can feel relaxed all together, and not heavy with unspoken imaginings of what the others are brooding over.

We laugh a lot. Rosemary and I have many things in common that Jim doesn't share, and there are things we all do. It's a great new warmth, to be able to make friends out of such a new situation in my life, and I hadn't thought it was really possible.

The more time I spend with Jim, the better it feels. It is extraordinary how alike we are in the hidden recesses of our selves, and very wonderful to be finding them out. In spite of the storms that blow by, unpredictably because we don't know each other that well, and the short times we're together, the sense of having always been close grows stronger.

Beatrice's relief shines through this letter. She genuinely liked Rosemary – but did Rosemary like her? Could she? Beatrice was slow to perceive the effect she often had on others. In this case, however, it seems that the two women really did develop respect and liking for each other.[13]

Working on what she and Jim called the 'Cosmology of Ancient Grease-Proof Paper' gave her particular pleasure. To special friends she told light-hearted stories of the fun they had while working on this, their next cosmology paper. Eating fried chicken, licking their fingers clean and jotting down the next notes gave the work its name. Nevertheless there were storms, sometimes low-pressure storms that she thought were born of Jim's depressions and withdrawals.

As for Rowena and herself, she continued:

How did all the urges to live complicatedly and energetically fall on us? Yes, what you say about falling unawares into some hole beyond the limit of one's emotional capacity is all too true!

I seem to discover each new limit that way, all my life, and fail to protect myself because I can't predict that a limit is coming.

She settled down in quite different mode to take her father up on a comment he had made after reading the report of a talk she had given:

It may be, as you say, 'bad science' to like the universe being open because it *feels* better, but there is in me a strong delight in that possibility. I think I am tied to the idea of expanding forever – like life in a sense – more than spatial infinity. In fact more complicated theories are possible, in which the universe is closed spatially, so finite in extent, but will expand forever. Eddington was wedded philosophically to that model. Currently James Gunn and I are working on the possibility that a model like this really fits the observations best. I'm afraid it's so complicated to get good enough data and bright enough ideas that we may never feel sure.

It amuses me a lot the way scientists' philosophical prejudices colour their arguments. Our friend Jerry Ostriker and I continually accuse each other of making biased scientific arguments because he says I really *want* the universe to be open, and I say he really *wants* it to be closed. Gets us nowhere, but makes good parties.'[14]

A weekend visit to Caltech in Pasadena, she told her father, had been part work and part party and part helping her hosts, the Gunns, to prepare and then clean up, 'but somehow plenty of science got wedged in as it always does in a rich place like that. It's awfully nice having friends among the astronomers all over the country. There are some great people among the best scientists.'

Naturally this was all she told him, but her letter was infused with the pleasure she felt in being in the company of both the Gunns. Perhaps it was her evident happiness that Edward Hill reacted to: women who leave their families should not be happy. Whatever the trigger, Edward again began confronting her with what he saw as her abandonment of her husband and children. Not that he wanted her to be miserable, far from it. But he was still ordained, still practising the Oxford Group's Absolutes. He was also receiving plaintive letters from Brian, looking for an ally. Beatrice could have foreseen all this, but it infuriated her.

Edward probably would have been satisfied that he had done his fatherly duty if Beatrice had sounded contrite. Contrite she was not. In a rather less coherent letter than usual, she opened her heart and let fly:

You ask and worry over again about the children and why I left Brian. How CAN I explain? It was and is grossly difficult for my own feelings, and to justify it as more right than any other cause, to leave Alan and Terry. They were the only solid reason I stayed (more or less) with Brian so long. I'd wanted to accept a much longer job than six months at Maryland in 1973 but he firmly wouldn't let me take Alan and Terry if I'd done so; that was when I had to come to grips with leaving them if I wanted to leave him, out of needs to be alone and to pursue astronomy as my heart was telling me.

I tried to kid myself (and Brian too of course) that it was only the work dragging me away, but now I've realised what a great negative burden it was being with Brian and what a great positive joy it is to find myself again. I don't know – and it's no use speculating – how differently I would feel if I had the children with me now. Certainly Brian always knew, and still does, that I would want them if he didn't. Now that they ARE with him, it seems easier on them not to have me near by.

My close friend Bea Wolf, Alan's piano teacher, says they seem full of joy and very undisturbed, which is not a trivial comment from someone who so dearly loves children.'[15]

The Sloan Foundation gave out about a dozen prestigious grants a year to young scientists in all fields, so Beatrice felt very lucky to receive one. It was 'a lot of money, $19,000 to be managed by Yale', she added, money which she could spend on research in any way she liked, such as financing travel to England in the summer. It would mean 'no haggling' with government grants and university administrations for funds for a while.

By mid-March a long stretch of free time lay ahead as Yale was not requiring her to teach until January 1976. For the first time in her life she had many months to herself – to research, to write up for publication work she had already done, and to think. And there was Jim, able to come and go, and with a bigger part in her life than ever.

She celebrated the beginning of this stretch of free time by writing at even greater length than usual to Rowena. Like her sister, she had been surprised by what an uprooting the previous year had been for them both. She felt she had been naive not to have expected this. Her life was now very full of good things but she thought often of her children, and dreamed about them. Aware of Rowena's interest in the interpretation of dreams, and in this case fairly sure herself what they were saying to her, she gave some details:

> The other night I had an extraordinary series of dreams (waking between them) in surrealistic colours and clarity, of the children playing, all feeling intensely happy – and so was I in the background.
>
> Then the dream image became just a roll of typing eraser paper, the kind you type white letters over black like white carbon paper. In the dream I pulled out several inches and they were all used, covered with scattered black letters, so the whole roll was useless and nothing could be erased. Took me several days to get right away from that image. It comes back now as I use the stuff.
>
> But basically I value my solitude more and more. After three nights with Jim recently I woke the morning after he was gone feeling Ah! there's just me – even though I was totally happy with him and couldn't have mentioned any way in which it seemed like an imposition on me when he was there. That must be why I don't get involved with anyone else closer; sometimes I think it would be good to have a lover whom I saw more often, but I seem not to care

enough to do anything about it.

For now it seems exactly right for me to have the sort of relationship I do with Jim; being far enough away geographically that I have space to sort myself out after so many years of suffocation. I could be in danger of getting lost in some relationship built on dependency again, and never come to grips with existentialist aloneness. Not that it feels lonely; and there was no creativity in clinging or being suffocated.

I think I'm doing better science now for psychological reasons, as much as because there are fewer practical distractions.

Beatrice in this letter then returned to the sessions which she and the Gunns had had together with the psychologist, O, in Pasadena. Rowena had earlier expressed reservations. Beatrice said she had felt exactly the same before getting to know O, 'the shrink' – reservations such as what good could it do, would they be told what they should do and think, would it become a blood-letting, and so on:

> But O never gives advice or makes judgments. He will ask how you feel about some situation, then say, 'I smell problems or worries in that remark,' and draw out what was behind it.
>
> Then one tries to be very guarded, and say what one thinks one should feel (or hopes the other people listening will believe), and O is likely to come in with 'Seems to me you are saying ... ' and pour it all out in very blunt language, mostly true, though we all argue with some of it. The result is we expose all sorts of things to useful sorting out, things that otherwise we wouldn't have thought worth the bother (or dared, or have been able to verbalise). My feeling is that after a couple of hours of this sort of help, I get the point, and find it easy enough to carry on a valuable open relationship with Rosemary, and with both of them – don't see any point in going on back ...
>
> We're doing a lot of fascinating new science, and the 'Cosmology of Ancient Grease-Proof Paper' might turn out to be quite fundamental. The possibility it discusses is fundamental, but we don't seem to have enough data or theory to pin things down yet. We're writing it up to publish as far as it has got, and nobody else will appreciate the full aptness of a quote at the end from Timaeus (where Timaeus apologises to Socrates for not explaining everything in the universe quite consistently yet).

She wrote on about the troubles she got into when editors of science journals sent her papers to referee. This meant 'offering the editor advice on publication or not, or giving constructive criticism, or rude rejections to the authors'. Inevitably, some papers were by people she now knew. One scientist had written 'a foolish paper', misunderstanding part of the subject. She was rueful because, although in principle one could be, and should be, an anonymous referee, she thought the latest such paper she had commented on would have clearly carried her style, *her* remarks, and the author would have realised it was her verdict. This astronomer was acutely sensitive, 'and I fear the worst.'

She might have been startled had anyone suggested that here she had taken over her parents' creed of Absolute Honesty, and was applying it to her scientific work; but this was what she expected of herself and of everyone else. Nothing less would do. She could not compromise.

It was one thing in her family letters to fudge the truth by omitting anything which could pain them while selecting and highlighting what could bring pleasure. Science was altogether different. She had no mercy on herself when she made errors, and none on other people. To correct an error was one step nearer truth. It sometimes seemed to her fellow scientists and students that she expected them to be pleased when a mistake in their work was pointed out. This had been her expectation in her first confrontation with Allan Sandage, at Austin, and she was taken aback by his reaction. Beatrice was kind, but her overriding mission was to find scientific truth, and an error detected was a step along the way.

Her life was very peaceful and happy, she told Rowena. She was grateful to Brian for pictures of the children 'looking perfectly beamish' in a freak fall of snow. They would come to Santa Cruz to stay with her in June. And she ended: 'I know Daddy is upset about me, so thanks for trying to explain. He has said he will stop asking me why it happened!' [16]

Her months of free time – without students – were punctuated by working visits from collaborators and speaking engagements in many parts of the US as well as the symposia she held at Santa Cruz. She also joined Sandy Faber and one of her graduate students, David Burstein, in considering a much talked-about thesis by a Caltech student, Donna Weistrop, which her adviser was calling a major accomplishment.

It claimed there was a much larger population than had been expected of M dwarf stars, red stars that are tiny and which dominate the mass of stars in the neighbourhood and affect the calculation of gravity. Over about three years there had been some 30 or 40 references to this thesis in other papers, the bandwagon effect. Astronomer Kyle Cudworth suggested to Dave Burstein that he should look at this hot topic again. Sandy agreed. Meantime, Beatrice had wondered whether in the thesis there was a flaw in the brightness measurement of the stars. The three of them set to work, with Burstein the instigator.

He did the nitty-gritty work of photometry, measuring the stars' brightness, and contacted Ivan King at Berkeley for some independent information. One day at Santa Cruz the three of them talked until late one afternoon. Sandy and Dave went home to their families, and when they showed up the next morning Beatrice had it all done.

'It was galling!' Sandy has said. 'It was the first taste of what it was like to work with her. It was sheer brilliance and effort and volume. I envied her the freedom to concentrate on work, but then I wanted other things besides work.' Beatrice, indefatigable, had computed models overnight and worked out how an error would translate. The flaw was such that the thesis had greatly under-estimated the brightness of very red stars, indicating that there were a lot more of them than there really are.

Sandy wrote the resulting paper. She considered Beatrice the leading light, as she had done all the calculations, but Beatrice insisted Sandy be the first author as she had put it all together. 'She was very generous to me. I desperately needed papers then.'[17]

They corresponded with the author and felt for her, but were of course glad to have put the record straight, or, as Sandy said, 'We had raised grave doubts. To the author's great credit she observed again after our paper, 'Rediscussion of the Local Space Density of M Dwarf Stars', was published. Donna was a good person.'

This 'awful story' underlined to them all the dangers of publishing without first thoroughly checking one's facts, something they henceforth emphasised to their students. It also alerted them to the great responsibility of being a thesis adviser.[18]

Jean Audouze became visiting professor for six months at the McDonnell Centre for Space Physics at Washington University in St Louis, Missouri. This also gave him the chance to work with Beatrice at Lick on a big paper she had invited him to join in writing, 'Chemical Evolution of Galaxies'.[19]

Beatrice also enjoyed talking more personally with Jean Audouze. She told him about decisions she had come to concerning her life – why she had decided to divorce, and the places she wanted to give in her life to her adopted children and to her research. Jim too came up from Pasadena when he could, and they worked on two large papers and part of another.

Edward Hill had found a way he could pay the Tinsley children a grandfatherly visit in Dallas. Through his continuing Oxford Group connections he had been invited to a congress in Brazil of the World Anti-Communist League, and then was able to visit his elder daughters. Edward had been important in Brian's life since they first met. For his part, Edward at that time thought that Brian had been hard done by, and was determined to let him and the children see that their grandfather would be a constant in their lives. Brian made plans and lined up numerous people for Edward to meet, so much so that Beatrice found she had to cut short the time her father could have with her. Still, she was pleased that her children would be able to be with their grandfather. It fitted in with her picture of presenting them with a continuing family background.

She had had to look hard at dates and her invitations to meetings in many parts of the States, so that she could be in Santa Cruz and free to leave her office and do nothing except devote herself to her father for his short visit. Of course her concept of doing nothing was fairly elastic. Edward Hill, in *My Daughter Beatrice*, recalls his amazement when he got up early on the last day of his visit. Beatrice was driving him to San Francisco airport and they needed to make an early start:

> When I got up I found her at the dining table with sheets of paper in front of her covered in obscure mathematical calculations. She had been sent an article to review for a scientific journal and had spent over an hour proving that the mathematics on which the author had based his conclusions were inaccurate. It was an astounding display of mental energy, especially as we had been up late the night before.

Beatrice's diary for April 1975 included visits to the University of Colorado in Boulder, to Sonoma and San Diego in California, to Princeton and Harvard and, she hoped, Yale for a couple of days. She had also to host an English astronomer visiting Santa Cruz from Cambridge, and there was a big American Physical Society meeting in Washington. Most of these engagements resulted from the publicity their cosmology paper had got in Dallas the previous year 'People want me to go and talk about it.' Incidentally she noted, 'The APS chose me rather than the other authors because it was organised by one of the most outspoken feminists in the university world!' This schedule of engagements came within six months of that casual put-down by the man who had not got around to answering her application to head the astronomy department at the UTD.

The Washington conference was enormous. Beatrice's talk generated much interest, and also a press conference. The most valuable aspect was being able to talk to some physicists afterwards. She was able to revisit some of her old haunts at Maryland, too. All in all the month of visits had made her realise what a good centre the Lick Observatory was, not to mention Yale. 'I feel very lucky not to be stuck in some second-rate place.'

The highlight of the month had been Princeton, both scientifically and socially. Beatrice had been invited to give an informal seminar at the Institute for Advanced Study, but instead found the event rather formal, even formidable in view of a few dominant personalities there:

> I decided to jump right in and talk about the cosmology on my mind, which is evidence that Einstein was wrong – right there in his own place. The result, as I hoped, was a lot of very interesting suggestions and discussions.[20]
>
> It's nice to be beyond the stage of being treated as a crank or an upstart.'[21]

Sandy Faber was to say that Einstein gave everyone a new way of looking at the universe, while Beatrice used her knowledge of stellar evolution and applied it to his work. 'She was the first to imagine that the life histories of galaxies could be modelled just like the life history of individual stars.'[22]

Edward had not taken the opportunity for a face-to-face discussion with Beatrice about her children. Now back in New Zealand, he wrote her a long letter. During a weekend when Brian had no child-minding assistance, Edward had been dismayed by what he considered Terry's attention-seeking and even violent behaviour, and inability to concentrate. Alan apparently was stoical, while Brian seemed greatly overburdened.

Beatrice at once phoned Brian to discuss the letter, then wrote at length to her father:

> I asked Brian directly on the phone if he thinks Terry is troubled or seriously hyperactive, and he told me that neither the doctor (a child specialist) nor her teacher thinks she has problems. In particular she isn't a problem with her behaviour at school, and is doing very well in her work – which she wouldn't be if it were a clinical case of hyperactivity. Terry has always been excitable and nervous and prone to react violently, so perhaps the problems aren't as worrisome as they look.
>
> We intend to arrange for Alan and Terry to visit me separately sometimes, which should be very good and helpful for them.
>
> I wonder very much how Terry would have seemed to you say a year ago, when I was there. She probably wouldn't have been your ideal of a little girl then, either, and I'm afraid I've never tried to tell her that girls should be ladylike. I also wonder how much both of them are suffering, consciously or not, at my departure. They seemed very accepting in December when I left, but the reactions are bound to be very deep.

No doubt life is difficult for Brian, but your description of his problems sounds exactly like my life was for years and years: couldn't open a briefcase from Friday afternoon until Monday morning (even if I was free and I did, Brian resented it!), couldn't rest without being interrupted, and had my career impaired by distraction and tiredness.

It must have been pretty unpleasant for you, visiting two halves of our broken family. I'm not trying to duck responsibility for the difficulties you describe at Laguna Drive, but perhaps you can understand that I believe that if I was still there, all of us would be a great deal more unhappy.

The children went to pieces over Brian's and my problems, and at least since I moved out in August I've been able to feel peaceful in their company. I wish they'd had an example of how to live harmoniously.[23]

Beatrice went on to thank her father for his frank opinion, which she appreciated, adding that it would be good if he lived closer to them all.

It is a sad letter in more than one way. Both parents cared very much for the children, but neither – through nurture, temperament or experience – was really able to perceive what was going on in the home, something that was obvious even to Edward. As for Beatrice's reaction to her father's retailing of Brian's difficulties and inability to get any time to himself, she must have taken a certain wicked pleasure in being able to describe so exactly how it would be for him. She did not remind her father that she had wanted to take the children but that Brian had steadfastly refused to let them go. In fact, she most probably reprimanded herself for letting the ghost of a grin appear while she was writing; she would despise herself for gloating, but, all the same, it was interesting to see the tables turned ...

Beatrice was pleased by the write-up she was given in *Science News*, the periodical read mainly by scientists so they can catch up on other fields, but as usual her next letter home made no mention of her main concern over the past weeks. This was her relationship with Jim. When they came together it was still often too much like lightning on a mountain top. Everyday living, quiet appreciation of each other, did not seem possible or did not last except when they were working together.

Then they went away for a weekend. Beatrice expected it to be a quiet, contemplative time. Instead, Jim was wracked with depression as he looked at his life. One factor in this was discovering that he had been wrong in aspects of their joint project. In science everyone's good work is built on the good work of others. In theory, scientists should rejoice in each new discovery. In practice, scientists find it no easier than other mortals to discover they are wrong. Beatrice was one of the very few who could delight in scientific discovery even if it proved her wrong. At that stage in his life Jim reacted differently.

As he was to explain it,[24] they came to realise that galaxies can change in two ways. Galaxies are made of stars and many galaxies interact, in the process exchanging stars, gas and dust. This process, called dynamical evolution, was effectively demonstrated in a thesis by Scott Tremaine, a student of Gérard de Vaucouleurs. Jim had invested great effort into looking at different elliptical galaxies, and now it was found that part of this work was faulty. 'Our results in several papers were severely compromised, although we learned a lot along the way.'

It was by no means just his work that was troubling him, however. All the uncertainties and strands of unhappiness in his life, dating from his father's death and what he saw as his inability to cope emotionally, now brought him to a point where he declared he was no use to her. This was the end.

Beatrice was devastated. The break was happening so quickly, and she had not seen it coming.

These two intellectual giants, as their contemporaries saw them, had never come to terms with the emotional scarring from the losses in their childhoods. Their relationship had never been grounded in the actual, down-to-earth world, and part of Beatrice knew that what Jim was saying was true. Part of her refused to accept it. Pride kept her silent. Many years later Jim was to say:

> We had an intense intellectual relationship both before and after a brief intimate one. Science was more important to Beatrice than sex. Doing science was riotous living for her. Our close relationship wasn't sustainable at heights and at close up. Beatrice was never possessive, never. She never argued, never reproached me. But she never levelled with me. She didn't tell me how she felt.[25]

She would not, and could not. Later, in her turn, she was to talk to Rowena about 'all that angst' she had gone through before coming out the other side into true friendship with Jim. There was no question of her confiding in Sandy at that stage. Bea Wolf back in Dallas was the only person she might have talked to, had they been together. Somehow she had to absorb this new trauma, and carry on with her work. Part of her dreaded the imminent arrival of Alan and Terry. Would she be able to hide her distress and act normally with them? In fact she felt they helped her with their happy talk.

At this reunion with their mother they were very proud of themselves

Brian and Beatrice on their wedding day in 1961. Beatrice wears the family's wedding veil.

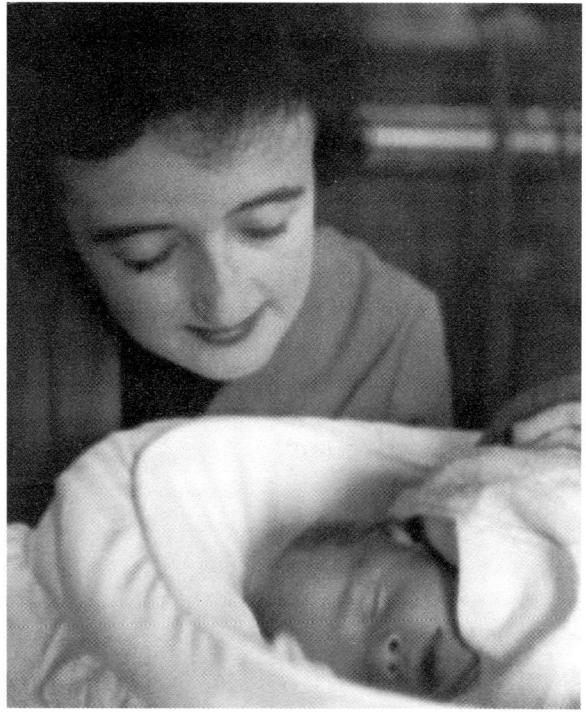

Beatrice and adopted baby Alan, in Wellington, 1966.

Beatrice and Alan at a picnic.

Family reunion in Forte de Marmi, Italy, in 1967. José, Rowena, with their two children, Cecilia and Andres (Lila and Sito), Beatrice, Theodora and Brian holding Alan.

Beatrice with her children
Terry and Alan

Beatrice's studio portrait for her family.

At Yale, photo for a magazine article.

Fred Hoyle (front, centre) and Virginia Trimble (right) at lunch during Hoyle's 60th birthday celebrations in Venice in 1975. Rear table: Beatrice (centre) with Margaret Burbidge (opposite).

AIP Emilio Segre Visual Archives, Clayton Collection.

Beatrice with astronomers Einasto and Ambartsumian, Tallin, September 1977.

Richard with Terry and Alan, exploring.

Beatrice with Wellington Harbour background, 1980, as photographed by Edward Hill: 'My daughter, the professor.'

> Dear Daddy and Mattie,
>   This is
>   HAPPY BIRTHDAY
> to both of you. I know D's is on the 30th, so the card should be (!) too early, but I have no record of M's birthday except early march. And I don't have dramatic episodes to recall as D. did for me! Anyway, I think of you a whole lot, not only on Birthdays, and wish you strength and happiness in the coming days. I honestly don't think the length of life is important.
>   Very much love,
>   from
>   Beatrice

Beatrice's last hand-printed letter to Edward and Mattie, March 1981.

Terry and Alan at their mother's grave.

At Austin in 1985, from left: Rowena Hill, Jim Gunn, Jill Knapp.

Linda Stryker.

Richard Larson.

Sandy Faber.

Alan, Jacob (Jake) and Teresa (Terry) Tinsley, July 1997.

Teresa Tinsley and her son Jake.

for flying from Texas unaccompanied. They were now nearly nine and seven years old respectively, and woke up at 6am with Alan saying he felt like climbing a mountain. Beatrice was able to write home that she was overjoyed by their zest for life and seemingly inexhaustible inventiveness. For children who had grown up in Texas and so far from the sea, the biggest excitement was the Californian beach. Terry rushed into the water, shoes and all. Beatrice had made a list of things to do and explore, and organised swimming lessons. She wrote that she enjoyed showing them off at barbecues and a departmental picnic:

> They both seemed to me very healthy and a lot happier inside than I'd expected (more so than they were) for which Brian is to be congratulated.
>
> They talked very freely about their feelings for me being away. Alan says in his philosophical way that about half his class have divorces – no doubt a help to them.
>
> I was talking to my neighbour whom I call William the Witch-doctor[26] before they came – problems I expected, and my feelings about it – and he was exceedingly helpful. Funny how wise some people are. It isn't a common quality of professional psychologists, either.[27]

Sandy had helped by arranging for her student Dave Burstein and his wife Gail, who had a two-year-old son, Jonathan, to look after the children for some of the time during the week while Beatrice worked. 'A happy arrangement for everyone', Beatrice called it. It was not. The Bursteins were distressed by what they saw as the children's unhappiness.[28]

Their picture of the children was about as different as it could be from Beatrice's account in her letter home. Gail Burstein had figured it could be an 8am to 5pm job on weekdays, but the children were left with her till 7-8pm. On their last night, when Brian arrived and came with Beatrice to pick them up, it was 9pm. Each day, after Beatrice dropped them off in the mornings, Alan would go straight to the TV set and was reluctant to budge, even for lunch. Terry wanted to sit on Gail's lap, pushing her own child away. Gail became very worried:

> I got Dave to come home mid-afternoons and he helped me walk them. They needed exercise. Alan started opening up with Dave.
>
> Jonathan had been ill earlier in the year and we'd got a book called *PET*, Parent Effectiveness Training, to help us. It said to watch out for 'the following danger signs', about 20 in a checklist. And these kids had all these problems!

> I was a young mum, and Dave a student and in awe of Beatrice. We didn't feel we could discuss it with her. Her children had come out to Santa Cruz to be with her, but she didn't seem to understand what that involved. Nearly all the time she seemed to want to be off on her own, working.[29]

Nobody at Santa Cruz was to know that Beatrice was using her work more than ever to blot out her distress, to fill up her mind so that there were few chinks left for slivers of pain to slide in. She intended doing her parenting job well and efficiently, in the same way that she did other jobs. It would not have occurred to her that parenting is not exactly a job. As she did not have instinctive empathy with her children, she could not have said what was going on in their heads. Alan was eventually to say, 'Both parents loved us and they tried all they knew how – but –'[30]

Just before the children had to go back to Dallas, Jim returned to work with Beatrice. He came with the Faber family to join them all on a last picnic at a sandy beach by the river in the redwood forest, 'an idyllic picnic', Beatrice told her family. Sandy remembers this differently. No doubt the children were keyed up by the impending departure, but they ran around out of control, with Beatrice becoming more tense and snapping at them. In Sandy's memory the picnic was far from happy.

This is another instance of Beatrice's having an entirely different view of an event, as with that excursion long ago when Alan was a toddler refusing to leave the beach, and when she summarily declared the outing over for everyone and brought them all home. She needed the framework of the orderly, the rational, the planned and expected. Underlying everything was the stress of trying to come to terms with what had happened with Jim, and settle into a new relationship with him. A picnic setting, with Jim present, must have been a near-intolerable strain. The children would have picked up on her emotional state. Rampaging and shouting kids, taut with the sentence of banishment back to Dallas, would have set her nervous system jangling.

Almost certainly Beatrice was not deliberately fabricating the scene when, writing home, she used the word 'idyllic'. She had persuaded herself that the picnic was what she had planned it to be, not what in fact it was. Overlays of her own childhood remained potent. She had been infected with her mother's seeing things as they should be, not how they really were. When Beatrice became the scientist again, however, fact was fact, white could never be confused with black, and there was no compromise.

The three astronomers at this farewell picnic talked about their current work, with lawyer Andy Faber looking on. Beatrice showed only polite interest in Sandy's family, which saddened Sandy and slightly annoyed but amused Andy. Beatrice may have been jealous of Sandy's family interests. Nothing in her experience had helped her understand how basic those interests and bonds were. What she wanted to talk about was their shared enterprise in science.

There were under-currents in the relationships of the three scientists. By this time Beatrice and Jim had several different papers and projects under-way and at last they seemed to have one complete and accepted for publication. This was the first in what was to be their series on evolutionary synthesis: 'Evolutionary Synthesis of the Stellar Population in Elliptical Galaxies. 1. Broad Band Colour and Infrared Features'. Published the next year, 1976, in the *Astrophysical Journal,* it was 'a long and very serious paper', Beatrice said. Another just about ready to send away and which she described as 'short and rather riskily unorthodox' was 'An Accelerating Universe?' *Nature* published this later in the year. That same year, too, they hoped to finish two more papers when they were in England, at Cambridge, but this would depend on whether they had time among the many other projects with the astronomers who would be gathering there during the summer.

Much later Sandy Faber was to say, recalling this picnic, that she was envious of the collaboration with Jim Gunn, he who had perhaps the single most creative, practical and far-ranging intellect in all their field. While responding warmly to them both, she envied them the time they could devote to papers they were working on together, and because Beatrice was single, unencumbered (except temporarily) and able to focus entirely on her work. There was the fact, too, that theoretical astronomers can make great leaps by using the observations of others. Jim was and is both observer and theoretician. Beatrice did virtually no direct observation. Sandy herself did both.

But Sandy had always known that she could not be, and would not want to be, entirely focused on her work, as Beatrice now seemed to be. Sandy wanted to lead the full life of a woman with husband, children, home and friends, being open to the world as well as doing her utmost to advance her area of cosmology. 'To understand the universe is more important than what I contribute to this understanding. I want many different experiences. Otherwise I would have guarded my time and not taken on strange collaborators or given so many talks or helped build telescopes and so on.'[31]

Of all her women friends at that time, Beatrice probably felt she had most in common with Sandy Faber. Now Sandy was 'bravely expecting again' – Beatrice's words – in January. Her first child, Robin, was two years old. Beatrice admired Sandy's careful and confident planning. Sandy intended, as before, to stop work for a month, take another month working half-time, and in other ways carry right on at 'her usual admirable pace'. Some people, Beatrice wrote in her farewell letter home from Santa Cruz, took on twice as much work as others, and then took on more.

She did not say that almost certainly this is what she herself would have done if she and Brian had had children of their own, and if she had had a job at the time. For the present she was glad, very, that her own days of intensive care of infants and toddlers were over. She could see, although not yet quite believe, that she would be able to move on to a different relationship with Jim now that her dream of a future close to him seemed to be over. They would still collaborate, anyway, and now was the time to be her new self. So much was at stake. So much time had to be made up. So much lay ahead.

CHAPTER SEVENTEEN

GOING SOMEWHERE

'Science is not a heartless pursuit of objective information. It is a creative human activity, its geniuses acting more as artists than as information processors.'
STEPHEN JAY GOULD

Beatrice set out on her long and challenging drive to New Haven. When she crossed the Californian border she had had the spontaneous feeling that she was going somewhere, rather than leaving somewhere, she said in a long letter home.[1] This was enough to give her more confidence in her undertaking. She was 34 years old, and she knew that a new life was under way. But she had much to do before this life at Yale could begin.

In Chicago she had two days of rest, staying with Hungarian astronomer friends, Jeno and Madeline Barnothy. They were full of talk and ideas, ideas which most astronomers considered quirky or even crackpot but which they presented with great warmth and enthusiasm to their 'little Beatrice'. She must have been one of the very few who always listened seriously to them, sifting out the good grains and letting the chaff fall where it would. They had a unique way of looking at things and sometimes found chinks in conventional explanations, which Beatrice found exciting. As Jim Gunn was to say,[2] 'She didn't care that the establishment thought them oddballs. Beatrice was after the truth, wherever it came from.' Indeed, the Barnothys were to be among the first astronomers whom she invited to speak to her students at Yale.[3]

Richard Larson had been the first person she sought out when she reached Yale because she had been so attracted by the papers and contributions he had made at meetings and seminars. He gave her particular help in finding an apartment in tree-lined Whitney Avenue, only 10 minutes' walk from the Astronomy Department. She made friends with the Italian building superintendent and his wife, the Armatinos, who lived in the basement, observing that their 'countless children' would be great for Alan and Terry to play with. Everyone helped her settle in before she had to travel on, to England. The chairman of her new department, Pierre Demarque, brought a ladder and fixed light fittings for her at her apartment. Many of her new colleagues, too, invited her to their homes. Officially she was now on the Yale faculty, as an associate professor, from 1 July. 'I feel very happy,' she wrote home.

A blanket of happiness and anticipation, in fact, began to settle over her and ease her underlying distress. Now, in the same way as she had helped Brian pack for yet another of the conferences or scientific expeditions he had flown off to over the years, she was the one setting off. Not only the intellectual feast of Cambridge awaited her, but, in a signal honour, she was to join the select band of world astronomers invited to Venice to celebrate the 60th birthday of the great Sir Fred Hoyle, on 24 June 1975. Nor was she to be just another guest. The organising committee had asked her to give the scene-setting address, a review of cosmology.

She would have been deeply moved by this honour, anyway, but she had a profound feeling of a most important circle being closed when she returned in her mind to her schoolgirl days and her first excitement at reading Hoyle's *The Nature of the Universe*. The deep sense of awe and of infinite possibilities which he had conjured up had never left her. She was already receptive but Hoyle was mainly responsible for opening her mind to the brave new world of astronomy and cosmology. In some way he had personalised it for her, letting her see how she too could find a place in this ultimate territory. Now she would be able to say something of what he had always meant to her. She was also to give a talk on chemical evolution and its effect on galactic evolution.

Jim could not get away from the US in time for this birthday celebration, but he had often talked with Beatrice about the tremendous influence Hoyle had had on him, too, with his books, his papers and then in person. He has said:

> Fred was always tremendously open and inventive, fun to be with. To Fred the universe was a fantastic game, an absorbing puzzle. Jerry Ostriker and Fred were similar in this way. Fred could come along one day and expound on something, but be proved wrong. Ten minutes later he could come up with a new explanation. It was a mental game to him.
>
> But Beatrice was more serious. She was after the truth. She wanted terribly to be right. It wasn't a game or a puzzle to her. She wanted to know if the observation was correct. I'm more like Beatrice. I'm probably much too sceptical. Beatrice was open to ideas. But it's a tightrope – you must look everywhere for the truth. Fred Hoyle was

a gadfly, the opposite of Sandage who has privileged information from God.⁴

Beatrice continued working on her papers for Venice during her plane trip to England.

Members of her parents' families were keen to see her when she arrived, but time ran out and she soon hurried off to Cambridge to install herself in a flat lent to her by Malcolm Longair, the Astronomer Royal for Scotland, who would be out of town. He had insisted there were plenty of astronomy books and gramophone records to entertain her and had also told her to use his food supplies. She summed up the latter as consisting 'mostly of Scotch whisky, Scotch shortbread and Scotch marmalade'. She had also been instructed to use her friend's extraordinary bicycle with miniature wheels, and this took her safely into the town in five minutes in one direction, and to the Institute in five minutes in the other. She decided she was very well set up, although to her amusement she found she had to get used to the light switches being the wrong way up, and to the fact that her vocabulary was reverting to the English way of saying things.

The meeting in honour of Fred Hoyle was held in a conference centre on the extraordinarily beautiful island of San Giorgio, the first island that the visitor to Venice sees, fairytale-like with its ancient buildings and church spires. Participants were spread among several hotels and met up each morning on the boats going over to San Giorgio. It was a great social as well as scientific occasion, with Beatrice's contributions being particularly well received. It was to remain one of the outstanding meetings of her life. Many of her friends were also there to honour Hoyle, all agreeing that this was the greatest place to celebrate a 60th birthday, any birthday.

Astronomers who go to so many meetings around the world find details blurring as to who said what, and when and where. Papers can be referred to later. It is the talk over drinks, the exchanges at meal times, which often stick in the memory. And the subjects do not have to be astronomical. Beatrice and Virginia Trimble⁵ were at the same table during a long lunch on San Giorgio when the subject came up of a woman's right to choose – abortion on request being one of the planks of the new women's movement of the 1970s. The two women were forthright about their own experiences. In Virginia's recollection, Beatrice described her earlier situation when she had felt that neither her marriage nor her career needed another child. What she did need, in Texas, was her husband's consent before she could have a legal abortion out of the state. In Virginia's case, she said, it had been 'extreme ignorance' of what did and did not constitute effective birth control; she had been 19, not married and with no desire to be. The two women, different in their personalities, had respected each other and got on well from their first meeting. They continued to speak openly whenever they met.[6]

In Cambridge Beatrice was still trying to finish the papers she had begun at Santa Cruz. In spite of the pressure of work, Cambridge was the ideal place for this as there was never a shortage of people to read the drafts and give useful criticism and new ideas. After her years of being shut off from the mainstream, this was what she revelled in perhaps most of all: the immediate, face-to-face reaction and debate with her peers.

Beatrice must have felt some apprehension when Jim and Rosemary Gunn arrived, but all those who had worked throughout the summer felt they deserved some fun, so the Gunns were swept up into group outings. They and Jean Audouze joined her in visiting a fair on Midsummer Common where they chased one another around in dodgem cars. Audouze insisted he did not enjoy it – too much like Paris traffic. Jim seemed in the grip of depression, with sudden mood swings. Beatrice found that mostly she could look at him squarely, with old affection yet without feeling entangled.

That last meeting had generated publicity which had even reached Der Spiegel in Germany. It had also resulted in the *Scientific American* asking for an article from the four authors of 'Unbound Universe?'. Beatrice thought this journal excellent so she was happy to spread their ideas by collaborating again with the other scientists. All four of them happened to be at Cambridge at the same time, the first occasion they had all been able to get together at the same place. By pouncing on whatever hours they could spare from other work, Beatrice shepherded the other three astronomers to look again at their original paper and produce a popular article called 'Will the Universe Expand Forever?'[7] As they were all so busy they were lucky to get what Beatrice described to her family as 'literary help from Rosemary Gunn'. Because it seemed to be a good chance 'to teach our stuff to the general public', Beatrice volunteered to be the editor.[8]

She had one last week of trying to finish things off before people disappeared. In particular she wanted time to do more work with Jim, but, as she said, they kept having new ideas and not finishing, so there would probably still be loose ends.

On her last evening in Cambridge she was taken to dinner at Trinity College by a visiting fellow, the American Richard Gott, one of her three co-authors of 'Unbound Universe?' She described him as 'a remarkable young and brilliant astrophysicist'. The formal meal at the high table was followed by the 'passing the port' ceremony upstairs in the elegant Combination Room. She met some famous old people who lived there as life fellows, and thought it an impressive way to retire after a career as a distinguished professor. Richard Gott gave her a tour of Trinity, including echoes in the cloister where Newton first measured the speed of sound.

Settling in at Yale gave Beatrice just as much pleasure as she had expected. Because she had already met so many people in the department she was able to feel from the beginning that she belonged there. To her father and Mattie she wrote on 14 September 1975:

> Yale is proving a very pleasant place to work. There are really nice people in the astronomy department here, and an extraordinary absence of bad feelings amongst them.
>
> This year there are two new young people at junior lecturer level.[9] They and my old colleague Dick Larson[10] and I are forming a gang that I think will do much useful astronomy.
>
> Another fun thing is that this year, that is till next June, I'm responsible for getting visiting speakers each week, which means I can organise visits from people I want to see. It might be quite a burden on my time, entertaining them, but I don't mind if I choose the people!

'And did she go at it, at once!' Richard Larson has said. Where most people would have taken a while to think about such an opportunity, weighing the rival merits of possible speakers, Beatrice was decisive. Invitations flew out, and she quickly had an impressive list of visiting astronomers drawn up.[11] The students, sensing that this was a real shake-up for the department, talked with excitement among themselves.

This was also the faculty's first taste of Beatrice in action. They were delighted to have her. As the department's Bill van Altena later put it, 'Beatrice was the best person in her age range in her field in the world. It was also helpful that she was a woman. Yale was in deep trouble because it didn't have enough women lecturers and professors.'[12] Helpful, yes. But her gender was scarcely the main reason for her appointment (although a few male astronomers continued to put this about). A significant grouping saw that she would both unify and electrify the Yale department scientifically by bringing together two of its strongest areas, stellar evolution and galactic studies.

Beatrice asked to be put on committees which could help women. One of these aimed to advance women's status on campus and give them opportunities, but she made it clear she always looked for excellence. Her high standards applied equally to women.[13]

The faculty noticed that she sometimes seemed suspicious, fearful that people would steal her ideas. Some of her experiences in Texas and at Caltech had obviously hurt her deeply. Richard Larson particularly noted her cynicism. Pierre Demarque observed her initial lack of ease with them, 'as if she were ready for dirty tricks'. Sometimes, for instance, if Pierre saw her sitting working with Richard or Gus Oemler, and approached them, she would say something like, 'This is private', as if she were afraid of intruders: 'I felt a little excluded.'[14]

Astronomers at that time tended to regard Caltech in particular as something of a cut-throat place, certainly not always friendly and not somewhere for people to discuss their research before it was published. Her colleagues decided that something in her experiences – although she had made some deep and lasting friendships there – was why at first she wanted to keep her work to herself. Pierre and the rest of the faculty had never known the open, trusting, co-operative young woman who had been the Beatrice of only a few years earlier, but before long they did come to understand something of what had affected her

Looking back years later, Pierre said, 'We thought the world of Beatrice. We are used to dealing with very bright people but she was outstanding. She was a very private person, not speaking much at all about her private life. She was also tormented – this was clear to us.'[15] The torment was rooted in the loss of her children, and, at first, the aftermath of what she was to call 'all that angst' with Jim. She was determined not to let her real feelings be seen, even going to the extent sometimes of answering enquiries about the children with glib replies. The pioneer woman astronomer Dorrit Hoffleit ('I was named for Dickens's *Little Dorrit*') was one of those whom Beatrice shocked when she spoke in this way.

The two women from the same department had not begun to know each other properly when they met in the town while doing Christmas shopping. Beatrice was buying gifts for her family, and they stopped to talk. 'My children were interfering with my career,' Beatrice said. Dorrit was chilled. 'We hardly knew each other, and she said that!'[16] It took some time for Dorrit to get to know the real Beatrice and to see that this was a foolishly offhand remark to conceal pain. Then they became friends. 'We didn't ask too many questions,' Pierre said, 'and her suspicions faded. Yale was very good for her.'

And, as a number of her friends were later to say, Beatrice was very good for Yale. 'Yes, Yale saved her, but Beatrice put Yale on the map.'[17]

Beatrice was almost certainly unaware of her colleagues' first reactions. Yale to her felt like coming home professionally. From the beginning she experienced the great joy of good teachers, who know when their teaching is resonating in bright young minds. Beatrice knew that the Latin *educare*, to lead out, is the foundation of education. Her students have never forgotten how they responded to her. They in turn to a large extent became her family.

She was concerned in case Jim Gunn might not continue collaborating with her on their previous level. This thought proved quite unfounded, but it did add to her delight in quickly discovering the pleasures and satisfactions of collaborating with Richard Larson. Bill van Altena was to say:

> Richard is a pure theorist. Beatrice took his theories, made sense of them and drew inferences. Her ideas sparked and fertilised his. It was a most fruitful collaboration. She would say, 'If this, then that – '. The Tinsley-Larson models are still used.
>
> Her thoughts have fertilised more people's ideas than anyone I can think of.
>
> Beatrice egged him on to get more involved and deal with truly relevant current problems. He still would have been a great astronomer but Beatrice encouraged him to broaden his horizons.[18]

From the beginning she felt at home with Richard scientifically. She also talked freely to him about her own life:

> She saw her parents as famous people – high and mighty, the movers and shakers of their day. Beatrice didn't speak of her mother but I learned she was an author and cellist, and her father had been mayor of a city – both were so distinguished, and she grew up in an atmosphere of expectation. She said she felt intensely driven from the very beginning, driven by an intense desire to excel, to win fame.
>
> Beatrice said to me, 'I worked like anything to do as well as Rowena did. She did very little study, but excelled. I don't think I was exceptionally gifted, not like Rowena was, so I had to work extra hard.'
>
> I could see from what Beatrice said that Rowena's salvation had been to become a rebel, although she felt Rowena had never been in a position to make the most of all her talents.'[19]

At intervals in her crowded days, Beatrice tried to get on with furnishing her apartment. A furniture salesman amused her by saying, 'They've found the universe is even bigger than they can ever measure, or something.' She realised his remark arose from the considerable media publicity given to one of her own lectures, when the American Physical Society had put out a press release. She agreed with the salesman without mentioning her involvement.

Her group of like-minded astronomers – Richard Larson, Gus Oemler and Chris Wilson, who made up the 'gang of four' – drove with her up into Vermont one weekend to enjoy the maple trees in every imaginable shade of red, purple, orange and yellow. The autumn weather and the trees reminded her of her student life in Christchurch, and its English-style parks. 'It's strangely like a full circle in life, starting a new long-term life here. I have a sense of hope and power over the future that had escaped me for many gloomy years,' she wrote to her family.[20] The next morning she was off to Philadelphia to give another lecture to a group of physicists and astronomers. 'Life is very good.' So good that she wrote a poem as well as writing about the day in her diary. Beatrice recognised Rowena as the poet in the family. She herself turned to writing poetry mostly as a release from great stress and unhappiness, the reason for keeping it as a private part of herself, but happiness also moved her. She called this short poem 'Autumn, Again'. It was written in October 1975:

> Summer's over, the stifling stresses past,
> The future's open, too infinite to know,
> My mind expands in the brightly coloured air,
> As it did in those autumns half my life ago.
>
> The time is new again, despite the years,
> And I am young again, so my music flows.
> There are stars in my sky again, becoming ideas,
> And love is unbound again, so the universe grows.
>
> Days of delight are back, and the peace they bring.
> I've hope again, for autumn is my spring.

She told Richard that she liked to write down what she saw and thought. He noted that she had a time each day for writing in her diary. Every minute had its purpose, even the purpose of recreation such as a walk on trails in the woods. He liked to wander and stop and look around. She had a path, and a goal, 'organised and purposeful and scheduled'. Beatrice did not like it if he took a different path through the woods. 'Once we got separated and she was kind of upset. She'd had a plan, and I hadn't, I guess. She wasn't obsessive, but there was always some goal.'[21]

By now Beatrice was practising her violin and had embarked on chamber music again, mostly Bach fugues arranged for violin and cello, which she and Richard played together. To have this close rapport with Richard in music, too, was possibly what put the seal on their relationship. His calm reliability and pleasure in interacting with her ideas meant she had a sure foundation to her life. As ever, playing music as well as listening to it seemed to spark off new astronomical ideas, ideas likely to turn into lengthy projects, as she told her family. Even if the problems turned out to be frustratingly difficult, the process of contemplating them was always enjoyable.

When the university president gave a speech on the status of women at Yale she was not greatly impressed. 'He said all the right things in a little too polished a way, so it all seemed rather condescending,' she told her family. Nor did the atmosphere in the university as a whole seem particularly friendly to women. It was going to be hard to think of good solutions to the very real problems women had. The situation was not made any easier by militant feminists or by suave conservatives. She also realised there were legal requirements that often hindered any real effort to find, or end, sources of discrimination. These requirements and time-hallowed traditions did not help everyday situations, either.[22]

She was amazed to learn that women had not been admitted as undergraduates to Yale until 1969, nine years after she herself had graduated BSc in New Zealand, although Yale had granted a PhD in English language and literature to Elizabeth Deering Hanscom in 1894. As for Harvard, it had been even more conservative. Learning that women had had to sit outside lecture halls to listen to Harvard professors did not sit easily with this Canterbury graduate. But whatever had happened in the past, Beatrice felt sure that at Yale, anyway, she would not have to contend with the sort of hostility to women that she had encountered in Dallas.

One day in the autumn of 1975 a couple of carloads of people from Yale's Astronomy Department set off for Boston for a two-day meeting at Harvard. Jim Gunn was to be a special guest. Among the 'Yalies' was a graduate student, Jim Rose who was apprehensive as this was his first public presentation of his research. He was there because of Beatrice. She had singled him out, encouraged him and arranged for him to present it. 'It gave my confidence a sudden enormous boost to realise that people were interested in what I was working on.' He was incredulous when Beatrice invited him to dinner with Richard, and Jerry Ostriker from Princeton. 'It was unbelievably inspiring to be able to listen to three such intellects sorting through a problem. Beatrice had a totally refreshing approach towards us grads, and what a difference that makes when there's good rapport between faculty and students.'[23]

Because Edward Hill always asked for news and liked to pass on her accounts of scientific meetings, particularly to Theodora and David Lee-Smith, Beatrice tried to keep a stream of information coming. On this occasion, rather than take the time herself to reduce the complicated proceedings to ideas basic enough for lay people to grasp, she sent her father a report of the meeting, a lengthy *New York Times* article by Walter Sullivan.[24]

At Harvard, as she and Jim were as usual working on something new, one of the population synthesis papers, they snatched what time they could to exchange more ideas. By now Beatrice had put most of the distresses of their previous relationship well behind her, and was back to the intellectual attraction to Jim which she had felt from the beginning. Now, too, there was the firm foundation of a special friendship between them.

About this time Sandy Faber observed that Beatrice was becoming increasingly devoted to Richard. 'There was a deep emotional involvement and a totality of professional interest and managing students together.' [25] None of this appeared in Beatrice's letters home. Instead she wrote of Jim

> One of our recent efforts has caused a spate of notoriety to which Sullivan added, including quoting some Plato that we used to preface our paper.[26]
>
> Jim escaped from the Harvard meeting straight up Mt Palomar, to go observing, but I've been besieged by phone calls and so forth, including a telephone interview with the Canadian Broadcasting Company.
>
> Funnily it's by no means the most significant piece of work I've done, but it captures the public imagination more than most things. Fun being famous for a little while!
>
> Sometimes I remember in a rather startled way how I used to read the encyclopaedia as a kid, and wish I could understand and contribute to cosmology.[27]

The publicity kept growing. The *Christian Science Monitor* covered their paper, too, and a science reporter from the Voice of America interviewed her on tape. She also enjoyed talking about new developments in cosmology to a conference of science writers, with emphasis on one of her enthusiasms, interpreting science for the non-scientific.

One week the visiting astronomer was Jerry Ostriker whom she described as 'another of my close friends and colleagues'. Earlier he had become close and comforted her after the break-up with Jim, so that now they had a special friendship. To her family she wrote that they had fun of the kind they liked best, vehement scientific arguments that ended with everyone realising the situation was more complicated than anyone had thought. 'Arguments that are exhaustive and exhausting are usually the most stimulating.' [28]

Alan and Terry arrived for a full week's holiday, to take the place of a reunion at Christmas. Beatrice reported to her New Zealand family[29] that the children were in extremely good health and spirits, and that they had thoroughly explored the area, often with 'the three bachelor astronomers'– as when they all drove to Mystic Seaport, the reconstructed whaling village. The six children of the Armatino family who lived in her basement joined her two to present a floor show on Thanksgiving Day, 'pretty entertaining, more so than they had intended'. Alan made himself the boss or producer, and Terry with her great sense of theatre was undoubtedly the star actor.

Richard Larson had not known what to expect during the children's visit. He had some young nieces and nephews in Canada but saw them only at Christmas family gatherings. It was obvious that Beatrice was rather tense during expeditions with the children, but he and the other members of the 'gang of four' enjoyed them, and were sensitive to their situation. Alan was comparatively calm and obedient, but Terry, who from the beginning fascinated Richard, was neither, although she was certainly beguiling.

Beatrice wrote to them in Dallas frequently, and treasured the notes they sometimes sent her. Her letter-writing was far more prolific than anyone, even her family, ever realised. Only rarely in her regular letters home did she mention her other correspondence. It ranged from loving letters to Nanny Gullidge in London and occasional catch-up letters to old friends, such as the Metcalfs in Christchurch and the musicians Penny Saunders in London and Pip Jackson in Sydney, to some of the most eminent astronomers worldwide.

She also tried to keep up with the output of younger astronomers and send even the briefest note of encouragement, with perhaps ideas for further work. But being true to her own high standards was an article of faith. She was discriminating. Anyone who had clearly put in serious work but who had made miscalculations or who, in her view, had drawn faulty or inadequate conclusions could expect to receive a lucid exposition of the matter, with encouragement to go further. Anyone who presented sloppy work might also hear from her, and Beatrice in attack mode – which she seemed to relish – was formidable. It was as if a combative streak in her, dormant in her earlier years, found full expression when she could tell herself she was promoting scientific truth.

As ever, not everyone appreciated this. 'Some didn't take her comments kindly, especially if they were older than Beatrice. She got into a number of fights this way,' Richard Larson has said.[30]

In the world of astronomy most people try to excel, but they are also mostly – although by no means always – collaborators in the search for new knowledge. Nothing excited and delighted Beatrice more than another significant discovery or theory, regardless of where it came from, and she was loud and generous in her praise and exhortations to continue. Meanness and jealousy, not unknown in this as in any other small world of experts, surprised as well as saddened her.

A quality she lacked, however, was tact. Sometimes it seemed to her friends that she thought being tactful was pussy-footing. It was her old 'black and white and no shades of grey' approach to life, her parents' Absolutes. But Nelson Caldwell, who was to become one of her graduate students and to work closely with her, came to believe that one of the most striking characteristics of this 'most striking of teachers' was her ability to change and grow, particularly in the way she perceived others. He, a young student changing rapidly himself, had observed her whole-hearted esteem for the great scientists she was now meeting so often. 'In her mind they were entirely great, unblemished by any trace of common human frailties.'

It so happened that at a conference Nelson had innocently shown some of his work to one of Beatrice's greats. This great man later 'tried to take unto himself' this particular work. Beatrice saw the evidence. She still very much admired him and kept up a correspondence, but he was no longer shiny white. Rather he was a being in shades of grey, like most of humankind.

Nelson observed other incidents with colleagues and students where Beatrice was able to reassess her first opinions, upwards as well as downwards, and see humour in the shortcomings of others as well as in her own behaviour. 'One doesn't expect people in their late 30s to change, or want to, but she did broaden her views and come to a more complete understanding of people.' [31]

The more Beatrice's own work progressed, however, the more it became apparent to others that science to her was a pure, abstract activity, with no room for emotion. Or at least this was her ideal. As Richard said, 'It was all logic, and she did carry this to extremes. She certainly didn't think she was rude in her reactions to some work – rather she expected everyone else to share her purely rational response, and she was surprised, genuinely surprised, when some people responded in a visceral way.' [32]

Something Beatrice often thought about was how to simplify for non-scientists, without over-simplifying, some of the basic ideas and their background that she and her colleagues were working on. When the *New Haven Register* sent a reporter to interview her, it was good practice. It turned out to be a better than usual collaboration of reporter and astronomer, she said.

An undated clipping of the interview, with a photograph of Beatrice and headed 'She's Astronomer Without Telescope', was passed around the family circle:

Beatrice Tinsley views galaxies from the end of a lead pencil. From this vantage point, the Yale theoretical astronomer-physicist has produced, for an upcoming issue of *Scientific American*, enough data to indicate that our universe is expanding and will continue to expand forever.

'Just in the last couple of years we've put these things together,' said the petite New Zealander with a pronounced English accent, an associate professor of astronomy.

The scientific conclusion, which is one of the ultimate tests of Albert Einstein's theory of relativity, resulted from her work and that of three colleagues – Dr J Richard Gott 3rd of Princeton, Dr James E Gunn of the California Institute of Technology and Dr David N Schramm of the University of Chicago.

'It's like the problem of shooting a rocket' she explained. 'If it leaves the ground with insufficient velocity it will fall back to earth. If the velocity is great enough it will escape the earth's atmosphere and go on forever. It's essentially the same question.'

Analysing the total density of the universe in grams per cubic centimetre, using orbital calculations, ensuring velocity and looking as far back into time as she did into the future, Dr Tinsley concluded that the collected facts weighed heavily on the theory that this universe is behaving like the rocket shot through the earth's atmosphere.

'We don't rule out that we can change our minds with new facts,' she said. The alternative theory states that the universe may have reached the peak of its expansion and is retracting, eventually to collapse some day.

'We're fairly sure if you just take a bit of the universe and average the particles together, the density would be below the critical value.' (The critical value is the density necessary to draw the galaxies back together.)

There is one hitch. Dr Tinsley said, 'The biggest test comes from those galaxies billions of light years away – determining how fast they're going. We're not sure. Our only handle is how it looks. The luminosity of stars, and of the galaxies made from stars, changes with time.'

The ultimate destiny of these spinning orbs won't affect anyone alive today. 'If the universe goes on expanding forever, we will just see an awful lot fewer galaxies. Before that happens, billions of years from now, the sun is going to burn out, anyway.'

Dr Tinsley, who started scientific star-gazing in high school, does these paper and pencil exercises for 'philosophical curiosity'.

She is a proponent of unmanned space flight, for economic as well as scientific reasons.

'I think the unmanned space probes are more useful to science. There's so much extra cost involved in sending people up. It's for adventure. The value of having a man in space is fixing something that goes wrong. The scientific returns from unmanned probes are much greater than manned, cost-wise.

'There's so much astronomically you can do beyond the earth's atmosphere,' she said. 'The air around the earth just gets in the way. It's partly dirty and emits light.'

This newspaper interviewer, like many others, raised the subject of astrology. Beatrice was definite in her reply. The daily addiction of millions of people to astrological forecasts was a strange, disappointing ritual, a sad reflection on how little people understood reality. 'If people were aware of the real information about the evolution of galaxies,' she said, 'they would find it every bit as intriguing.'

In the run-up to New Year 1976, Beatrice prepared lectures, played chamber music, and had her friend Bea Wolf from Dallas to stay. They visited New York and local galleries, and talked frankly about what had happened in their lives and how they now felt. By mid-January the astronomers' chamber music duo had become a trio. Playing tentatively at first with Richard had given Beatrice much pleasure. Now they felt good enough to expose themselves to others who were recruited for trios or quartets. Her own playing was not yet up to Richard's, she declared.

She certainly did not tell her family this, but for some time now she and Richard had been in a close relationship, although she was adamant she would never consider marriage again. They would continue to live separately, keeping their own apartments, but as their world saw it they were a couple. Richard in particular was a private person, and few of the faculty and students commented in any way, but there seemed to be universal pleasure in seeing together two such special people who had perceived each other, and who complemented each other in so many ways.

Beatrice continued to get great satisfaction from her duty of selecting, inviting and entertaining the faculty's visiting astronomer each week. Everyone particularly enjoyed Alar Toomre of MIT. He was a superb lecturer, putting on a staged performance with what Beatrice described as 'fantastic wit'. She also enjoyed his fascinating calculations used to show the way galaxies form, and movies of his computer results, which she considered unique.

She had begun teaching a new course for undergraduates, and liked watching their exposure to the visiting lecturers. 'What a pleasure after the University of Texas undergrads!' In Texas almost anyone was admitted who could get through high school, and many took astronomy as a supposedly easy way of fulfilling science requirements. At Yale, however, the undergrads were a highly selected bunch. She was teaching the most stringent of three introductory astronomy courses so she could be certain they were all genuinely interested in science and that they enjoyed mathematics, and accordingly were a pleasure to teach.

Nevertheless it was her graduate students who particularly absorbed her. Her first PhD students were Barbara Anthony and Bruce Twarog, who married. Barbara talked to Beatrice about what name she should take or keep or use. Beatrice warned her to think most carefully about it; she should be prepared to stick with her decision, no matter what, as it was too hard to force a transition on the name in one's publishing record. Barbara opted for a hyphen between her and her husband's names.

Beatrice was a very important person in their lives, and remains so. Bruce Twarog put it this way:

> Beatrice obviously had a dominant role in my career, but she also had a major, continuing influence on me personally.
>
> Of the many things that made her special, perhaps the most important is simply the way she treated the students. She clearly had more talent than all of us combined, but still treated us as colleagues rather than underlings and, more important, she meant it.
>
> As a result of her own experiences early in her career, she treated those around her, no matter what level they were at, with respect and courtesy, and attempted to involve everyone in her love of the field. It made up for a lot of issues that arose while we were at Yale, and has always been the basis of the way we treat our students.[33]

Barbara Anthony-Twarog remembers how Beatrice trusted her thesis students to make their own decisions:

> In retrospect, and after so many years in academia, I am astonished at how much freedom she gave me! My project was observational, and she trusted me to pick the best targets and methodology. She did need to correct my work with respect to statistics and some aspects of mathematical physics that were weak, but she trusted me, and Bob McClure, to make the observational decisions.
>
> She was so quick, mentally, and such a fast talker which isn't necessarily the best characteristic for a classroom speaker. She was much better one-on-one. Once someone described her lectures as like a stream of dried peas on a window pane.

In spite of such quips, Beatrice was very much appreciated and respected by all of us. We were curious and shy about her divorce and children, about her relationship with Richard Larson (whom we also admired very, very much) – a variety of nearly adolescent things like that. In the years since then I have learned for myself how wildly inaccurate students' speculations are about my personal life, so I suspect ours were silly, too.[34]

To the students it seemed natural, inevitable, that so many of the brilliant astronomers they saw were attracted to one another, loved one another. In many ways it was a golden age for astronomy, for the study of galactic evolution, in America. Rare and remarkable intellects were at work in astrophysics. Electricity was in the air. They perceived and appreciated one another, and sometimes this spilled over into physical relationships too. But, regardless, they seemed to move in a circle of glowing warmth and intellectual excitement, and drew the students into their orbit. They were grateful. As Barbara Anthony-Twarog said:

> Two things all of us were especially grateful to Beatrice for. One was that so many fabulous people came through our lives – just because she was there. The list is extensive. Gunn, Knapp, Faber, Penzias, Peebles, Ostriker, Gott, van den Bergh, Rubin – these names are just the first to come to me.
>
> The other thing was our strong conviction that she was genuinely protective of us as students. She was wary of answering too many questions about our work when they were posed by some people she suspected of being just a little too interested in an ongoing project. The point is really not whether she was right or wrong to be suspicious, but that she was, and on a student's behalf.[35]

Her interest and concern, too, stretched to include the wives or husbands of her students, a number of whom had married young, some to ease their financial circumstances. Beatrice noted this wryly, contrasting students with the faculty in various departments, many of whom seemed to be living in *de facto* situations.

Suddenly, without warning, there came terrifying news from Dallas. Alan was very seriously ill. A tumour had been discovered in his chest, he had problems breathing, and he was in hospital for tests. Jill Knapp happened to be with Beatrice at Yale when the news came. 'She was completely shattered – distraught, and terrified for Alan.'

Brian said she should not fly to Dallas until they had definite information, as Alan could be alarmed by her sudden arrival at the hospital. Strangely, she went along with this. The excuse most mothers would give – 'How lucky I just happened to be passing through Dallas' – either did not occur to her or she spurned it as a lie.

She sat down on 1 February 1976 to tell her father and Mattie her fears. If it was cancer, such a tumour would most likely have spread the malignancy, and be inoperable:

> I've agreed with Brian (against my instincts but in accord with rational judgement) that I shouldn't go until something is more definite, or he'll be frightened. Brian sounds terribly distressed. No wonder. Terry thinks Alan's just having a bad cough checked up and is happily telling me riddles on the phone.
>
> Even if it is a benign tumour the poor kid will be in for a lot of drugs, to shrink it, and a long illness. I don't know if Brian has told his parents so you shouldn't contact them (I don't know if you would, anyway).
>
> Life seems to have changed colour since I heard about Alan. I'll probably go to Texas about next weekend anyway; it wouldn't be too surprising for Alan to have me visit him after a week in hospital. I'm sorry this is such an awful letter.

But six days later, in Dallas, she was able to write with good news. The tumour was not malignant. It had been a major operation, and Alan was still in intensive care, but the danger was over and Brian was quite cheerful. They were all able to sit beside Alan's bed.

Alan as an adult was to say that he became so ill because he had not told anyone about his accelerating problems in breathing until after he collapsed in class in fourth grade. School had become so important to him that he did not want to miss a day. This was Hamilton Park School, the first integrated school in the Richmond School District in Dallas. Half its students were now black. 'We were bussed in – big deal. The teachers were very good and there was a waiting list.' Then one day he woke up in an oxygen tent in hospital – 'and there was Mum. She could hug me but not kiss me. Actually she didn't usually go in for kissing or touching though she was very caring.' Even as a young child, Alan had noticed that Beatrice avoided intimacy. 'With Dad and later with Richard she didn't show affection or seem to need it, or attention.' [36]

Richard noted that Beatrice, back in Yale, spoke 'really well and sympathetically of Brian' because of his fear and distress when Alan was so ill. Perhaps this shock about Alan would bring his parents closer together. Beatrice for her part hoped this happier state of affairs would continue and that there would be no more sarcastic and wounding remarks. Before too long she was able to say that Alan's illness was like a bad dream, now over, although she told Theodora that she felt emotionally and physically exhausted herself. Everyone was astonished at how quickly Alan had got better.

The Yale faculty enjoyed a three-day visit from Richard Gott, now, at an extremely young age, on the Princeton faculty. Gott could talk for hours on end, literally, putting one idea after another on cosmology, she wrote. home. 'Very fascinating, but exhausting for the less mentally energetic!' He and Richard Larson were friends and rivals, both being in the business of theories of galaxy formation: 'somewhat rival theories, of which I think Larson's are far better, but they can discuss them for hours'.

Prospective graduate students quite often visited the department and were invited to listen to any visiting speaker. One of these prospects was Nelson Caldwell who happened to coincide with a talk by people he called 'those curious friends of hers', the Barnothys:

> I couldn't understand a word of the talk, and still can't figure it out, and neither did any of the grad students. Beatrice somehow thought these people were interesting, and they did have some wild ideas which later became of use, such as gravitational lensing, but it was hard to sort the good from the wild.
>
> I suppose I felt that if this was how Beatrice thought, it was pretty hopeless for me to ever be able to talk to her. I later learned that I could talk to her, though she was certainly pretty intimidating for a year and a half.[37]

Nelson Caldwell came to decide that this openness was something else which made Beatrice different. Most astronomers who seemed intimidating at first meeting remained that way: it was their way of dealing with the world, a practised method, he found. Beatrice was open, without barriers. It was the power of her intellect that was intimidating, not Beatrice herself.

In fact, she was the one to whom the poor Italian family, the Armatinos, kept bringing their stories of victimisation. She was enraged. Anger when someone else was the victim she expressed more easily than anger on her own behalf. In any case, taking up causes on behalf of individual families as well as movements such as Zero Population Growth was part of her life. Thus pages in her letters home over months were headlong blasts of indignation on behalf of the Armatinos.

Good things were also happening. She and Richard were engaged to teach at an astronomy school in the Swiss Alps. They drove around Lake

Geneva, drinking in the scenery, to the small town of Saas-Fee, 1800 metres up in the mountains of the Canton of Valais. Beatrice was overjoyed, calling it the loveliest place she could imagine visiting to teach astronomy, 'not to mention astronomy is the nicest thing I can think of doing while not walking in the Alps'. Altogether Saas-Fee was an idyllic experience. Richard had never seen her as happy – ecstatic, in fact. 'Fancy being paid to go and see all that,' she wrote.[38]

Each day she woke very early and walked up mountain trails while the rest of the school were still asleep. It was more than the environment, however, which was stirring her blood. The experience of being away with Richard, their first big excursion together, made her realise what a central position he now had in her life. How easily it could have been otherwise, she was to write to Rowena. After all, she had first been attracted to him because of his brilliant contributions at astronomical meetings, but not always, as she had discovered, did brilliant minds mean the kind of person she could wholeheartedly feel for, day by day. Richard was later to remark dryly that at that stage she was still mainly in love with the Swiss Alps. Beatrice thought otherwise.

In all she gave 10 lectures and spent hours talking with the students individually or in groups, using mixtures of English, French and Italian – she had given up on German. The students had to cope with her New Zealand accent, which by this time was becoming overlaid with American. She also made some conscious effort to slow down but continued to find it almost impossible to control the way she spoke. The accents of the others, Australian Ken Freeman and Canadian Richard, gave the mainly European students some further problems. But the tutors had all arrived with extensive lecture notes that they planned to turn into a book, so the students had written material to turn to for help.

This book, *Galaxies*, by Freeman, Larson and Tinsley, was published later in 1976, by the Geneva Laboratory. Beatrice's contribution, the longest of the three, she called 'The Evolution of Chemical Abundances and Stellar Populations'.

Soon after they were back at Yale, Rowena's visit at last materialised. Her main purpose was to catch up on what they had done, thought and felt, and what they now wanted to do with their lives. Not that they forgot Theodora, by now awaiting the birth of her first child, and seeming to be blissfully happy and settled with her husband David. But what they saw as the smooth sequences of her life seemed such an enormous contrast with theirs.

One evening the astronomer Alvio Renzini was having dinner with the sisters when Rowena light-heartedly asked him to explain just why they were doing astronomy. Alvio said astronomy was fun, a challenging intellectual game, a chance to visit places and to meet interesting people. Beatrice would not accept this. 'She was much disappointed by my uninspired view of the job and reprimanded me, saying, "We are doing it for no less than understanding the universe!" '[39]

Rowena with her fluent Italian then tried to trace their different attitudes back to Beatrice's Calvinistic background and to Alvio's pagan Mediterranean attitude. Alvio found her interpretation plausible, 'possibly because it satisfied a narcissistic side of me.' Thinking about this in later years, when he 'had matured somewhat', Alvio realised that his research of that time was on a very small scale compared with the breadth Beatrice's research had already achieved. 'She was so much farther ahead in her scientific work. My work was indeed like playing games – compared with hers.'

The students were permitted to invite two special visitors each year. Jim Gunn was now their choice. Beatrice was doubly pleased that she was able to look forward to his arrival with equanimity, but still with intellectual excitement at the prospect of more work discussions. She happily hosted a large gathering of students, Astronomy Department members and physicists who were old friends of Jim's. 'Not much effort,' she commented in her letter home. No, because her idea of entertaining had changed very little since her student days. She was quite aware that others regarded party-giving as an art, and was able to laugh at herself but be entirely serious when she said, 'My style of party is to get wine, beer, nuts and cheese, and let people swallow them while talking.'[40]

She had a crowded schedule of meetings at Austin and was a PhD examiner there, so could see the children for a day as she passed through Dallas. Brian's parents were staying with him, and she gathered the children were having a great time. Alan was almost at full strength again, and Terry full of bounce as usual. Beatrice then wrote home unguardedly:

> I didn't see Brian's parents or go near the house because he was adamant that I should not meet them, even for a moment while picking up the children. Strange, but I can't be bothered asking or thinking why – there are enough past and present hurt feelings without scratching for more. Brian was quite friendly when I visited Alan in February but this time he didn't even see me – left the kids at the kerb and asked my friend to bring them back after I had left. Anyway, Alan and Terry are beautiful and loving. Only 10 days now until I see them in Colorado when there should be time to catch up on them a little better than one day of fun.[41]

Beatrice was actually holding back in this letter, although she was more outspoken than usual when in family-writing mode. In fact she talked angrily about what she called Brian's cutting and vindictive behaviour. She should have known that their coming together when Alan was so ill was something that would not last. Never, ever, would she marry again.

An outsider – such as Bea Wolf – could perhaps see that Brian was also full of pain, outraged too at being the one who had been left. It was only human for him not to want his parents to see a bright, fulfilled and happy Beatrice.

Because she could have the children with her in a holiday atmosphere, Beatrice had agreed to give lectures at Aspen, Colorado. She arranged a number of treats and people to look after them during her astronomy sessions, and thought it had all gone well. Richard saw rather more. When it was time for the children to say goodbye, Alan took it bravely but Terry was in floods of tears. She could not understand why her mother had left the family in Dallas to go away to work. 'Couldn't you work in a pizza place or something?' Beatrice was distressed but also her logical self, saying to Richard, 'It is always a wrench when the children have to leave.'[42]

Their department had just been moved to the Gibbs Laboratory at 260 Whitney Avenue. The rooms were now much bigger and looked out into the tops of huge old oak trees where squirrels lived. The astronomers had one very long floor, the others being occupied by physicists and biologists. Everyone struggled to settle in.

One day, while they were doing this, a young woman from the astronomy Department was mugged near the Gibbs Laboratory. She fell heavily, concussed, and was taken to hospital. When she became conscious she began speaking in French, her childhood language, and mentioned Pierre Demarque. He was summoned, but the young woman did not believe him when he tried to explain she was not in France. He told Richard, who decided to wait a few hours and then see if perhaps he could make contact.

When Richard met Beatrice in passing and told her what happened, she was appalled, possibly recalling her own ordeal in the university carpark in Austin. She nevertheless came with him to the hospital in case she could help. They found that the victim was still not speaking normally. Her husband was with her and Richard also talked to her, hoping to strike a chord, but Beatrice in her distress and indignation seemed stunned speechless.

Then the young woman said, 'Oh, I know you! Let me hold your hands.' She said later that the two astronomers were the first people she recognised, and she went on to recover, except for memory loss of the two weeks before the attack. Richard did not forget what he saw as Beatrice's over-reaction.

After setting her new office to rights – everything functional with little decoration except for large colour photos of mountain scenery – Beatrice hurried off to the big American Astronomical Society's meeting in Haverford, Pennsylvania, where she gave one of the invited lectures, which she thought was well received. Again she met up with old colleagues from all over the country.

She must have been a travel agent's joy at this time. After the AAS meeting she flew to England on 1 August, to Cambridge again for a busy summer. She took time off to fly to Scotland to visit her astronomer friend from Maryland, Jill Knapp, of whom she had become particularly fond. Jill, whose parents were in Edinburgh, recalled: 'I think she was a bit taken aback by the way my father greeted her. He made up a parody of that old song, "I was seated one day at the organ" and sang lustily, "I was seated one day at the telescope." Her kind of astronomer spent time at computers rather than at telescopes.' [43]

Jim did not come to Cambridge that summer. Rosemary had at last told him to leave, and he was in Chicago on sabbatical with a visiting professorship. Since her Maryland days, Jill Knapp had been working at Caltech as a post-doctoral student. After her marriage to Steve Knapp ended, Jim had sought her out with some material they could work on together. This became a three-way paper with Scott Tremaine. Jill and Beatrice could now talk together with real understanding of what it was like collaborating with Jim –'exhilarating, though I couldn't work as quickly as Bea could.' [44] For her part Beatrice felt both pleasure and relief that she could be so genuinely happy at the thought of two of her dearest friends together.

After the International Astronomical Union cosmology symposium at Cambridge, there was another big meeting at Grenoble, the IAU General Assembly – 'thousands of astronomers, mostly anonymous French ones'. One who was far from anonymous, of course, was Jean Audouze. At Grenoble, Audouze organised a special session on 'Carbon, Nitrogen and Oxygen Isotopes in Astrophysics', for which Beatrice wrote a paper.[45] She also gave another talk, 'Masses of Supernova Progenitors', as part of a special session organised by Dave Schramm. Everyone noticed how stimulated Beatrice was by this gathering.

It was in France that the young Scott Tremaine had a meeting with Beatrice that he never forgot:

> I was just a post-doc and kept my head low while the great and famous astronomers spoke. Beatrice, however, sought me out at one point and invited me to join a group that was splurging at an upscale restaurant. There I found myself in the midst of people I had heard

of but never met and I probably learned as much at that one meal as I did at the rest of the conference.

'I have often thought how unusual and thoughtful it was for Beatrice to make that effort on behalf of a young colleague whom she hardly knew.'[46]

With the beginning of the 1976–77 academic year, Beatrice had a bunch of new students to get to know, as well as catching up with the previous year's, and delving into research projects again. To her family she wrote, 'I seem to have spent the summer talking about the last set of results, and now I want to find out some new things!' She was happy to get ideas from anywhere and anyone, and to pass on her own insights to anyone interested. Yale's convivial atmosphere had long since obliterated the suspicions she had shown at first. Teaching was proving most agreeable, and standards were high.

One student who had decided he wanted to do research with Beatrice and Richard Larson, even before he applied to the Yale graduate school, was Curt (Curtis) Struck-Marcell.[47] (He was to drop the Marcell from his name some years later.) As a senior undergraduate at the University of Minnesota he had done some research with another galactic astronomer, Jay Gallagher, who advised him about graduate schools. Jay had worked with Jill Knapp, who had passed on glowing accounts of Beatrice and Richard at Yale. 'I learned that they worked as a team (it took me a while to figure out that there was more to their relationship than that), that they were bright stars in this field, and that I would be doing very well if I got the opportunity to work with them. They were extremely important to me in the five years I was at Yale for my PhD.'

Curt Struck did not have a course with Beatrice that first term, but was very much aware of her as a luminous presence in a generally welcoming department. For the new students, staff held a party in the tower across the courtyard, where the library and cafeteria were, and Curt spent much of the time talking in a group around her:

> Like many times later, she was a focus, a centre of energy. Even though she was very much farther up the academic ladder, Beatrice seemed much more accessible than most of the faculty. It was possible for a first term grad student to feel that she was more nearly one of us, and to hope that things went well for her.
>
> She always seemed to be the 'tribune' of the grad students, in the Roman sense of spokesman, and representative.[48]

When she was working with a group of these students or giving a talk, there was a charge in the air, expectant, confident. The lecture room seemed filled with an unseen force. Beatrice was doing what she most wanted to do, with years of work and research in prospect.

Nelson Caldwell, another of these new graduate students, was also struck by this same apparently inexhaustible energy, and the way she seemed always to operate on a higher level than anyone else. 'She never sauntered down the hall. It was always at breakneck speed, often speaking out loud what was on her mind, her dark eyes blazing.' [49]

Very much on Beatrice's mind at this time, quite apart from all her other responsibilities and interests, was the fact that she had been put in charge of organising an important international conference to be held at Yale next May. It was taking considerable thought. 'Only about 100 people, though,' she wrote.[50] That 'only' in her letter home may have been unconscious, a way of reassuring herself. In fact, the organisation was such a big job that the senior faculty had begun discussing it, putting Beatrice in the role of planner and organiser, even before she was formally appointed. She seemed the ideal person to ensure a memorable meeting.

To hold an international conference on evolution of galaxies and stellar populations was Pierre Demarque's idea. He had discussed it with Richard Larson and Bob McClure. They all knew it was a burden to put on her, a big job of organisation, but thought she was the natural person to do it. Richard had been deputed to sound her out one time when he was to give a talk at Santa Cruz, and Beatrice had picked him up at Berkeley and driven him there. 'We talked about the idea of the conference. She asked a few quick basic questions regarding its scope, and said yes.' [51]

Now Beatrice was on her mettle. She chaired a scientific organising committee comprising Sandy Faber, Ken Freeman, Jim Gunn, Richard Larson and Martin Rees,[52] and checked out some of her ideas with them. Getting in touch with the very best people possible in the field, discussing their papers and trying to ensure the optimum shape for the conference meant that she was spending her evenings at the computer, usually till 1am. She took time off for a reception to welcome some new staff members, 'bright and interesting new grad students', and also went up to Hartford to give a lecture on cosmology to the local Sigma Xi, the scientific society of doctors, industrial chemists and similar professional people. 'Quite fun,' she told her family. Otherwise it was head down, working.

Bob McClure saw her as 'a real whirlwind'. Well before he left Yale for the Dominion Astrophysical Laboratory the next year, he had seen enough of Beatrice to realise she was a great complement to the rest of the department. There was one possible problem, he noted: she worked so quickly that she left others behind. 'It would have been very difficult working in the very same field as she would be publishing papers way before I would ever be getting started.'[53]

In November the children had a whole week with her including Thanksgiving, and joined the Armatino children on various activities. Richard observed that Beatrice was particularly happy to see her children horsing round with the Armatinos, riding on one another's shoulders and so forth. 'I never had that when I was a child,' she told Richard. Nor had Terry and Alan before the divorce.

Again there were the three bachelor astronomy friends, Richard, Gus Oemler and Chris Wilson, for a Thanksgiving party and picnics in the state parks. Writing to her family, Beatrice again described idyllic times. Richard, becoming increasingly fond of the children as well as close to Beatrice, had reason to wonder about the actual outings because Beatrice found it hard to relax with them and accept what he considered normal childish behaviour:

> There was so much bottled-up tension in Beatrice. She just couldn't take things in her stride. She couldn't compromise. When the children came to her at holidays it brought so many of her own tensions to the surface.
>
> Her impulses were always to do good, hence the Armatino family, but she wasn't emotionally involved with them. With her own children she was uptight about whether they were, or were not, doing something or other. She had a short fuse – she lost her temper rather often with them.
>
> The two kids would fight. You couldn't put them together on the same car seat. She'd say, for instance, 'I've had enough. Shut up. That's it. We're going home.' Once when she said 'Shut up' to Alan he bravely said, '*You* shut up!' She really exploded then, made them get out of the car and walk until she picked them up.[54]

'We walked for quite some time,' Alan was to recall dryly.[55]

Parents are often most angry if their children do something they themselves were not permitted to do as children, so, as Richard noted, when Alan answered her back 'she was the angriest I ever saw her.'

He also noticed that if something happened at home she would shut them in another room. This usually was not for long and they did not appear to remember it, but it was clear they were used to it This was how it had always been in Dallas – and in Beatrice's childhood.

She was to refer to these traumatic occasions as 'squabbles', saying dismissively that she remembered doing the same things with her sisters 'incessantly'. What she may have written in her private diary is another matter. It does seem, however, that when it came to her relationship with the children she often unknowingly followed her own mother's mode of denying reality, and keeping at an emotional distance from them. Among onlookers in the department, Gary Steigman, who greatly admired Beatrice, was one who felt that Richard was more 'maternal' with Alan and Terry than Beatrice was; Richard appeared to be the parent in the group.[56]

Besides teaching, Beatrice had been researching one paper and writing another for 'a monstrous meeting' at Harvard of 600 people. This was the follow-up to the huge gathering in Dallas just before she left there at the end of 1974. Though held in Boston, it was still called the Texas Symposium on Relativistic Astrophysics, the fun name dreamed up between drinks by that group of Dallas scientists beside a swimming pool, but a name now esteemed in astronomy worldwide. Beatrice's part, as she told her family, was to give a talk, 'a very popular and inevitably "gee-whizzy" review on cosmology' – the State of the Universe address, as it had been called by a reviewer when Jim Gunn gave it. This time he was away, observing.

For once she felt she had had her fill of conferences and needed to limit her time there, so decided to drive up to Boston and back, three hours each way, in one long day with a student to keep her company and keep her awake. Beatrice's planned stay fitted in with the talks Curt Struck particularly wanted to hear. He was delighted though a little over-awed to be travelling with her, and during their journey asked about her talk. She went over it with him, and indirectly expressed something of her nervousness about talking in such distinguished company.

They both had reasons for anxiety at that time. She was still a new junior faculty member at a place very famous for not keeping junior faculty. He was one of seven first-year graduate students in a department that did not like to admit more than three or four a year, so there were rumours about how many of them would fail, either in classes or in the PhD qualifying exam.

Curt Struck found the Harvard meeting extremely exciting. He vividly remembers many of the talks, particularly Beatrice's, 'The Cosmological Constant and Cosmological Change'. Her conclusions summarised the cosmological status quo at the time. Other memorable contributions included Stephen Hawking's on what has come to be called Hawking radiation. This is his discovery that black holes, especially microscopic black holes (if they exist), can radiate energy (and particles), as well as suck energy and matter in, and will eventually dissipate. Vera Rubin, too, was memorable as she discussed Virgocentric flow, the 'flow' of galaxies located on the outskirts of the Virgo galaxy cluster.

At the banquet, where Curt sat at Beatrice's table, he listened to Jocelyn Bell on the discovery of pulsars. For a student to be included in proceedings along with so many renowned astronomers was something he had never dreamed of.

'I was very proud of her talk and pleased that it went well, and was well received.'[57]

It felt great to be from Yale, and to be associated with Yale's Beatrice. Everything was going so very well.

CHAPTER EIGHTEEN

YEAR OF THE GALAXIES

'Galaxies are the footprints of the Big Bang.'
SANDY FABER

Arrangements for the big galaxy conference in May 1977 continued to occupy Beatrice. Pierre Demarque did the local organising such as arranging accommodation and coffee breaks, but the weight of the meeting fell on her:

> When I'm not teaching or researching, this conference keeps me hassled (more grey hair every day). The great and famous speakers are largely temperamental prima donnas, seems to me. I hope I never become one!
> Richard pointed out that very famous and busy people can be choosy – and they are.[1]

Beatrice was particularly sensitive to giving people their due, and ensuring that nobody was missed out when credits for work were called for. This was especially true if students were involved and she suspected they were being exploited by famous names (something not unknown in university circles worldwide). She was still incensed by what had happened to her when she worked at Caltech. In her ground-breaking thesis published in 1968 she had presented results which she believed to have been rediscovered by Searle, Sargent and Bagnuolo and republished in 1973 without any reference to her thesis. They – very much the seniors – were at Caltech when she was. Beatrice had heard about their work and had told them about hers, showing Leonard Searle her thesis. There was still no acknowledgment of her work. 'She was extremely sore. More than anyone else would be,' Richard Larson observed. 'There could have been an explanation such as their paper being already in press, but we knew of no explanation.'[2]

Students vividly remember the story she once told them about the difficulties she had had getting her PhD thesis published, and getting it discussed, let alone generally accepted. On one occasion she had presented some of this work at a conference that Allan Sandage was chairing. She expected everyone at least to be interested in her conclusion that the then-popular method of determining the acceleration of the universe, a method used by Sandage himself, was seriously affected by galaxy evolution.

At the end of her talk, Beatrice said, Sandage had got up and said something like 'Okay, it's break time.' There were no questions. Nothing. It was as if she had not spoken.

When it came to arguing and disputing details of one another's work, everyone expected Sandage and de Vaucouleurs to disagree. They were of an age, and their sparrings and confrontations were like the steps of a formal dance, expected and acceptable, seemingly even to themselves. They were part of the landscape. But for Beatrice to clash with Sandage, that father figure in astronomy, was almost unthinkable. She was young, she was unheard of – and she was a woman. 'Most of all,' as somebody murmured, 'she was Beatrice Tinsley.' There was nobody like her.

It had taken Richard some time to accommodate himself to Beatrice's style, although he had begun to do this when they first collaborated in 1973–74. Now she was impatient, bubbling with ideas. Once she happened to say to him that when she was walking to work she was thinking of what she would put in a referee's report, and when she was walking home she would think of a paper. He worked at a milder pace. It was hard to digest all the ideas she tossed out so he tried to select among them. Usually she accepted his revised version, and maybe added to it. Trial and error was not her way.

Richard decided that her skills were needed to help him with a paper, and prodded her into something she was not totally confident about. They eventually called the paper 'Star Formation Rates in Normal and Peculiar Galaxies', and it became famous when it was published in 1978 in the *Astrophysical Journal*. It showed that peculiar galaxies that were just merging likely underwent star formation triggered by the merger. The paper began with the idea of trying to find evidence of really young galaxies; perhaps a catalogue of peculiar galaxies might give clues. The idea was to age the galaxies, taking observed properties, such as colours, and see if they could be explained by a mix of red (old) and blue (young) stars, or see if, for instance, they were all young. Richard asked Beatrice if she could model the colour of the peculiar galaxies, hundreds of them. He has explained:

This – modelling – meant getting a mathematical representation of the observable properties, colour in this case. Stars have different ages, masses and chemical composition. Actually only elementary bookkeeping is required. But the art comes in finding the right ingredients to mix together – how to put together the best information we have at the time. In this work a great deal of theoretical astrophysics and observational astrophysics are brought together.

Beatrice gathered up information as if she were a magnet. She was like a vast, intellectual vacuum cleaner. She gathered it from the literature, from anyone she could talk to, and from her own ideas, bringing everything together. And everything was to answer the big questions about the origins of the universe.

Her synthesising method maximised scientific inputs from everywhere, and minimised the guesswork. To her 'the revealed truth' was what was being revealed by scientists on a continuing basis – that is, 'state of the art' truth, which changed, and will continue to change as more is discovered.

She was always alert to whatever was going on, and how it might connect. Most scientists are interested only in small, defined areas. Beatrice put it all together.[3]

Her students, coming from widely separated universities, were quite unaccustomed to being taught by anyone like Beatrice. Curt Struck had never encountered anyone remotely like her before. As he said:

> At this point I needed a lot of explicit direction, but Beatrice was not one to take a thesis student who needed direction all the way through. Guidance yes, direction no.
>
> The project was in the very heart of her own work, actually a straightforward extension of work she had just done with Richard. There was lots of 'insider' stuff that I didn't understand at first, but she knew intimately.
>
> I began by helping her update some of the stellar input data, putting it into the form used by her colour evolutionary computer models. I have many pages of the boring tables in my files – the tables may have been boring but I bought into her enthusiasm that the results wouldn't be.
>
> As to my notebook of hand-drawn coloured-pencil graphs I made once we got into the colour modelling – these days my daughters would be ashamed to hand in such crude work to their high school teachers, let alone to the world's leading expert on the subject.

But they were quick working sketches, and sufficed – or not exactly, at first. There would be a scene something like this. 'Well, let's see, you've drawn the axes backwards,' (astronomical colour graphs go from positive to negative numbers), 'so this must be blue and this red.' While I turned very red.

She was very patient with my first fumbling attempts, and I got better. In the end, as expected, the infrared colours we were studying proved very sensitive to stellar population ages, and were a great complement to the optical colours studied previously.

So we wrote a paper and I turned aspects of it into a thesis. Of course I had to draft that paper myself, using the previous one as a guide. Beatrice did not do her students' work. My first drafts weren't good, and resulted in more embarrassment for me and more patience on Beatrice's part. I am not a great writer in any sense, and beyond that basic fact I had a devil of a time picking up the technical writing style. There were no manuals or tutorial sessions in those days. Beatrice was a master of the style.

In fact Beatrice was legendary for the way she wrote papers. If a paper was ready to be written she was known to be able to go home at the end of the day, work through the evening (not all night) on her typewriter, and have a complete draft ready the next morning. Not a perfect draft, of course. She was a sharp editor of her own work as well as that of others. She was also fearless about having lots of people read her drafts.

Much later I would do a lot of editing for her, including on her big chemical evolution monograph, 'The Evolution of Stars and Gas in Galaxies', for *Fundamentals of Cosmic Physics*, her swan song in 1980. One of my most important editorial assignments was to point out what a typical grad student might not understand. This is something which all lecturers should do, to remind themselves – but few ever think of checking up. I also got formula checking duty on the *Fundamentals* paper. This paper, 'The Evolution of Stars and Gas in Galaxies', has 580 citations in the technical literature (from the Astrophysical Data System, which often misses some); 26 of those citations are from this year. By the standards of most researchers, that's pretty amazing longevity.[4]

Beatrice, back in her 'travel and talk' routine, went first to Maryland to give several lectures at the Goddard Space Flight Center, then went back again to Johns Hopkins University in Baltimore, followed up by a two-day informal conference at home in Yale. Called the Eastern Astronomers' Neighbourhood Meeting, it was one she was taking part in but did not have to organise. Yale, she was discovering, had a number of good traditions centring on sharing discoveries and work in progress.

Gary Steigman revived the tradition of meetings between Yale and Harvard astronomers, taking in astrophysicists from MIT and the area. He recalls the organisation as being very informal, with those who were interested pitching in to help, 'although a few of us took overall responsibility to see there was a programme.'[5] Beatrice was quietly amused when one of the students told her that organising this conference had been exhausting.

Immediately after this came a two-day meeting of a small team working on a special project for the Hubble Telescope. She explained it to her family[6] as being a camera project for a large telescope to be flown as a satellite, 'some time in the political never-never'. Describing her role, she said she could not design circuits but she was 'supposed to have ideas for things to work on'. The team of which she was a member was concentrating at this time on the high-pressure system competition for an experiment on the forthcoming space telescope. It was not scheduled for launch until 1983 and Congress had not even authorised the funding as yet, but some centres had already spent tens of thousands of dollars developing instruments, 'electronic detectors like fancy cameras in effect'. The astronomers were now busy proposing a series of observations they wanted to have done. She had been persuaded to join what she considered a very competitive team: 'fun, win or lose, but the tight deadlines are killing.'

Beatrice, never interested in instruments or telescopes as such, apparently did not play a major role apart from suggesting and debating scientific priorities for the space telescope. She was much more involved in the preparations for the galaxy conference on 19–21 May.

This turned out to be one of the most memorable conferences in the lifetimes of the participants. Yale's International Conference on the Evolution of Galaxies and Stellar Populations was held in the Kline Geology Auditorium, a compact room next to Gibbs Laboratory. A deliberate decision was made to keep the conference small and select, just the outstanding people in the field worldwide. Numbers had to be limited because there were only 120 chairs in the auditorium. Too many astronomers applied to attend. Beatrice had to keep declining some who persisted. She was upset and felt bad about it, but stuck to her guns. To be as fair as she could, she did her homework on people, such as asking a referee what a particular person could contribute. Eventually about 130 came.

She remembered so well the intense excitement of the first meetings she herself had attended, when she was the lowliest of students. Now here was an opportunity to enable as many of her own students as possible to see and hear proceedings. She would make them 'student workers'.

As she explained, their main job would be to run. At the end of a talk, when questions were asked, they were to run – or hurry discreetly – to each questioner with a form on which they were asked to identify themselves and write out the questions they had asked, or the main points of their contributions to the discussion. This was to facilitate inclusion of the questions and answers when the proceedings were edited and published, as happened with IAU meetings. All sessions were to be tape-recorded, but in too many conferences it was found that some parts of the tapes, such as questions from the floor and the ensuing discussion, were inaudible. Hence the runners.

As a low-level junior student worker, Curt Struck found the conference exciting from the outset. Thus his experience of the evening reception in the Hall of Graduate Studies:

> Very neo-gothic, lots of dark wood and stone, warm, yellowish art-deco lighting. I kind of hovered on my own, before Beatrice dragged me into a conversation she was having with Martin Rees and Virginia Trimble. They were legendary figures even then, of course.
>
> And this was so very typical of Beatrice – drag the poor, ignorant grad students in, and get them involved. I don't think there was anything too special about the conversation. It was ordinary details about travel to New Haven, and meeting arrangements and happenings. But I was there – with them.[7]

When Beatrice stood up on the dais the next day to open the meeting, something extraordinary happened. As Ivan King of Berkeley recalled, 'She was greeted by a tremendous, prolonged ovation – quite unusual.'[8]

He himself then got proceedings under way. He believed that Beatrice had originally asked Allan Sandage to be this first speaker, and that when Sandage had declined she had turned to him. In the event, everyone seemed to agree that King gave an exceptional talk in a number of ways – that it was a brilliant notion for him to get up in front of such an elite audience and tell them, in effect, that their whole field was a complete mess, in a state of ferment, that here was a list of 12 of the worst problems, and what were they going to do about them?

This was Nelson Caldwell's first real conference, and he was immensely struck by the way both Ivan King and then Sidney van den Bergh of the Dominion Astrophysical Observatory set the tone by pointing out questions which needed addressing. As he said:

> Some of these questions were actually answered during the meeting,

and, as is the case in astronomy, some of them were answered by new and more interesting questions.

King made one of his classic, provocative statements when he said too many astronomers had wasted their lives studying quasars instead of more tractable problems. Yet here, sitting in the front row, was Maarten Schmidt, whose picture had once been on the cover of *Time* magazine for his discoveries in quasars. And, later in the meeting, Wallace Sargent presented the first data on using a quasar as a light source to study intervening material, a method that has produced literally thousands of papers since then, and much real hard science.[9]

From the outset the conference was about attacking the big problems of extragalactic astronomy. A sure sign of a memorable meeting was the upbeat corridor conversation – that time between talks and sessions when pent-up reactions and ideas newly ignited burst out in a hubbub of talk.

Not that every session seemed so eventful. De Vaucouleurs's talk was not new, and not sparkling. The corridor consensus, as gathered by some of the graduate students, was that the great elder just had to be given a chance to speak. Whether his long-time rival, Sandage, would have had anything new to contribute had he decided to attend was an unanswered question. Leonard Searle's talk about the relatively new Searle-Zinn idea of how the Milky Way formed was important and well received, although it was familiar to some. Much of the material presented at the conference was very new at the time although some Yale people had the advantage of having had previews of some talks, such as Larson's.

Of the famous allies, the three S's – Sandage from the US, and Searle and Sargent both Britons long resident in the States, with the first two father-figures in astronomy – Sandage had vacillated, accepted, then changed his mind. The organisers did not know why. Searle had also vacillated but then agreed to talk.

By the time of this conference, Beatrice and Richard's paper, 'Star Formation Rates in Normal and Peculiar Galaxies', had been rejected, twice, by a referee for the *Astrophysical Journal*, (it eventually published the paper the following year), with remarks such as 'Not enough new and original material'. Beatrice and Richard suspected the referee might have been one of the authors of a 1972 paper by Searle, Sargent and Bagnuolo. This paper had presented some similar models but did not address the same applications. Now Richard presented their paper, Beatrice not wanting the extra involvement of presentation while she was concerned with the conference overall and a chamber music concert that evening.

Most of those present saw that this was ground-breaking work. An entire industry is based on that work now.[10]

The last to give a talk – apart from Jim Gunn who summarised the conference – was Wallace Sargent. He spoke on 'Quasar Absorption Lines as Probes of the Past Intergalactic Medium'. Sargent was a noted speaker, with the cutting sarcasm of many of the English. As he was on his way from the floor up to the dais to speak, he erupted so that everyone could hear: 'Whenever Searle and I write a paper, it seems we do so under the pseudonyms of Larson and Tinsley.'

The whole auditorium rippled.

He later apologised to the two Yale astronomers.

'What had happened,' Richard Larson was to explain, 'was that I had failed to mention in my talk the work of Searle, Sargent and Bagnuolo. Of course we had done this in our written paper. We'd made ample reference to their work.'[11]

The audience turned to look at Beatrice. She had sat quietly during Sargent's attack and had laughed nervously, as everyone else did, but she turned beet red. Her position as conference host forced her to try to look composed.

Nelson Caldwell, looking back, summed up:

> From people's expressions, it was plain they thought it was simply nonsense, this implication that the Larson and Tinsley work was somehow derived from that of Sargent and Searle.
>
> 'It was utter rubbish, as a reading of the papers can easily show. We students discovered that it was not Beatrice's way to argue in public in that manner, and I myself certainly never heard her criticise her competitors in public in such a way, although there were times when she would make quiet remarks in colloquia about things she disapproved of.
>
> It also became well known to us that at meetings she made a habit of asking at least one question to which she was sure the speaker knew the answer – and she knew too – just to make that speaker feel comfortable.[12]

Comfort was not the aim at the galaxies conference. Discussions were lively, verging on aggressive at times, and caustic comments often drew laughter. Witty asides were welcome. Thus Sargent said the only thing he had learned at this Yale meeting was not to eat Greek pizza before giving a talk.

The Sargent-Larson-Tinsley spat apart, people had other opportunities to see the great and the good and the would-be's in action. Curt Struck drank it all in:

Another thing that was neat about this meeting was that, as well as the major talks, little mini-contributions were snuck in around them. Personalities were on display, and there were great opportunities for people to make their mark, some quite aggressively. At the end of Sandy Faber's talk, 'The Chemical Composition of Old Stellar Populations', Dave Burstein made quite a substantial contribution. I realised later that this was the time in his career when he needed to be noticed.[13]

Curt was beginning to learn the politics of astronomy.

Something else which stayed in the memories of participants was another lively discussion on the Hubble constant,[14] with proponents for each value. This had been voted on at several meetings. In the end, at this Yale conference, there was a vote of sorts, resulting in a compromise between the then-entrenched positions of 55, Sandage's view, and de Vaucouleurs's 95–100. Beatrice's researches took her to a value of 70. The result of the discussion was 75, not far from today's generally accepted value of 72 km/sMpc.

To the young Struck, the summit of excellence among so many excellent speakers on the last day was Alar Toomre from MIT, who spoke on 'Mergers and Some Consequences':

> To use a very American metaphor to describe the importance of Toomre's talk to the field of galaxy evolution, this was like the reading of the Declaration of Independence to the Continental Congress.
>
> Although Toomre conveyed some doubts himself about how far his hypothesis might extend, and how exactly it worked, overall he was at his most magisterial.
>
> Beatrice and Richard played an important role in the discussion, which included the beginning of the Ostriker-Toomre debate about whether mergers can make ellipticals. Nobody wanted the discussion to come to an end. It was electrifying for anyone with an interest in the field.[15]

Looking back many years later, Ivan King also considers Toomre's talk the high point of the meeting: 'Toomre's suggestion, that elliptical galaxies were made by collision and merger of other smaller galaxies, was novel at the time but is now fully accepted.'[16]

Ken Freeman agrees, but also puts at the head of the list the Searle-Zinn paper about the formation of the galactic halo by accretion of small galaxies:

These ideas really changed the subject. I think most people at the conference realised that history was being made. This meeting is known now as the Yale Conference. No date is needed. I've been to more than a hundred international conferences now, and this was certainly the meeting of a lifetime for me. In fact it turned out to be one of the most memorable astronomical conferences of all time.[17]

Participants generally agreed that they now had clear, defined lines of enquiry in extragalactic astronomy, such as galaxy evolution in clusters, galaxy collisions as a means of galaxy transformation, stellar population variations among galaxies, chemical evolution at different redshifts, and detailed galaxy structure.

Beatrice, the meeting, and the way she had organised its components, all won loud praise everywhere. It was a galvanising event for her, for the Yale department, and for extragalactic astronomy. It was a fortuitous time for such a conference, and she had triumphantly used her personality and connections to bring together a large number of people who had only just begun to define extragalactic astronomy and where it would go in the coming years.

Almost everything had gone well, 'a smashing success', one of the best ever, participants said. The atmosphere was informal, relaxed. 'Beatrice had no time for British or any other kind of pomposity. She wanted every minute to count.'[18]

Jim Gunn, giving the concluding talk, had ended it by saying, 'Perhaps there will be another seminal conference, some day soon.' This had been a special time, with special people and special ideas. As he was to say much later, the conference was a genuine watershed, a testament to Beatrice's work and her vision, and a pointer to where she was going.

Nearly everyone, he said, had come along to Yale holding the view that galaxies were individuals in the universe whose lives were entirely influenced by their own internal dynamics. Everyone emerged from the conference knowing that galaxies lived in a world full of other galaxies, and that their interactions among themselves were every bit as important as their internal interactions.

'It was the single most important galaxy conference in the history of the subject. There has been no other similar conference on galaxies since then. There is no Beatrice.'[19]

When the last participant had departed, someone asked Beatrice if she would do it all again the next year. She groaned. But at once she threw herself into producing the book of the meeting. She wanted it to be the best possible record of everything that had been said and argued. Richard was co-editor. They were helped by their librarian, Delores Campbell Gehret, and by Barbara Anthony-Twarog.

Everyone who had talked had done so on the understanding that they would give a written version by a set deadline. Nearly always, at astronomy meetings both then and now, participants talk from notes or an outline, and use transparencies. Rarely does anyone come along with a written-out talk to read. Their later written papers, therefore, can quite often differ from what was actually said at the conference, particularly if the authors absorb and incorporate some of the points made in discussions.

Wallace Sargent did not front up with a written copy of his address by the time of the deadline. He had of course been taped, and Barbara Anthony-Twarog was deputed, somewhat to her alarm, to write up his talk from the audio tape: 'Since I didn't know enough about quasars to rewrite anything he had said – and wouldn't have dared, anyway – it probably is one of the best matches between oral and written talks in history.'[20]

By 1 July 1977 Beatrice was at last able to write again to her father, who, with Mattie, was by then staying with his sister Rosamond in England. She underlined that she had been extremely busy and that the conference had left her collapsed. In some ways, though, the really solid work had been putting the conference proceedings into a book. It included about 20 hours of taped discussions, which needed a lot of editing and checking with the participants. She was particularly pleased, she said, that discussion was part of this book as it often contained some of the best points made at conferences yet was hardly ever included in reports of what had gone on. By this time the book, *The Evolution of Galaxies and Stellar Populations*, was very nearly ready for the printer. Her goal was to finish everything connected with it, then get back to doing some original science again instead of 'having to be so absorbed in other people's work'.

Alar Toomre of MIT, that 'summit of excellence', was her very last contributor. He had personally handed his paper to her only the evening before. She had had to do rather more than sit and wait to receive everyone's talk. Beatrice had badgered him, among others, saying there was a deadline for his paper of 15 June. When he still had not delivered she kept at him, saying, 'Okay. There's an absolute deadline of 30 June.'

Beatrice in badgering mode was formidable. Richard said she kept 'beating on him', and on 30 June Toomre had his son drive him to New Haven from MIT in Boston, he proofreading in the car to get to them in Yale by 5.30pm. The four then went out to dinner to celebrate.

Alar Toomre has said that nobody before or since has succeeded in getting a manuscript from him so promptly. He added that Beatrice and Richard had done him a great service by making him put down his ideas in a new direction.

The book was published by the Yale University Observatory in just 11 weeks from the date of the conference, apparently a record for such publications. Not all editors are willing or able to continue 'beating on' their contributors until they come to heel. It ran to 450 pages, double-spaced on an electric typewriter. Computers at that time were used just for number crunching, not word processing.[21]

Richard, in the preface, wrote that the book could 'perhaps serve most usefully as a signpost of changing directions in galactic research'. To open the preface Beatrice supplied this quotation from William Herschel, taken from that favourite book of her student days, *Theories of the Universe*:

> If we indulge a fanciful imagination and build worlds of our own, we must not wonder at our going wide from the path of truth and nature ... On the other hand, if we add observation to observation, without attempting to draw not only certain conclusions, but also conjectural views from them, we offend against the very end for which only observations ought to be made.

When she next wrote home,[22] she had nearly finished the mailing out and the handling of accounts for the book's supporting advertising. As she had been appointed director of graduate studies in the department, a post involving a lot of time with students together with handling mail and general paper work, she had to be extra careful in organising her days and nights so she could carry on with her own new research.

The children had spent 12 days with her in July, 'a great joy', she wrote. They had had a few days at Nantucket Island staying with the 70-year-old astronomer Dorrit Hoffleit, whom Beatrice liked and admired. Dorrit, who had long since changed her first unfortunate impressions of Beatrice while never forgetting what had caused them, had now officially retired. She directed the little observatory on Nantucket but still returned to work at Yale in the winter months. When she was at Yale, students such as Curt Struck sometimes helped her in her work on her Bright Star Catalogue. Now she had invited Beatrice to give one of the summer public lectures on Nantucket: 'She was brilliantly successful at connecting with her audience.'

One morning Alan came quietly downstairs very early, and told Dorrit a great deal about his family and the arguments leading to the divorce. 'He was the first to give me background to what had happened. He seemed very mature, and appeared to understand why his mother had had to leave.'[23]

Beatrice thought Nantucket idyllic. She took the children to see a remarkable house built in 1790, the birthplace of Maria Mitchell, America's first woman astronomer. 'The children are fun to do such things with.' When she drove them up to their summer camp, she left them feeling confident that everyone was happy about the arrangement, a feeling confirmed on visiting days.

Edward then sent Beatrice the typed manuscript of his memoirs, 'Half a Life', with its poignant story of a lonely boyhood and then a tortuous account of becoming virtually taken over by the Oxford Group. The manuscript made a deep impression on her, but not the one Edward intended. She found it insufferable. To Richard she said, 'I hate it.' [24]

With her deep feelings for her father, Beatrice was in a conflict of emotions. She despised the English class system and Edward's proud part in its upper echelons, and noted how her father theoretically wanted equality for all while knowing he was not equal. He recounted his childhood and young manhood in detail for this memoir. It was not hard for her to see how his upbringing had led to his immersion in the Oxford Group, contributing to his never having found a true and lasting vocation.

Edward planned a second volume and intended the title of this autobiography to refer to the first part of his life. Beatrice reflected sadly that in some ways the title could be taken literally. She mourned the waste of his abilities, and the fact that he had not lived the full life that could have been his.[25]

Beatrice wrote to her father with care, but chose to be indirect:

> It gives me a lot more understanding of your life and the enormous contrast between the world you lived in at my age, and the world I live in! I wonder what I'll be adjusting to at your age?
>
> Another contrast is between yours and Mummy's very sheltered early childhoods; the less sheltered time you could provide for us in New Zealand; and the almost unprotected exposure to the evils of the world that my children face. With public schools, neighbours and television, evil would be hard to avoid these days.[26]

'Sheltered early childhoods' is certainly one way of describing how both her parents were brought up, insulated by wealth and privilege. The 'less sheltered time ... in New Zealand' referred to the Hill children's attendance at public or state schools and their friendships with children from a wide range of backgrounds. The Hill parents were mostly successful in their aim of keeping the harsher facts of life, such as reports of violence and crime, away from their daughters' eyes and ears. Beatrice told friends she had seldom seen newspapers. New Zealand's state-run radio service would rarely include such things in its news reports (the Hills did not listen to commercial radio), and there was not yet television, so she grew up not knowing about quite a lot that was happening in her community. It seems not to have occurred to her that her own children would be more affected by open unhappiness in the home than by any evils of the world as seen on television.

Once, when the children were staying with her in New Haven, Beatrice found them reading a newspaper feature about a years-old murder on East Rock, where she had often taken them to play. She was horrified. Children should not know about such awful things. She pounced on the paper. 'Anything like that had been hidden from her in her childhood,' Richard Larson said. 'She grew up unable to accept the existence of crime and poverty. They shocked her. She couldn't deal with them. Her reactions were over the top.'[27]

Beatrice tried to put her father's 'Half a Life' out of her mind but she could not let her feelings for the Oxford Group stay altogether unexpressed, particularly its enduring influence on her parents' lives. She declared to her father: 'The Oxford Group – Moral Rearmament – is an almost incredible phenomenon to me. Did Frank Buchman's character change? Or was the original emphasis, on personal religion, a sham?'[28]

Further than this she did not wish to go. There was the old fear of hurting him, so it was best not to get involved in discussing religion. Disagreements and feelings of alienation too often resulted. But these old childhood attitudes and concealed rebellions coexisted in her even as she tackled an expanded workload, more ideas for new research which had germinated since the May conference, and a determination to seek refreshment through more music in her life.

Chamber music, baroque and classical music remained her great love, either listening or playing herself. Her personal music situation now seemed even better. One of the new graduate students in 1977, Linda Stryker, played the viola and violin, and was very keen to join them. They had already managed a couple of evenings of string trios and had plans to expand into piano quartets, she told her family. First the original trio began tackling the Mozart Divertimento in E Flat which she had so enjoyed playing some 18 years earlier with Glen and Wal Metcalf in Christchurch.

Linda Stryker, a few months older than Beatrice, was to become one of her closest and most trusted friends. When Linda was nine she had vowed to become an astronomer. Her teachers regarded her as a child genius in science, mathematics and music. When she was 13 she had been taken to Caltech to meet Edwin Hubble, Horace Babcock and J. S. Adams. Linda herself was never so sure of her abilities. Her parents were dismayed at the thought of her becoming a physicist or a mathematician; any girl like that would not get a husband. If she went into those fields, they said, they would not support her at college, but they would if she chose music.

Linda had won awards and a scholarship in music, so, knowing she was pleasing her parents, she went in that direction. She made a career as a classical guitarist and teacher, playing and teaching many instruments such as the viola and violin after obtaining an MA and part of a PhD in music. But her old longing to know more about the universe resurfaced so she began again, getting a BA in physics and an MA in astronomy. Then, after having been accepted by several universities, she chose Yale for her PhD: 'My mentor at Kitt Peak, Tom Kinman, recommended Yale as they did good solid astronomy there. As for Yale accepting me – I've always thought the viola may have been the deciding factor for at least some of the people there.'[29]

For the next few years, whenever she was at Yale, Linda was to play at least once a week with Beatrice and Richard. She had switched from violin to viola a few years earlier, but felt she had never really mastered it. Richard was the best player of the three, she thought, in that Beatrice had not practised enough over the years, 'but we were a passable little trio. It was truly fun and a great break from the departmental work.'

Although Beatrice was cautious about being drawn into administrative roles which took time away from her own work, she was pleased to be appointed to the committee that suggested who should receive honorary degrees from Yale. Already, after its first meeting, she found the committee interesting but was thankful that it met only four times a year. Some academics seemed almost to make a career out of sitting on committees, she noted. She did not intend becoming one of them.

Brian had met and planned to marry a Dallas woman called Yvon de Guider who had a daughter, Courtney, aged eight. 'I really hope it all works out. It would be best for everyone if Brian remarried happily' was Beatrice's comment to her father.[30] When Alan, now 11, and Terry, two years younger, arrived for their usual Thanksgiving holiday the marriage had already taken place. The children had not been present. 'We came home from school and it had happened,' Terry said indignantly.[31] Beatrice in her letter added that she was naturally curious about their new stepmother. The children's only reply was 'Very nice.' This was a great relief, and they were cheerful. Terry even seemed considerably calmer than a few months earlier. 'Brian's life should be a lot happier and easier again,' Beatrice added.[32]

One of the guests at the special Thanksgiving dinner was Dorrit Hoffleit, their summer hostess on Nantucket Island. Dorrit enjoyed the children's company. She had been walking along Whitney Avenue one day when a car drew up beside her with Beatrice and Richard, and Terry in the back seat. They chatted, and Dorrit told Terry she would like to take her and a friend to lunch one day. Terry pointed to Richard. 'He is my best friend.'[33]

Alan starred one evening when he gave a travelogue-type 'seminar' in the department, showing some of Brian's excellent slides of the family's trip to New Zealand, and talking about them. Many from the department came to this evening. Everyone wanted to be supportive, but were also pleasantly surprised by the little boy's lively talk. Beatrice was proud. She was much happier saying goodbye to them now that they had a 'very nice' new stepmother.

Instead of flying to an astronomy meeting in Washington, she decided to drive there and back, 650 miles, with a passenger, because a student on a zero budget wanted to go to the meeting too. This was Linda Stryker. Beatrice told her family she liked her a lot: 'an extremely capable and pleasant person almost my age, having been a professional musician until she went back as an under-graduate in science some years ago'. Beatrice so enjoyed talking with her student passenger that she noted in her letter that she felt she was making more and more friends of the kind who were life-long.[34]

For her part, Linda felt she got more from her conversations with Beatrice than she did from the meeting:

> Beatrice impressed me as being very kind and thoughtful, unprejudiced – which I really appreciated – very knowledgeable, and fun to be with. I felt there was good rapport and was grateful, as I was still nervous at being new to grad school, being back East, wondering if I would be able to cut the mustard ...
>
> She introduced me to many astronomers. We stayed in a hotel near the White House. As she was preparing a talk to give at the meeting, I tried to give her space to work on it. On the way back we talked about the meeting – she filling me in about various people who were there – and about guest speakers at Yale, her work, my interests ... I felt I had been very lucky.[35]

The year was nearly over. It was traditional for the students to put on a skit, a gentle or not so gentle satire of the faculty and notable visitors, which was performed at the Astronomy Department's Christmas party. The revue this year was a long take-off of *Gone with the Wind*, this time called *Gone with the Stellar Wind*.

It was written at what the students called 'informal nocturnal seminars', not quite parties, with most students contributing, although the main writers were Barbara and Bruce Twarog, John Giuliani, Jas Smith and Peter Stetson. A mock Beatrice was the star.

Starlett O'Tinsley was the southern belle played by Barbara Anthony-

Twarog (largely, she has said, because she had a suitable dress). Starlett spent the evening either trying to preserve, or find her version of, the cosmological constants. She kept after the value of q_0, the cosmological 'deceleration parameter' which measures the rate at which the expanding universe is slowing down or speeding up. Each time Starlett carried on, her leading man moaned, 'Frankly, my dear, I don't give a damn.'

Curt Struck, suitably padded, had the character role of Ajar Tummy, aka Alar Toomre. Students played other galaxy conference delegates, including a number of those who had been collaborating on papers with Beatrice. Some could well have been disconcerted if they had been present to see their foibles and peculiarities seized on and presented through the students' eyes. The students knew, however, that they could go a long way in their satire, providing they were even-handed. Beatrice summed up in her Christmas letter home:

> I was Starlett O'Tinsley who had to preserve her plantation in order to find out whether the universe would expand forever. Quite a role to live down! The rest of my colleagues were cast as other combinations of themselves and characters from the book and played by various students.
>
> At the end 'I' won a lot of money and announced that now I could get famous solving the problems of cosmology. But the rest of the cast shouted back in chorus, 'Frankly, my dear, we don't give a damn!

To Linda, watching her first of these pre-Christmas revues, the students were brilliant both as scriptwriters and actors. She had declared herself far too shy to act, and when she was called on to hold up a sheet as a prop, she laughed non-stop and could not hold the sheet steady. 'The audience literally fell over, they were laughing so hard.'[36]

It had been a great year for astronomy at Yale, a great year for Beatrice. In many ways she seemed the focus of a renaissance, herself embodying a professional excitement which pervaded the whole department, faculty and students.

Curt Struck, looking back, put it like this:

> She was so obviously ecstatic about her work, so wholly engaged. It was contagious! It was well known that she had felt stifled in Texas. Yale was her professional nirvana, as she made clear at many different times. And because one of her great talents was synthesis, pulling all the clues together and solving the mysteries, she made it plain she felt in a perfect place for her work. She had inputs

from Larson, dynamical modeler and general theorist; Demarque, stellar evolution theorist; van Altena, astrometry; McClure, stellar observation; Oemler, extragalactic observation; Steigman, cosmology; and Eardley, physics. She loved all their inputs, and not only did she seem able to put all the pieces together but she also did it very quickly. Her speed of thought was incredible.

Beatrice's very positive attitude was contagious. Everyone could feel it and respond. Even from some outside points of view the Yale astronomy department at that time was viewed as Camelot.'[37]

Her department now planned to make another tenured appointment. Its chairman, Bill van Altena, had made preliminary enquiries, asking others who chaired astronomy departments for names of astronomers whom they considered highly qualified in theoretical astrophysics, particularly in cosmology and galaxies. Now he wrote again for further opinions.

For Beatrice, the new year of 1978 had begun when she managed to have a day in Dallas with the children en route to an American Astronomical Society meeting in Austin. She again stayed with her friend Bea Wolf, catching up and talking music, and then had a whole day with the children. She was also invited to meet Yvon and her daughter Courtney, 'both of whom I like a lot. It looks like a happy scene,' she told Bea Wolf. Relief and gratitude that everything seemed to be working out so well in Dallas energised her.

In Austin at the big meeting of astronomers, the best thing for Beatrice was that she was able to move quickly from one person or group to another, and talk to people individually.

In her correspondence she almost never mentioned buying new clothes – actually, it was something that did not happen very often – or say anything about her own health or appearance. But here she added a cheerful postscript saying that she had improved her hair a lot –'I think' – by having the front cut in a fringe and the sides trimmed. 'It's less stern.'[38]

It was as if she were beginning to take at least some degree of interest in her appearance again, instead of rejecting such matters as part of the trappings of marriage, the state which she continued to despise. As she actively disliked being photographed, 'before' and 'after' photos do not exist.

Beatrice was busy, happy – the two always went together – and as full of plans for collaboration and lines of enquiry of her own as any single astronomer could possibly hope to accommodate. Altogether 1978 looked as if it could be even better than 1977, and that had been her best year ever.

CHAPTER NINETEEN

TIME OF CRISIS

[A scientist makes science] 'the pivot of his emotional life, in order to find in this way the peace and security which he cannot find in the narrow whirlpool of personal experience.'
ALBERT EINSTEIN

Not long into 1978, Beatrice noticed a smear of blood on her towel when she was drying herself after a bath. She felt the back of her left leg on the outside and about two inches above the bend of the knee. Blood was coming from a rough, crusty spot. She slapped a Bandaid on it. As it continued to bleed after she bathed or showered, she got hold of two mirrors so she could see more clearly what it was. She saw a black, raised mole.

Beatrice had always taken sensible care of her health. She had regular medical examinations for possible breast and cervical cancer, took vitamin supplements, and enjoyed brisk walking. But those who live alone have nobody to tell them if something untoward develops on parts of their bodies they do not normally look at, or cannot themselves see.

Ironically, this was the mole, then quite small, which Brian had pointed out to her, years previously. He had said it looked ugly when she wore shorts, so she should have it removed. Beatrice had shrugged this off.[1] What she looked like did not matter. She was not concerned about inconsequential matters of personal appearance and women's magazine exhortations to 'make the most of yourself'.

Most lay people in the 1970s did not think about such everyday things as moles. Certainly they were rarely written about other than in medical journals. But Beatrice did know, although Richard did not, that moles which bled or changed appearance should be treated at once. Later she was to tell Richard that she had waited for about a month, hoping the bleeding would stop, before seeking advice. When Richard asked her why she had delayed, when she knew the bleeding was a warning sign of cancer, she said something to the effect that she 'wasn't in the habit of going to the doctor with every little complaint'; she had decided it was just a minor nuisance. He was left wondering what that delay might have cost her, because the doctor on duty at the Yale Health Center at once realised that the bleeding mole must be removed immediately, for biopsy.

This was done under local anaesthetic that same day, by a woman surgeon. Beatrice was taken aback by the professionals' reactions, but she apparently decided there was no point in worrying about anything until she knew the results of the biopsy in a few days' time. Meanwhile she would carry on with her life as usual.

Back for the results, Beatrice realised that the news was serious because the nurse in the office with the surgeon was crying. To this surgeon Beatrice began by presenting herself as coolly scientific, insisting on all the facts. She was then given them, straight. The mole was a melanoma, a particularly intractable form of cancer – and, as Beatrice was to discover, with no known cure once it had spread from its original site.

As soon as possible she must have a much bigger operation to remove a much larger piece of tissue around where the mole had been. The lymph nodes at the top of the leg must also be removed. Skin to cover the excised part would have to be taken from the top of her other leg.

When she made her way back to the department and to Richard's office, she shut the door and sat across his desk from him, scarcely able to speak. Her face was blotched with purple. 'She looked and sounded totally anguished and devastated.'[2] Never had they thought of anything like this.

She was referred to a well-known specialist surgeon at Yale New Haven Hospital, Dr Stephan Ariyan. He was developing an experimental technique for treating melanoma, involving aggressive treatment of the affected area with chemotherapy. The scientific aspects of this slightly consoled her. Whatever happened, she would be part of a scientific experiment.

Again asking for all the facts, she was told that her cancer was serious, very serious indeed. On a scale of one to five, with five being hopeless, her melanoma was at stage three, possibly four. The probability of a cure, based on previous experience, was 50 percent. Dr Ariyan told her there was a possibility that his new treatment could increase her chances, but he was careful not to make any claims that were not supported by experience. She must have a major operation just as soon as she had recovered from the side-effects of the tests they had done in hospital, she told Richard, giving him her chances of survival.

He immediately took the best possible view: 'I thought of her as having a 50 percent chance of surviving. She thought of having a 50 percent chance of dying.'[3]

To Beatrice, what she had been told was equivalent to a death sentence.

All her great dreams, all her ambitions and the things she knew she could do, all her ideas were going up in smoke, into thin air, just like that, and just when a full professorship and a new life lay before her. Her children, all her friendships – she had had the whole of her life planned ... Now this.

To Richard it was as if she took the news in the sense that she now had to prepare for the idea of dying soon. From then on he gave her whatever support he could. As he saw it:

> I am a 'bottle is half-full' kind of person. To Beatrice this bottle was half-empty – but she was the one with the cancer. I clung to my view of her situation and continued to be optimistic.
>
> Never did she show emotion like that in public. To me, privately, she did for a time, and at times. After that she just did everything possible to survive the cancer while fully preparing herself for death. She was her old, highly planned, self.
>
> After the initial shock we were both preoccupied much of the time by mundane concerns about how to keep living from day to day when life kept throwing obstacles in the way. Life just became more difficult, and pretty soon there wasn't any energy left for emotion.[4]

On 25 February 1978 Beatrice broke the news of her forthcoming operation to her father and Mattie, who were asked to tell Theodora and David. With Richard as her lightning rod, she had begun to come to terms with what lay before her, raw though her emotions were. In her letter she downplayed her feelings and brought her rational self to the fore:

> I'm in for a rather unpleasant week, sorry to say. A lump on my leg has turned out to be skin cancer, and they want to excise a rather larger piece than can be done under local anaesthetic. So I'm going into hospital, and will be operated on.
>
> It's going to involve a skin graft so I expect to be limping for a while – the place is a few inches above my knee. About a week later I'll be out of the hospital and in the Yale Health Center, a nice place on campus where my friends can visit, for about another week; then I expect to be home and I'm assured I'll be okay enough to look after myself.
>
> That's hardly even necessary in view of the marvellous friends I have here. This all happened suddenly, but everyone has taken over my teaching and other duties, filled in various lectures, offered all sorts of help, and been human support. A friend will be picking up my mail, here at home and at work, so you can write as usual.
>
> I'm not ringing you in advance to tell you about this because it

would be needless worry for you. Yale-New Haven Hospital is the best place in the world to have this problem taken care of. I expect to be back at work by 1 April at the *latest*. Meanwhile we played lovely trios last night, Mozart transcriptions of Bach fugues for violin, viola and cello.

Two days later, she made a 'living will'.

THE LIVING WILL

To my family physician, clergy, attorney, or medical facility:
I hereby request that all those responsible for my care and knowledgeable of my condition be completely honest with me in the event of a terminal illness, that I may make my own decisions and preparations as much as possible.

If there is no reasonable expectation of my recovery and I am no longer able to share decisions concerning my future, I ask that I be allowed to die and not be kept alive indefinitely by artificial means or heroic measures. I ask that drugs be administered to me as needed to relieve terminal suffering even if this may hasten the moment of my death. I am not asking that my life be directly taken, but that my dying be not unreasonably prolonged if my condition is hopeless, my deterioration irreversible, and the maintenance of my life an overwhelming responsibility for my family or an unfair monopoly of medical resources.

This request is made thoughtfully while I am in good health and spirits. Even if this document be not binding legally, I request those who care for me to honour its intent, which is in part to relieve them of some of the burden of this decision.

She had it witnessed, and signed it on 27 February 1978. Let everyone think it is just an ordinary skin cancer, she had decided.

Her unemotional letter home, factual as far as it went, like so many others of her letters to New Zealand had not begun to tell the full story. She read whatever she could about melanoma. It made her feel more in control. To Richard she poured out her feelings and frustrations, always coming around to the fact that she had 'a whole world of work to do'.

Her operation was delayed until 13 March. Each day of waiting was anguished. She tried to work normally with her students. The best thing would be to treat the whole matter calmly and scientifically, and work on as usual. She told Linda Stryker alone in her office, saying flatly that she had a 50–50 chance of survival. Linda, stunned, blurted out something about how glad she was that Beatrice had such a good chance. That earned her the retort that it was 'a horrible chance'.[5]

Sandy Faber was in bed with a back injury when she received a letter from Beatrice. Sandy called her at once. 'She put on a brave front and was so solicitous about my problems. A big plant arrived from her the next day.'

A week before the operation in New Haven Hospital Beatrice wrote to Rowena:

> I'm feeling very hopeful, most of the time, because there is some chance that I already have no cancer cells left in me, since a large chunk was taken out with the mole; and there is a 50% chance of the operation being a total cure. Perhaps this time I can be one of the lucky ones, as I have been so far ever since the first bad news.
>
> Sometimes I lose all courage and collapse, as I did tonight after supper, but dear Richard picked me up, as usual, and by the time he left we were even laughing at things. He has proved, more than I could have found out without such an awful test, to be an absolute gem of support. He's there when I need a motherly hug, and he can provide all the lines of thought and pleasure that will keep my mind off this thing and on to living while it is possible.
>
> I think my perspectives in life, and probably his too, will be altered forever by the time we were wondering if I might have only a few weeks to live. Every day I wake up and find I'm at home (even if with a headache and a sore leg in several places) I feel full of joy just to be alive; I think I'll feel that way when I regain consciousness after the operation, unless the cancer chemicals that were fed to my leg have escaped enough to make me sick (a possibility, but not for long).
>
> All the things I read or hear, and the people I talk to, seem sharp and important. Music is fuller, science is more fascinating, people are more lovely and accessible. My friends are being marvellous, too, not just Richard.

She told Rowena about some of the long-distance phone calls she had been receiving, after breaking the news to special friends. Jill Knapp, whom Beatrice reminded her sister was 'Jim's girl friend – the one he won after all that strife' – had phoned earlier, 'and talked for ages'. The two women, who had liked each other from the beginning, when they had first met at Maryland, had kept in touch.

Jill's version was that she had happened to phone about some astronomical problem. Beatrice had answered at length, being extremely enlightening. Jill said she would see her soon, at a forthcoming meeting. Beatrice said no. She had to go into hospital:

Beatrice said this in a straightforward manner, without drama or self-pity. I made an equally casual response, thinking it some minor operation. Then she said she had cancer. I was absolutely knocked by this. Beatrice comforted me. In fact she was the one who got all her friends through her cancer. She made it possible for us to accept it.[6]

For her part, Beatrice rejoiced that she could be genuinely happy that Jill and Jim were together. Jim had been deeply shocked when he learned of the cancer and had called her up for a long chat, eventually sounding more cheerful. Telling the Barnothys in Chicago had been difficult, but Madeleine had written such a heartening letter. 'I can rely on those two to face anything with spirit, somehow looking for all the possible lines of good and help.' Beatrice had to let so many people know what was happening. They were all offering 'whatever the real equivalent of prayers is', and their support was certainly a mental help.

As for her graduate students, she told her sister, she had tried not to worry them, fearing for their morale as she was in charge of them. But they and the staff were offering all kinds of help for the time when she would not be able to get around. And Richard and Bill van Altena had just appeared before a committee to present her case for promotion, but had to leave before a decision was made. It would probably take 'the Big Wigs of Yale' weeks to announce anything. 'Anyway, little problems like whether I get tenure this year don't seem so important now! I'll be happy if next year comes at all.'

The operation to remove more tissue turned out to be major. Beatrice had questioned her surgeon beforehand, so that she had at least theoretical knowledge of what the surgical team was going to do, and indeed find. The pain surprised her, although she had resolved from the beginning to be as stoical as possible. As soon as she was sufficiently recovered from the anaesthetic to be coherent, she asked for details. They had indeed found that the cancer was so far advanced that she had at best only a 50–50 chance of recovering.

The medical team put as encouraging a light on the facts as possible. Beatrice looked at them and through them, and put into words what seemed to be undeniable. Here was confirmation that there was at least one chance in two that she was going to die. Again Richard bore the brunt of what might or might not be her death sentence. When she saw how anxious and distressed her hospital visitors were, she did her best to play down her own feelings, and soft-pedalled the doctors' explanations. With Richard alone she felt she could be her vulnerable self. New treatment procedures, however, did give grounds for hope.

She was back in her apartment and able to climb up her stairs by the last week in March, she told her family, and felt sure she was improving every day, although she had a lot of fitness to recover. Sixteen days after the operation, she was even able to spend a few hours in her office. The bandage had come off her skin graft – 'the graft looks atrocious to me, although the doctor is very proud of it' – and there were only a few nagging sores from the donor area, which was on her other thigh, and a large incision at the top of her leg where the lymph nodes had been taken away. Her surgeon and his team told her they were very pleased with the operation.

Her responsibility of ensuring that her father did not worry was now to be tested because he wanted to fly over to visit her as soon as possible. Again she had to juggle possible dates as she had no intention of forgoing any of the meetings and conferences which had been scheduled for months.

As early as the first week in April she began teaching her students again, sitting in front of an overhead projector, sometimes in a wheelchair but always with her left leg up. Every day she found she could manage a little more activity before her body told her she must rest. She blamed this weakness on the total inactivity she had had imposed on her for 11 days, when she had been sternly told she must not move or it might affect her large skin graft. Because she managed to give the impression that the cancer and the operation were things of the past, and that astronomy was what mattered, even in the close graduate-teacher relationship it was possible for her students to tell themselves she had 'beaten the beast'.

She had worked out how much everyone – apart from her very close friends, to whom she owed the truth – should be told about her condition and likely future. Matter-of-factness and hope were to be the ingredients. Thus people gained the impression that, although it was not one of the least worrying forms of cancer, and although it had not been caught as early as the doctors would have liked, many treatment possibilities existed. As doctors in New Haven were actively involved in research and experimental therapies, Beatrice was pleased to be able to stress this. Optimism would be the best thing for everyone, particularly the students with examinations ahead.

Seven graduate students had their oral qualifying exam for PhD at the end of the 1978 spring term. The panel of five faculty included Beatrice, along with Larson, Demarque, McClure and Eardley. Curt Struck, who was one of the seven, remembers the ordeal clearly:

> Beatrice led the proceedings, helped me get started on several problems, and gave me a problem or two that she must have known I could handle. I left the seminar room at the end, and walked down

to the library to await the decision, feeling very dejected. Like most students I felt sure I hadn't cut it. I was about the fourth of the seven to undergo this ordeal, and, as I recall, at least one had failed before me. I think three failed by the end of the week. It was not an easy time.

Then Beatrice came down the hall and told me the committee had agreed that I had done a good job. I was in![7]

Linda Stryker found her final take-home exam very tough. This was for the galaxies class which Beatrice had taught, with her leg raised. The students had about two weeks to work on it. Questions and research were taxing. Her final oral qualifying exam was the big ordeal. 'I took along a box of Kleenex, just in case. Afterwards I ran to my office, and waited ... Then they said I'd passed. All the students came with me to pig out on vanilla milkshakes. When the boys passed, they usually went out for drinks.'[8]

While picking up the threads of the work she had missed when in hospital and then convalescing, something had suddenly occurred to Beatrice. She wrote to ensure that her father carried medical insurance each time he travelled in the United States.

> It is important. For my first three days in hospital – two weeks before the operation just for tests – the bill was $1001, about equally for 'room and board' (including nursing), and things like X-rays. It will be much more for the last two-week hospital visit.
>
> Luckily, Yale employees have a comprehensive coverage, and I won't pay a penny, as far as I know. But someone like you could be clobbered if there were an unfortunate accident or illness.

As it turned out the bill for the actual operation was more than $8000. Almost all of it was covered by the Yale employee insurance, but, for some reason she could not fathom, Beatrice was called upon to pay something like $8. Her second hospital bill, excluding doctors' fees, was $4715, but this too was covered by insurance.

By this time she had a number of detailed questions from her family to answer:

> It all looks pretty ugly still. Today I gave away all my dresses and skirts that don't cover the knee, because I'll always have a rather unsightly gash just above the left knee, although at least it will slowly gain a normal colour as the fat layers grow in. Wanting fat is an unusual experience! I've several pounds of weight to regain too

so I'm trying to eat all the right things.

This type of cancer, melanoma, is very mysterious. It's associated with places, Australia and New Zealand included, and with ethnic groups, with excess sunburn, and possibly other things, but the doctor told me it was a waste of time to wonder what might have caused a particular case. He says lung cancer is the only cancer whose cause they are nearly always sure of (smoking of course). I'm just lucky at the recent advances in the treatment of melanoma![9]

In spite of these reassurances, Edward arranged to visit her as soon as he could, arriving in early May. He found her not completely recovered from the operation, although she was determinedly cheerful. She was still able to drive her car for short distances only. Edward accompanied her to a small park that everyone hoped would be free of traffic noises, and where a television team recorded her voice for an astronomy documentary. Another time she drove her father up East Rock outside New Haven – scene of the long-ago murder whose newspaper account had so distressed her when she found her children reading it. There father and daughter had a superb view while they sat and talked among beautiful kalmia bushes for a long time. No doubt Beatrice did not want to linger on the subject of her cancer. In any case her father later recalled that she mainly talked about her children and her students and her concern about their differing situations. She spoke about the students almost as if they, too, were her children.

A visit she greatly enjoyed was that of the eminent astronomer Bill Morgan,[10] who had come from the University of Chicago to receive an honorary degree from Yale. Beatrice was among those on the special committee who had recommended this honour. 'He's an incredible person – treats science with the passion of an artist and life (at 72) with the intensity of a poet,' Beatrice wrote home. She saw how genuinely moved he was by both the degree and the citation. Bill Morgan had heard about her illness and gave 'quite a lecture' on surviving by the will to live. He told her he himself had been suicidal some 25 years earlier, but was very well out of it now.[11]

Rowena was the member of the Hill family who was most interested in body-mind relationships, the effect of mental or emotional stress on physical processes. Beatrice hitherto had tended to dismiss such matters as unproven. Now she became more open to the possibility, eventually telling Rowena she believed it was the trauma of leaving her children that had triggered the melanoma.

For the present, the operation had either cured her of her cancer – or it was in remission. She looked squarely at the chances, and packed as much as possible into each day. There was absolutely no point, she had decided, in behaving like an invalid or thinking of herself as one. Thinking-time needed to be devoted to science. This put everything else into perspective.

She habitually left her office door open, which encouraged diffident students to go in and talk. This office was across the hall from the room where Nelson Caldwell worked, and he could not help noticing the chunks of her time that she gave to students. He observed that one in particular spent many hours with her, talking over his difficulties and problems while she listened as an understanding friend. 'She genuinely wanted her students to succeed.'

Not just visitors cut into her time. Members of the public would often phone the department with questions or to discuss astronomical issues of the day. The department's secretary would pass on the call at random to whoever happened to be in at the time. But word must have got around; callers began asking for 'the woman astronomer'. Gary Steigman decided this must have been because 'Beatrice was more patient with them than most of the rest of us were.'[12] This patience in her dealings with almost everyone, her students in particular, pointed up the testy impatience she showed rather often in her interaction with her children, something that she did not realise.

Her illness and her father's visit had stirred up so many memories that she decided to visit New Zealand. She knew it could be for the last time. December would be possible, with an astronomy meeting in Wellington preceded by one in Australia. She had been invited to speak at both. In Sydney she looked forward to staying with her old friend Pip Jackson, now a professional cellist. She would be able to combine the two visits, including seeing Theodora and family in Invercargill, as she wanted to catch up with her younger sister again, and get to know her husband, David Lee-Smith. Then there was her new young niece, Charlotte, who was said to look very like Beatrice as a child. (Their second child was named Beatrice).

Beatrice had no intention of curtailing any of her usual travel to astronomy meetings, but realised she would still have to rely on friends for help. Richard drove her to a week-long meeting at the University of Maryland. More than 200 astronomers from as far away as Australia made for an interesting but exhausting event, with perhaps too much crammed into the occasion – 100 papers. But she delighted in being able to talk to so many of those attending, such as two from Estonia, the only Soviet astronomers who were able to come. One of them brought her gifts, and an amber brooch from a Russian astronomer who greatly regretted not being able to meet her again. Beatrice decided to let this Estonian choose records as her gifts to them both. She was amused to watch him choosing the Beatles and Bob Dylan for himself and a classical record for the Russian astronomer. He explained the difference in tastes as 'the difference between first- and second-generation intellectuals'.

She estimated she knew perhaps a quarter of those at the conference. Although she gave a talk she found she had to sit and allow people to come to her, instead of her usual routine of moving rapidly from one group to another to get maximum coverage. She was able to walk a little, however. After one lunch she escaped outside with Vera Rubin. 'Beatrice told me in detail of her plans for her children, and for the rest of her life. She already expected it to be short.'[13] The children's education, and what she saw as their needs, were often on her mind.

When they returned to Yale, Beatrice found that she had been officially promoted to a full professorship with tenure, effective from 1 July 1978. The committee had been considering the matter when Pierre Demarque, who was senior, said that Beatrice's standing as an associate professor was such that they should not consider just giving her tenure. She should be made a full professor with tenure. The others quickly agreed.

Busy though she was, she planned to take a couple of weeks off while the children had their usual summer visit with her. It went well. More and more clearly, as she wrote home, she saw the difference in their personalities although they got on with each other most of the time. Alan, reflective and with a wry humour, enjoyed quietly reading. Terry still had problems settling to any activity for long. Now 10 years old, she spent the morning of the big 4 July holiday by darting from one place to another, cutting out pictures, drawing and colouring in, and helping Beatrice cook, 'which I do a few times a year'. Beatrice nevertheless considered Terry noticeably older and calmer. Brian and Yvon were having a baby, and both children were excited about this event.[14]

Because of her father's earlier concern about what he saw as Terry's hyperactive behaviour, Beatrice wrote to him after the children had left, saying firmly that they had all had a great time, and that Terry was much calmer. Both were fun to be with as they were so witty and quick to pick up ideas. She and Richard had taken them for a holiday, driving as far as Canada.

After they left she tried to fit in a great deal of catching-up work, as well as preparing for a meeting at Santa Cruz for a full week. In the plane on 23 July 1978 she wrote a long letter to Rowena about what she said had been an entirely happy time:

> Richard loves the children enormously, and vice versa; they were both in tears saying goodbye to him, and Terry introduced him to a stranger on the trip as her 'sort of father'. I am increasingly devoted

to Richard! He keeps showing (unpredicted) marvellous qualities, and I can't help making contrasts with Brian, who can still be very cutting and abrasive, on the phone even.

Musing about their happiness on this 'family' holiday, she decided it was partly because the children were older and in great shape. Terry's problems were subsiding, and Yvon most probably played a part in this, she wrote hopefully. Mainly, though, it was because she herself had had no tensions with Richard or conflicts over what they wanted to do. Their relationship was very important to her, she told Rowena, although they both still preferred to live pretty much alone. 'Richard has taken some of my more unbearable characteristics, like worrying and nagging the children, and managed to make them very much better – to my great relief – without me feeling nagged or annoying to him. In fact I simply find him nice.'

She did not expand on these 'unbearable characteristics', or indicate what the others might be. As Rowena had much earlier observed,[15] her sister was not introspective. 'Nagging the children' could be translated as shouting at them, expecting instant obedience. 'Worrying' might well refer to being over-anxious that they should always do the right thing, or the right thing as she conceived it; her expectations were so high. If she had really listened to Richard's mild rebukes when she sweepingly condemned whole groups of people – politicians were crooks, businessmen were crooks, and so forth – she might possibly have realised that she had taken over her parents' habit of thinking and talking in what they called Absolutes, while applying them in her own way.

In this same letter, written as she flew across the continent, Beatrice spoke openly and directly to her sister, beginning with conciliatory remarks about one of Rowena's interests that she had previously scorned, mysticism:

> I have just finished reading the book you so much recommended, the *Tao of Physics*, and I agree it's a great book. Among many other things it showed me what should have been obvious – that scientists shouldn't disparage mystical ways of thought, or vice versa. Of course the science that I do is far from fundamental particles, dealing instead with very microscopic things that behave mechanistically; and the religion all around is very unlike Eastern mysticism! It [the book] possibly explains how anti-scientific Daddy seems to be. He sees science and religion as enemies, it seems to me. 'Intellectual arrogance' etc. You know.
>
> I've often thought before about life and other parts of physics; in particular, which isn't a parallel but an application, that such a

highly organised system as a human being is much too 'improbable' (thermodynamically) to survive, and must return in time to inorganic dust.

Cancer seems to be a way of entropy catching up. Death in general is inevitable, just as is the dissipation of spiral arms, and other beautiful shapes in galaxies. All the same, it is hard to retain for long a feeling of cosmic unity and acceptance of my eventual decay; the reality of being a conscious individual, and wanting to remain one, is just as strong! I am curious to know whether immersing yourself in Taoism changes that.

Consciousness has always been the ultimate mystery to me, since I remember having a philosophical thought. It still hits me from time to time as a totally unsolved problem, and I can't retain a state of thought in which it seems like a false problem – ie. in which being Me Thinking isn't the central aspect of reality, and an absurd one. I can accept the fact that I will stop being conscious one day, but I keep wishing I understood how I arrived in the first place.

Funnily, my illness has sharpened both senses: awareness of life as I perceive it (and valuing it more), and awareness that death and returning to dust are natural and in harmony with the universe.

Rowena in Venezuela received this letter unusually quickly. Perhaps equally unusually she replied at once, on 16 August, writing by hand and then typing her letter. The handwritten version has survived:

What I immediately wanted to answer in your letter is a question of a quite different sort of 'identity' – what you call 'the feeling of being me as a conscious being'. This is all very difficult to express but – it's clear from your way of putting the question that you equate consciousness with personal identity – which is a mistake.

It is certainly true that immersing oneself in Taoism gives Zen – and related – ideas provided one has reached a stage of experience and self-knowledge that allows one to receive them – changes one's whole vision and sensation of existence. When I try to visualise now what life is (and means, which is not a different question), I see a great dark sea with waves that rise and fall, and in the heights of the waves there is light ...

For her work at Santa Cruz in the last week of July, Beatrice stayed with the Fabers and their two small girls, Robin and Holly. She greatly enjoyed the whole experience. Their comfortable house was big enough to have both her and Sandy's father as guests at the same time, and a living-in housekeeper, the ultimate luxury. In what she told friends was 'the nicest way', she envied Sandy Faber her obviously intimate and fulfilled relationship with her husband, and the way they all lived.

At the end of the month she received a sad letter from Terry. Their dog, Rata, that they had all loved so much had died, apparently a casualty of the extreme heat in Texas, together with her age and chronic arthritis. Beatrice called the children to sympathise. As she told her family, Alan had said to her very philosophically, 'These things happen,' as though to comfort her. Beatrice did take comfort from what she saw as Alan's strength and growing maturity. When he wrote to her, he used the word 'fun' six times in four lines. She counted them.

The children had a new half-brother called Winston Walter. They were delighted and held the telephone receiver near him so that Beatrice could hear his cries when she phoned them. Brian would surely be very happy about this, she commented, and she hoped again that his feelings towards her would become more mellow.

In the new term she was not teaching a course but found herself busy with students and a raft of duties, whereas she really wanted to spend most of her time on research, as so many ideas were clamouring to be followed up. She had three PhD theses to supervise and two term projects as well as the non-scientific business of all the graduate students, whom she found returning to her with diverse queries and problems. But the beginning of the new term marked the last in a series of vaccinations she had been having so she was able to hope that she had seen the last of medical setbacks, at least for a while. There would still be regular X-rays and medical consultations for a year or more.

As 1978 was the 150th anniversary of Schubert's death, his music was being celebrated everywhere. Jill Knapp once heard Beatrice say categorically, 'Most of Schubert is better than most of Beethoven.' Schubert certainly had always been one of her favourite composers. She allowed herself Schubert concerts whenever she could, even though free evenings were usually devoted to research. Again it was maple-leaf season so she treated herself to another excursion to Vermont, which she recorded with a camera she had just bought. It was also for taking pictures of people, anyone but herself.

When the astronomer Sidney van den Bergh, whom she particularly respected, came visiting from Canada, she held a drinks party for him. One of the graduate students, Peter Stetson, has recalled how they were always most explicitly welcome, 'so we were there again, chomping down the free food and schmoozing with the great man.' Beatrice merely mentioned to her family that her small apartment was completely crammed before most of them went on to dinner at a restaurant. Peter Stetson was more explicit: 'Beatrice particularly liked Basil's Greek restaurant, which had great egg lemon soup, and she used to get enthusiastically involved in the Greek dancing.'[16] There was no dancing now, but Beatrice wrote that she had driven herself to the IBM Research Center in the countryside near New York, to give another talk. These talks were all eating into her time. 'Sometimes I think it's a wonder I do any research to give talks on.'

She had been dreaming of her return to New Zealand, feeling the need to be sustained by its mountains, lakes, bush and long green hills. Her family and the friends of her girlhood, too, had been crowding into her mind. First came the Canberra conference and a visit to Ken Freeman at Mt Stromlo, and then New Zealand. The IAU colloquium in December 1978 at the Carter Observatory in Wellington had her as a speaker. Frank Andrews, an astronomer who had been a younger student at Canterbury with her, sat next to her for part of the proceedings and has a vivid memory of her. 'She had a fierce interchange with another astronomer, tearing his logic and reasoning apart. "If you haven't thought about it six times, don't publish," she said.'[17]

She stayed with her father and Mattie, trying to make as much time as possible available so her father could talk with her. They drove out to a distant beach at Makara and Edward Hill took a photo of her against a background of rocks and Cook Strait. He was impressed by her promotion to a full professor. In *My Daughter Beatrice* he wrote: 'Her hair is blown about by the wind and only her glasses could suggest she was a distinguished professor.'

Then she flew south to Invercargill, where Theodora and David Lee-Smith had been teaching at the two local high schools. David had decided to become a librarian and was working as such at the Southland Community College while he studied for his New Zealand Library Certificate. When she visited David at this library she went straight to its medical section and looked up what she could find under melanoma. She talked to him about it in a matter-of-fact way, preferring not to say much to Theodora, the young mother. Beatrice did not put it into words, but David felt that her visit to them was to say goodbye.

She was back in New Haven before Christmas, for another medical check-up. In a special message to her father she asked whether he could lay hands on photographs of her as a child. Some New Haven doctors were interested in trying to discover whether people who were in danger of getting melanoma could be detected early. They had asked her for any early childhood pictures to see whether she had always had a mark at the spot where her melanoma developed. 'Have you any photos showing an inch or two above my left knee, on the outside? I have a photo when I'm aged about four, but it isn't good enough to say.' It turned out that none of the photographs showed anything. Beatrice observed that if she had to develop melanoma she could not have done so in a better place than Yale, though that was hardly a good reason for becoming a professor there.

The children were with her for Christmas and a round of entertainments. She was able to fly back with Terry to Texas as she was scheduled to speak at Houston – Alan had had to leave three days earlier because of the new term starting at his school – and noticed how easy Terry was to get on with when she was not being competitive or tied up in her tense moments. These were getting fewer, perhaps partly because of the very good dietary ideas Yvon had introduced, she thought. This time in Dallas she was pleased to be invited to spend a short time at their house and meet the new baby, now called Walt.

Early in 1979 she flew with Richard to London for just four days, to give papers at the Royal Society at a session chaired by Jim Gunn. Hers was on 'Detectability of Young Galaxies', and Richard's was 'Star Formation in Young Galaxies'. She wrote at length to Rowena on 4 April, particularly mentioning her pleasure at seeing Jim, 'quite out of his deep depression by now, and seeming like a real old friend'. She had returned to discover that the Armatino family was again in deep trouble. A criminal employer had got his clutches on the unfortunate family. Beatrice's pen almost tore through the paper in her anger as she detailed the injustices which had befallen them – some of the anger surely displaced from her own situation: 'This country doesn't treat its poor and unemployed in a dignified way at all.' She had lent the family more money, given them clothes and was still trying to find the father a job. Beatrice, so often channelling her own emotions through the trials of this family, did not see that they were not always the victims, or so some of her friends believed.

In her letters to Rowena, Beatrice was becoming rather more introspective. Reflecting on how well they had all got on together, and how she and Richard planned another week-long drive with the children in June, she nevertheless wondered how it would be if she had them all the time. Continuing her long letter of 4 April she wrote:

> Sometimes I miss the children a lot, but I don't know what resources I would have for them most of the time. Being chronically tired is a new frustration, and I can't tell if it's age, aftermath of the cancer, or – what I can't really face – possibly still coping with stray cancer cells trying to set in a tumour somewhere. I have to face the possibility, and not collapse if it happens, and, anyway, face the periodic trips

back to the hospital for examinations and tests. Perhaps I'm simply trying to pack more and more into each week as the demands of work get stronger.

Astronomy is wonderful fun, and I still have more ideas for research than there's time to work on. Some stimulating students, too ...

Jim has caused a big stir in US astronomy circles by deciding to move from Caltech to Princeton next year – quite a shift of power.

It was largely the doing of his great friend, (and only real peer intellectually, I think), Jerry Ostriker, at Princeton, who was once involved with me during the turmoil of splitting with Jim. Funny, they don't seem complicated to me now! Jerry was here last week for a meeting, and we all had a great time.

Daddy would call it intellectual snobbery, but super-intelligent people are the ones I enjoy by far the most, despite (or perhaps because of) all their eccentricities ...

Thank God this isn't last year! This morning, waking up, I dreamed of a beautiful blue sea that I was walking beside, and I laughed and said in the dream, 'I didn't die, did I?'

CHAPTER TWENTY

FACE TO FACE

'I shall die, but that is all I shall do for death.'
Edna St Vincent Millay

Just seventeen days after she had had her triumphant dream, Beatrice sat down to break her news to Rowena. She had had another of her medical examinations. A recurrence of the melanoma now seemed most likely. The previous week a small lump had appeared where her lymph glands had been removed: a common form of first reappearances. More X-rays and tests were to be done, and the lump was about to be removed and examined. Her doctors were predicting that they would then put her on long-term chemotherapy, with treatments every few weeks, and there was a strong chance of future tumours:

> I could live for many years, or for one, and they can't predict because patients vary so much, and the treatments are so new.
>
> Ever since the original tumour I've known there was a 50–50 chance of this happening, so in some sense I'm braced for it and I'm nowhere near as shocked as I was by the first news. All the same, it's a lot to adjust to.
>
> Richard is again an inestimable source of strength to me, and the poor thing has now a lot to face in his own life, too. We've spent hours and hours talking in the last few days, and I can't imagine how I would be feeling without such a person to sort it all out with.
>
> Already I can feel how a strong buoyant sense of the value and joy in life can completely overcome the disappointment – a genuine understanding that quality of life counts for more than its length.[1]

Such a ringing statement could perhaps be expected from a strong-minded philosopher in late middle age. But only three months earlier Beatrice had had her 38th birthday. Her doctors soon determined that the new lump was another recurrence of melanoma, as she wrote to Rowena:

> So it is probably though not certainly a systemic disease by now. The doctors intend to put me on chemotherapy probably monthly, and there I suppose I stay until the End.
>
> They say they've had patients on the new drug treatment for three years, and even though they get tumours from time to time (treated by radiation) they live and work normally in between. So I like to imagine I'm chronically ill, rather than doomed. All the same, it's hard not to see the whole of life in a different perspective.
>
> Poor Richard has taken this very hard. It stunned him all at once, whereas I was too dazed and the meaning is only gradually sinking in. He's a really marvellous friend and has kept me just as much as possible thinking of how to adjust positively, to make the most of every minute, and not to make life brooding.
>
> I will never know whether I have months or years to live (until they tell me it's months) but then I never did, but then I wasn't aware of mortality.[2]

Rowena responded by sending her some of her latest poems, written in Spanish but also translated. Beatrice replied at once: 'I can't get over your linguistic abilities – writing poetry in a language that isn't your own.' She gave a brief rundown of her current treatment: a number of drugs with annoying side-effects, two vaccinations each month which made her feel 'flu-ish' for a few days, and three nauseating intravenous infusions on three consecutive days each month. The treatments were tolerable as the time in between them could be lived almost normally. By now she had learned to accommodate to having less energy by cutting down on her walks, and getting more sleep.

Otherwise life was better than ever, she underlined, with her research steaming along. As for her deepening relationship with Richard, 'Whatever happens, neither of us will ever be the same again.'

Not for one moment did her standards slip where her students were concerned. No opportunity to teach should be missed. Nelson Caldwell was talking casually to Beatrice one day about a supernova in a galaxy he was to study later. He doubted that the supernova could have been a star that formed in the centre of the galaxy and then moved outwards to explode into the supernova. She quickly calculated, in her head, the time it would take to do that, and found it to be possible. Nelson was impressed by how quick she was, but, more important, it taught him the importance of 'really knowing' something, of doing one's homework rather than speaking without thinking.³

Beatrice, however, was always ready to put out a hypothesis. She did not hesitate to speak up. In Brenda Maddox's biography, *Rosalind Franklin, Dark Lady of DNA*, she shows that in Rosalind Franklin there were some striking similarities with the character, upbringing and experiences of Beatrice Hill Tinsley. There were also differences. The *Economist*, reviewing Maddox's biography, spoke of Rosalind Franklin as being 'the unacknowledged heroine of DNA, the Sylvia Plath of molecular biology'. Maddox tells how Rosalind from girlhood had been taught always to be entirely certain of all her facts before writing or speaking in public, and how this scrupulous restraint cost her a share of the 1962 Nobel Prize, or the prize itself, for the discovery of DNA in 1953.

Unlike Beatrice, Rosalind Franklin was not given to hypotheses, to jumps of imagination. She had not yet completed all her work when one of her colleagues, the New Zealand scientist Maurice Wilkins, showed a key Franklin photograph of the B form of DNA to James Watson and Francis Crick – doing so without her knowledge or consent. These two men added it to their own knowledge, and to that of Maurice Wilkins, to make the intuitive leap that led to the discovery of the double helix, DNA, the secret of life. The Nobel Prize is not awarded posthumously, and Rosalind Franklin had died of ovarian cancer in 1958 when she was 37. Nevertheless, not one of the three men mentioned her when they accepted the Nobel Prize in 1962.

Beatrice, like most of the rest of the world, was unaware of this work by Rosalind Franklin. Until comparatively recently, Franklin is said to have been airbrushed out of history. There are many parallels in the lives of these two women, but one of Beatrice's particular strengths lay in her ability to float hypotheses. She would consider, discard, and then seize other apparently contradictory pieces of the puzzle, and synthesise them into something nobody had thought of before. Richard Larson has remarked: 'Beatrice didn't really deal with "facts", in the experimental sense, but with logical arguments and mathematics. She was always supremely confident of her logic. Probably her logic was always correct, but logic isn't everything, even in science.'⁴

The time came at Yale for the student lecture. Junior graduates traditionally each presented a 30-minute paper at the end of their year while Beatrice and the rest of the class listened attentively. One student, Jim Schombert, launched confidently into his presentation on quasars, with focus on the variability of quasars caused by internal cloud motions. After a few minutes Beatrice stopped him, took over, and demolished his premise. There was nothing left for him to say, no way he could fill his allocated time. As he remembers it, the room was silent with embarrassment. But Beatrice knew her student. He was shaken but not embarrassed:

> I suspect she felt free to interrupt my talk since a lack of self-confidence was not one of my problems. For me it was a critical lesson in terms of knowing what material is proper, and how to ask yourself questions when researching a topic. I learned a lesson that has stayed with me forever.
>
> Her galaxies class was the most powerful and in-depth astronomy course I have ever taken. Tinsley was at the boundary of change in extra-galactic astronomy. Much of extra-galactic research before 1980 was morphological, that is, mostly about classification and categories. She was leading the transition into a more physics-based system, where physical parameters such as the mass, colour and luminosity of galaxies were more important than morphological information. Her galaxies course, for which I still have the notes, was the clearest outline of the revolution in our extra-galactic thinking. The class was full of the history of galaxy research as well as homework exercises that involved computer analysis of galaxy data – something new at that time. We no longer teach the history side of extra-galactic research, so I remember those classes with the fondness of an ageing astronomer.[5]

At the end of that same year it happened that Jim Schombert was to be one of a bunch of students stuck in town over the holidays. Beatrice gathered him up for Christmas dinner, along with close friends. He was appreciative but found her apartment very different from his parents' in that it was singularly lacking in books and pictures, and the talk, which seemed mostly about music, went past him:

> I think she was just not a people person, and perhaps didn't relate well to my generation. I was young – and the position of instructor and researcher makes it hard just to sit down and have friendly conversation with students. Besides, I was a typical young white male of the time, more interested in basketball scores than classical music.
>
> As director of grad studies for astronomy, she always said of my work that I was not reaching my maximum potential, which was frustrating since I believed (and still believe) that I was working at 110 percent of my skill level. To this day I can't figure out what she saw in my work, or whether this was a standard line to encourage more work from grad students. We were frequently at odds about

my work ethic. I remember one occasion when she told me I was spending too much time with my wife, which I thought was way out of bounds for a prof. On the other hand, we did later divorce ...

Tinsley was a great researcher and had a keen mind. However, she was in a field that was ripe for sharp changes. You often hear extra-galactic researchers today lamenting her days when 'all the good problems' were being solved. It was hard not to do ground-breaking work at that time since there was such a surge of new information about galaxies. But she was the person with a strong work ethic and the right mental toolbox at just the right time in extra-galactic astronomy. Her discoveries were key to what we know today.[6]

But back to the time of that memorable student lecture that did not come off. The students departed for the summer vacation which for most of them meant more work. The feeling seemed to be that Beatrice saw considerable possibilities in Schombert, and wanted to spur him on, although at that time it seemed he wanted to enjoy a normal social life and standard of living while a graduate student. 'It doesn't work that way,' one remarked darkly. 'Grad student days are not fun.'

Nelson Caldwell went along with this. It was hard grind, and always with the knowledge that one might not make it after all. Luck, good or bad, as ever had a place in the scheme of things. He and Gus Oemler had earlier worked on a project as part of the course requirement. It was finished, but not really in a publishable form. Meanwhile Beatrice and Richard had begun discussing a different scientific problem. When Beatrice mentioned it to Nelson, he commented on the similarities. It was apparent that Nelson's work, or an extension of it, could help to answer some of the questions that were arising. That was enough for Beatrice. 'Come and join us,' she said. So they combined in the project on the evolution of S0 galaxies in galaxy clusters.[7]

Nelson Caldwell has said: 'The resulting paper[8] became very influential. It gave me recognition that helped me in my subsequent career. And it was all because of Beatrice.'

With the students away, Beatrice wrote home that she had been having a great time talking with people in her field who had called in at Yale. Discussions had been most useful as well as enjoyable. Good, Edward Hill must have thought as he read, relieved. But then Beatrice carried on in much the same tone:

> I saw the doctors yesterday to find out about the cancer. As I had been led to expect, the lump they removed was a melanoma tumour,

which means that last year's operation didn't cure it. (It was only a bare 50 percent chance, anyway.)

There is now quite a strong chance of further recurrences so I'm starting on drug treatments that mean several visits to the hospital's cancer clinic every month. I expect to feel unwell for a few days, but all their two dozen patients on this regimen are working normally.

These doctors are involved in an intensive research programme on the treatment of melanoma, and I've been told (by outsiders) that it's the best place in the world to be. How ironic to be at Yale for this. Anyway, I have the utmost confidence in their skills and care, and in their willingness to discuss every aspect of the treatment and prognosis with me.

You needn't worry. My friends here are simply marvellous.'[9]

All in all, Beatrice felt she could still meet her own standards as a teacher because the drug treatments now under way were reasonable. One made her slightly feverish for a day twice each month, but the fever could be fixed by taking aspirin, and other treatment made her sick for a few hours, but only once a month. She could put up with this, knowing it would end in time, and always took astronomical reading matter with her.

Even though they were soon to stay with her, the children phoned on Mother's Day. They had given Yvon a present and said they were enjoying having two mothers. 'Yvon and Brian must be doing very well. No wonder I'm happy about that set-up,' Beatrice wrote home a week later. This was a rosier view than the facts warranted. It is hard to know on what she was basing her judgments of the Dallas ménage. Wishful thinking must have been part of it. It also seems possible that the children wanted to spare her their view of the situation. Almost certainly she would not have picked up on what lay behind the things they were actually saying. Alan's recollection of those years gives a different picture:

After the divorce we lived hand to mouth, with different people looking after us all the times when Dad was away – an explosive situation. I became introverted and a loner. Terry became too extroverted. Every emotion seemed out in the open with her. It was the opposite with me.

Our family supported a whole counselling industry. It became a game. I always argued with the counsellors: 'It won't work, because –' Terry always agreed with them but that didn't work either. It became worse when Yvon arrived with her daughter. Terry was displaced. At first she was ingratiating, then she ran amok. I was good, she was bad. It was awful. Then I withdrew when battles raged. Dad didn't seem to see what was going on. One good thing – when Yvon came, Terry and I became allies again.[10]

Although Beatrice's old energies had diminished, not for her the expediency of writing round-robin letters. Besides her research, she was still producing a prodigious number of mostly hand-written words each month. She sent notes and full-scale letters to her family and far-flung friends, to the big range of astronomers with whom she was so frequently in contact, to her many students past and present, and to people whose work in astronomy journals in some way caught her attention. Her usual mail had swollen, too, with anxious letters of enquiry and good wishes. These sometimes included accounts of cancer 'cures'.

Her stepmother sent her one such, about laetrile, said to have cured some people. Beatrice was forthright. 'Thanks, but nothing would induce me to take it. The doses people talk about lead to cyanide poisoning!' Her doctors looked into all the unorthodox treatments, but were convinced that laetrile could do more harm than good, and there was no statistical evidence for it. Occasionally there were spectacular cases of cancer cures when people were taking various substances, but that also happened to people on no medication. Beatrice added: 'It would be like saying that the owner of a collie dog had a sudden disappearance of cancer, so collies cure cancer. There is no evidence for a cause-and-effect relation. Cancer patients get so desperate they fall prey to quack treatments.'[11]

When insisting that her doctors be completely frank with her, Beatrice told them about the crowded schedule she wanted to maintain if possible, and the great importance to her of the new ideas and stimulation she received from her far-flung astronomy meetings. The doctors did what they could to schedule all her treatments so that her summer plans were very little affected. To her family she said merely, 'It's easy enough for me to escape from the doctor and nurses for a week because I go in normally once one week, three successive days the next, once the next week, and not at all the fourth. This goes on until another recurrence (or unless), which would be much more serious.'

After the children's summer holiday with her, she flew off for three days in the French Alps to a summer school at Les Houches, near Chamonix. There she gave a talk, 'Cosmology and Galactic Evolution'.[12] Jean Audouze observed her with deep admiration as well as sorrow:

> Her mind was still at the same level of intelligence, and her talk was as brilliant as all in her brief but splendid career. But she got tired very quickly and all of us were very sad to see her struggling against

a cancer evolving so tragically. I believe she was nevertheless happy to be with us. We considered her to be as active and sharp mentally as ever.[13]

Back in New Haven, Beatrice settled to begin a long letter to Rowena, in part describing the hectic schedule she had just completed but mainly talking about what was most on her mind, her research and her relationship with Richard.[14] The main project was the research which had been 'hanging over' for seven years. This was the work which she had begun with Jim Gunn in 1972 at Caltech. She explained that it had stalled in 1975 partly for scientific reasons and partly because he was in deep depression and could not settle to reduce the data. Finally, however, Linda Stryker had been able to spend the summer at Caltech, getting the data into usable form for the mainframe computer. Now that she was working on it with Beatrice, with Jim visiting to collaborate, 'it really will be the important piece of work it was intended to be.'[15]

Linda had earlier put in some five weeks at Caltech, on part of Beatrice's and Jim's project. This was still the era of computer punch cards, and Linda worked day and night to get the boxes of cards ready. Occasionally she had to confer with Jim, who was busy with other projects but who would always stop and clarify matters for her. Now at Yale the three scientists put in good work on galaxy population synthesis. Linda was appreciative that they listed the authors alphabetically, so it became the Gunn, Stryker, Tinsley paper.[16] The way Beatrice and Jim worked together always impressed anyone who observed them. As Linda said:

> They worked very hard all day, constantly evaluating and discussing and trying out different things, assessing them when output came from the computers, and seeming to spark off each other.
>
> I was made to feel I worked with them, not for them. If I said I'd go and clean up the mess, Beatrice would say, 'We'll all go.' These two towering intellects never made me feel I was just the grad student.[17]

Beatrice and Richard had gone to different summer conferences, and they had been apart for nearly four weeks. She told Rowena:

> I miss him a lot but the extra solitude is valuable too; for just a few weeks I get a kick out of realising all the time in the evenings is 100 percent my own, and I don't have to consider what anyone else wants to eat! But in general the way we live minimises the necessary burdens of having a close relationship.
>
> I resent having grown up believing that the only 'responsible' and

respected way of having a lover was to be married!' [18]

At the same time as this letter, she wrote a quite different version of her summer for her New Zealand family, detailing concerts and meetings and her work as her department's director of graduate studies.

Beatrice had to miss part of the big IAU conference in Montreal in August because of sickness from chemotherapy, but found a spare hour away from proceedings so she could write to Rowena again. Her sister of all people was the one who most understood how things were with her, she felt. It was a relief just to sit quietly and write after almost non-stop talking and listening for a week, and this, of course, had come on top of the non-stop work and busy-ness with Jim at Yale. That had been a very successful working visit, she told Rowena, in spite of the problems her chemotherapy had caused:

> Jim is an extraordinary person. I easily understand why I was so in love with him, but equally easily see why it didn't last!
>
> He and Richard are great friends (were students together), and watching them talking and being such different people, I realise how much peace and stability I owe to Richard.[19]

At a meeting such as this Montreal IAU she continually ran into old friends and acquaintances, with all manner of personal and professional connections, Beatrice continued. Even while writing this aerogramme she had been interrupted three times in her supposedly hidden-away corner. 'The number of people I know is huge!' So many kind enquiries were emotionally exhausting – word had certainly spread, and it seemed everyone was now intent on asking her about her health. It was essential for her to get away periodically and be on her own, or just with Richard. And then she paid him the ultimate compliment, knowing that Rowena – and Richard too for that matter – would laugh and see it as the compliment it was: 'Richard is usually as restful as nobody.'

More than most women, Beatrice was able to cope with one of the distressing effects of chemotherapy, her hair beginning to fall out. Her determination to eschew make-up and 'feminine fripperies' had left her (at least on the surface) able to laugh at herself. Altogether she felt well enough to be host to one of her English cousins, Matthew Hill, a law student who was visiting New York, and who brought two of his friends to New Haven to see her. He found her at home, between treatments, working from a reclining chair with the floor of her room covered with reams of computer print-out. She was philosophical and matter of fact about her cancer but very thin, and eating carefully, just a salad.

Beatrice was able to organise a walking tour of Yale for the group, and drove them to the British Art Center. With her cousin, she mostly talked music. When he tried to enter her territory by speaking of the great excitement people were exhibiting over the moon rock in the Smithsonian Museum of Air and Space, she was at her most definite, and said tartly, 'I'm not interested in flying geology.' [20]

At the beginning of the 1979 September term Beatrice was involved both with the graduate students, new and old, and with advising a few freshmen. People observed that she seemed more engrossed than ever in all her students, at all levels. Faculty member Gary Da Costa, who had been a student of Ken Freeman's in Australia, noted how approachable she was:

> You could take her a draft of a paper. You can't approach many people like this, anywhere. She'd say, 'Yes, yes, but have you thought about this? Or that?' She would be very precise, and cut right to the centre of everything.[21]

Nelson Caldwell had a number of interactions with Beatrice over his thesis, 'Star Formation in Early Type Galaxies'.[22] It had a chequered history. He had changed its concept many times in spite of Yale's requirement that graduate students submit a prospectus of their thesis within about two months of beginning work on it. Beatrice had badgered him constantly about this, but over some two years he had continued to make changes in its scope:

> When I finally typed it up, I marched into her office and presented it to her proudly.
> She was glowing with happiness – until she read the date I had put on it.
> 'Shit!' she said.
> I had pre-dated it by two years to make it seem on time.
> This was the only time I ever heard her use language like this, and it was done in laughter.[23]

Beatrice had instituted what she hoped would become a tradition, writing a letter to each graduate student at the end of the year to tell them how they were coming along. Those not doing very well were told so, and what they might do to improve, or if they should look for another career. In Linda's case she was told she should learn more physics. Curt was told he should talk more with Richard. Otherwise her letters kept encouraging Curt to work on and get done. He was there for five years, whereas Beatrice's ideal time for a student was between three and four years.

As a teacher Beatrice was more general than specific in her teaching method. Nelson decided she had the right idea: talk about the interesting points because the small details could put off a lot of students, and in any case they should go into them on their own if they were interested in the subject. Beatrice always regretted not having as strong a background in physics as, say, Jim Gunn and Jerry Ostriker, but Nelson decided that this in one way was a strong point:

> It was sort of like the story people told of Einstein. He thought he needed more mathematical rigour and so went off to do more study for a number of years, but then never did any more great physics.
>
> Beatrice wasn't burdened with the detailed analysis that a physics background would allow, so she could explore the general questions that astronomy is all about.[24]

A Yale tradition which greatly pleased Beatrice was that every year the department made funds available so that the graduate students could invite a distinguished scientist, not necessarily an astronomer, to a colloquium. The understanding was that this guest would come to Yale for two or three days, not just arrive and give one talk and then take off. The students took this opportunity seriously, debating among themselves and nominating their preferences. One year they had tried to get Allan Sandage, but this kind of student-organised affair was not for him.

For the colloquium in the autumn of 1979 the students invited the distinguished physicist Philip Morrison, from MIT. The faculty naturally made up part of the audience for these talks, but it was understood that they were organised mainly by the students, and for them. They were well aware, however, that the stream of distinguished astronomers who visited Yale came primarily because of Beatrice. Even if they were not giving a formal talk they liked to check in with her, and whenever she could she put on a special reception at her apartment, and invited students. The number and variety of people she knew, and who liked to come and visit her, the students thought extraordinary – people such as Alar Toomre, Martin Rees, Vera Rubin and Arno Penzias. As Beatrice intended, these parties enabled students to meet and listen to the great, and begin to find out what they were really like.

Linda Stryker went off to work on her thesis at Kitt Peak, first arranging to be told if Beatrice's condition became worse, and insisting that in any case she planned to return at intervals. She had found an excellent pianist, Leona Francombe, to join the chamber music group while she was away.

Beatrice was told that the frequency of her drug treatments was about to go down, a most welcome let-up as she had a new graduate course to prepare. Her days were becoming easier.

Then began a dramatic change in her life. It was heralded, suddenly and unexpectedly, by Terry's arrival from Texas to spend a long weekend with her mother. There had been a domestic explosion in Dallas. Brian's wife Yvon had said, 'It's her or me.'

The visit was ostensibly to see how things could work out in New Haven, although a weekend, out of the blue, was scarcely a reasonable trial. Beatrice gave no hint of any turmoil in her letter home.[25] Writing carefully, she downplayed this event by talking about almost anything except Terry, saying the plan was for Alan to do the same the next term, by himself. Terry's arrival happened to coincide with a visit from her mother's brother Jocelyn Morton and his wife Kate. Beatrice was able to laugh at herself by saying that, in honour of her relatives' visit, she spent what seemed to be a whole evening shopping, cooking and cleaning the house. As she reported to her family, one of her friends had said, 'It is just as well that relatives visit sometimes.'

In a phone call Alan had told her he was enjoying school and liking science and mathematics best. Maths was also Terry's favourite subject, but she did not concentrate enough to do nearly as well as she could. 'I hope that changes,' Beatrice wrote. 'I recall being very mediocre at everything until I went to high school.'

Beatrice could always find a subject to write about other than her main concerns – if they were not suitable for family consumption. Now she wrote on about what her sisters were doing. The letter shows her carefully temporising: how much shall I tell them about how things really are over here? The way she described the situation, next time she wrote home, put a very different gloss on what was actually happening.

She and Brian, she wrote,[26] had decided that Terry was to leave the family in Dallas and come and live with her for at least six months. Beatrice emphasised that she was 'very thrilled' because the two of them got on extremely well together. She worried about how big an upheaval it would be for Terry, particularly starting at another school, but felt pretty sure it would be a good thing in the end. She seemed to be doing quite well at her school in Texas, including getting an 'excellent' for behaviour and co-operation. She and Terry had talked over the move at length during that weekend visit, and Terry had sounded very pleased, with one proviso: she had to be able to bring her hamster, named Empress Brownie.

Beatrice had been writing too optimistically about the Dallas set-up. The blunt fact which she now faced was that Terry and her stepmother did not get along well together. Terry's non-stop exuberance and great need for affection – at times she was like an extra-playful puppy – did not sit at all easily with Yvon. It turned out that, in Terry's affectionate rough-and-tumble play with her young half-brother, his arm had been broken. Yvon had become hysterical, Brian distraught, and the upshot was his sending Terry to live with her. All this Beatrice had of course talked over with Richard, going into every detail they could imagine. Richard would back her in every way.

The main question for Beatrice was whether her health would be reliable enough to enable her to care for Terry. She was about to start a round of tests again. Her doctor knew what she was contemplating doing. They talked about the worst, and the best, courses of the disease, and he told her it was most unlikely that she would not be able to be responsible for an 11-year-old until the following June. He did, however, advise against making a final decision until he had seen her X-rays and other test results.

She had explained all this as best she could to Terry during the weekend visit because obviously her situation could change at very short notice. Terry appeared to take it all in her stride, Beatrice wrote. The child had responded from the beginning to Richard. Nor did she show any obvious signs of feeling rejected by her father and his wife. Brian was flying with his new family and Alan to see his relatives in New Zealand, but Terry seemed not in the least disappointed to be left out of this adventure. All this must have sounded reassuring. Richard, quietly observing, thought that Terry had most certainly been affected, however. To him, she did not seem her usual cheery self but was more tense and insecure than before.

Beatrice plunged into looking over possible schools and wondering whether her present apartment could be rearranged for Terry, or whether she should try to move to a bigger one, although New Haven had an extreme shortage of suitable apartments. She decided to wait and see.

The new regime began happily. She contrived a small bedroom for Terry by curtaining off part of the dining area. As she had to give a talk in California during the week of Thanksgiving, Beatrice arranged to travel back through Dallas to collect Terry and hamster and bring them home with her. She had discovered the Cheshire Academy, a small private school that seemed to have a much better atmosphere and better teaching than anything else she had seen, but Terry had to be interviewed first.

Beatrice then wrote carefully to her family, saying that she had been increasingly wanting to have Terry come and live with her, sensing that things were not the best for her in Dallas:

> Now Brian is very keen for me to have her. Yvon doesn't seem able to cope with her extroverted and rather noisy nature. But Brian

himself finds her well-balanced, helpful and pleasant company in Yvon's absence, and that is what she nearly always is when she stays here. (The difficult moments seem to be rivalry with Alan, but it isn't beyond normal childish behaviour.)

Terry definitely needs to be moved from her present situation, and thank goodness I can take her. I hope this interlude will be a positive thing in her life. I have friends and their children who are great friends of Terry's already, and 100% guarantees of help when I'm sick from medicines and so on – very supportive people around.[27]

When Terry arrived she was tested and observed at the Cheshire Academy. It was about 10 miles out in the country, and had only seven children in the sixth grade which Terry would enter. From the child's point of view the day was a great success. Beatrice could only take a deep breath and hope for the best. When Terry was accepted, everyone rejoiced.

But the child's home surroundings, such as that curtained-off space in the dining area, were still distinctly makeshift. Virginia Trimble observed, when she arrived at Yale with Martin Rees who was giving a colloquium, that Terry was sleeping on a mattress on the floor and the apartment in general was not well furnished. Beatrice had no privacy.

Virginia was perplexed by what she saw as Beatrice's deliberate downplaying of her own attractiveness. She was now wearing glasses but had chosen frames that seemed far too heavy for her face. The two women came together infrequently, but – both outspoken – had fallen into the habit of talking frankly about intimate matters. When they had first met, Virginia thought Beatrice looked stunning with her hair hanging loose and framing her animated face. 'Then, after she left her husband, all that changed. She decided she'd been playing a role, and part of it was her appearance, the way she had been making herself look.'[28]

As Richard Larson explained, 'She often said she felt she had been living a life, a specific social life in a specific context, laid out for her by her parents, and that she'd been suckered into marriage.' She had made the decision to eschew hair styling and attractive clothes. Appearance didn't matter. She wanted to be her unvarnished self, and this even extended to no make-up, with her hair tightly pulled back. 'There were to be no more roles.'[29]

On one occasion, Virginia – 'and I was a bit afraid of her' – had confronted Beatrice about this and asked her why she tied her hair back and now never let it hang loose when it looked so marvellous that way. Beatrice had replied with another question: 'Do you know the average annual wind speed in Wellington, New Zealand, is 40 miles an hour?' This was a specious rejoinder, to say the least, considering she had been in

Wellington only once, briefly in 1978, since she ended her marriage, but it was a reply that Virginia never forgot.

Now, in New Haven, and wondering if this might be the last time she would see Beatrice, Virginia could only silently marvel at their differences, and wish her well.[30]

Beatrice was happy when Alan came up to join them for a few days. The visit was arranged because he would be unable to have his usual Thanksgiving holiday with her, as that was the time when the family in Dallas planned to leave for New Zealand. She managed to have times alone with him, trying to think of things to say which he could hold fast to in the years to come.

It was as well Alan came when he did. Almost immediately afterwards, Beatrice had to go into hospital again. Again cancerous lymph nodes were diagnosed – still very localised, however. She was put on to a slightly different treatment which included a drug less nauseating than the one she had been previously prescribed. Many of her students came to hospital to visit her and they and older friends reassured her about Terry's care.

Back in her apartment, she had recovered enough by mid-December to be delighted to meet up with her old friends and mentors Wal and Glen Metcalf, who were travelling the world and were staying in Providence. She took them out for an early dinner together with Terry, and they talked hard, although Glen remembers that Terry seemed unsettled and rejected the meal that had been ordered for her. Details of their short visit remain vivid:

> Terry's neatly foil-wrapped meal, all ready to go in the oven to be heated up, was given to Beetle as we left the restaurant. Doggy bags were a novel concept to a Kiwi then, and this was an elite specimen of the genre. Beetle was obviously concerned that Terry should feel welcome and feel she belonged with her, even though the apartment had no second bedroom.
>
> When we got back to the apartment, Richard was there with his cello. We played for an hour or so, Mozart piano quartets. It was wonderful, especially for me who had had no access to a piano for a considerable time. And it was magical to hear Beetle again in full flight. She was radiant that night, as beautiful as I had ever seen her – eyes sparkling, cheeks bright, all bounce and vivacity. I realised too, later that evening, that Beetle loved Richard and that Terry's presence meant that she was sacrificing needed privacy not just for thinking and working but also for the expression of that love.
>
> How could I believe her when she walked with me back to the train for Providence and told me she was dying?

What she prayed for, she said, was that she'd live long enough to meet Terry's needs. She also told me of her love for Richard. And then the men joined us, and she said no more.[31]

To her family Beatrice said, 'We played quartets together, very reminiscent of old times. The Metcalfs are such lovely people.' As for Terry, she wrote, she was in heaven because she had discovered ice skating at the Yale rink. This was also a favourite sport for Pierre Demarque's children, and they would all skate together. Terry seemed to be coping with regular homework, which was reassuring; the child was doing far more advanced and self-motivated things than she herself had done at the same age, she declared. No intimations of death tinged her letters home. Beatrice continued to look determinedly on the bright side of everything for her family, including Empress Brownie: 'The hamster's adorable. Most inspiringly energetic and happy.'[32]

Just before Christmas Beatrice felt well enough to have the whole Demarque family for lunch, after which the children all went skating. Terry also enjoyed the department's annual pre-Christmas party when the students did what Beatrice considered one of their best-yet skits. For 1979 it was a take-off of *Star Trek* with all the faculty somehow involved and with what she described as some hilariously good impersonations. 'I still giggle at the thought.'[33]

More than one student has admitted to having clearer recollections of these revues than of some of the classes they took. They were an important part of the Astronomy Department's life at that time, and were looked forward to all year. Because of the close, even affectionate, relationship between faculty and students, something that Beatrice had largely brought about, the students were emboldened to do and say things that some other departments considered outrageous. The 1978 revue had been fun but had disappointed because it did not include the usual impersonations of faculty members. As Nelson Caldwell said, 'Even the faculty complained. It seemed they didn't mind being satirised as long as they could get a few laughs at the expense of their colleagues. So in 1979 we threw in as many personal insults and exaggerated mannerisms as we could think up.'

Beatrice was Colonel Tinbeam, tinbeam being her computer name, and Bill van Altena was Buck Bill. A typical scene aboard the Star Ship Enterprise had Colonel Tinbeam, played by the post-graduate student Angeles Diaz, speaking in very rapid Spanish, with Buck Bill saying, 'I couldn't understand that, but then I never could understand Colonel Tinbeam.' This brought appreciative laughter.

There was also a running joke about the cosmological parameter called q_0, pronounced 'q naught', the amount of deceleration the universe is experiencing, something that crops up in some of Beatrice's papers. The earlier triumphant skit, 'Gone with the Stellar Wind', had a line that went, "Well, why not? Or should I say q naught!" This became a favourite, inserted into every revue, although not everyone got the point. On this occasion came the line, "Well, why not?" and from the audience came Beatrice's voice, '*Or should I say, q naught!*' As Nelson Caldwell noted, the point was made at last. Everyone laughed.[34]

Beatrice insisted to her family that she felt in normal health again, and that the new chemotherapy was relatively painless: two pills at bedtime with a little nausea in the morning. Her latest surgical scar was healing very well and her stitches had been taken out. With Terry she had been able to do a lot of walking around doing Christmas shopping, and just talking together. Years later Terry was to remember this time vividly:

> Something changed in Mom after the divorce, and specially after she got sick. She reached out to me more. I'd been living in a silent world. We both needed to talk.
>
> My school had sent home a note saying we and our parents should watch a TV special on sex and drugs. Mom and I talked about it first, then we watched, and then discussed it. It was really neat we could talk. She was much more open and ready to talk than Brian was. She talked to me about walking by yourself, and walking together with someone, and pointed out that she and Richard had independent lives but they also walked together. Even as a little kid I could see they meshed constructively.[35]

On the morning of Christmas Day, before Richard arrived for lunch and they opened their presents, Beatrice sat down to listen to Bach's B Minor Mass while writing to Rowena in one of her rapid-fire letters. She began by declaring that Terry, who had woken her early with her stuffed stocking, was settling in beautifully, and that they were nearly always very happy together. Her teachers seemed marvellous. They expected children to behave well and work hard, in an old-fashioned way, but the lessons were most creative and interesting. It was a long day for Terry, with the bus at 7.15am and home about 5.30pm, but she was managing:

> She can be very difficult, but nothing in her behaviour is abnormal for a lively and highly-strung child; I'm sure the disaster with Yvon must mean that there are serious problems.[36] The longer Terry stays, the better for all three of us, and Richard has shouldered most of the responsibilities of fatherhood, for which I am indescribably grateful.
> As for the cancer, I agree it's like entropy catching up! Was ever

anything as improbable as the perfect working of human cells? ... Somehow I find it comforting to think of decay and death (at 80 or whenever) as a natural process akin to processes on all scales up to the entire universe.

This was the first family Christmas she had celebrated for five years, because Richard had always visited his own family in Canada. She persevered on the busy telephone line until she got a call through to Edward and Mattie in New Zealand, and was surprised and amused to discover that Brian was actually there when she called, and that he had taken Alan and Yvon and her children to visit, too.

Her last family letter for 1979 ended on an upbeat note: 'I keep assuming that the treatment will be 100 percent successful. There isn't much point doing otherwise.'

Whether or not she should have radiation treatment was discussed in the New Year. Beatrice did not leave the arguments for and against this treatment entirely to her doctors. She read all the research she could find, and then had extensive consultations with the medical team. In Richard's words, 'She bombarded them with questions.' Finally she decided to take the course of radiation:

> I get the jitters at the thought of lying under a 4 MeV accelerator. But the idea is that if the cancer is confined to that area (by surgical scars), the radiation should finish it off. The problems are that the melanoma cells could have got away, in which case it's locking the stable door after the horse has fled, and melanoma is very very resistive to radiation.
>
> The doctors are admirable, and a very nice group. They have discovered how to use certain high doses to shrink melanoma tumours. I don't have a tumour, so it is sheer conjecture that the same dose could kill off a small number of cancer cells – I think it would be silly not to try.[37]

She carefully boiled down her life to the essentials: Terry and teaching, Richard, the treatment her doctors ordered, her post-doctoral researchers, music, friendships and always her own ongoing research. In the icy winter conditions, both she and Terry got influenza. 'The hamster is sometimes the liveliest creature about the house.' Others on the department staff became ill, too, but in spite of hospital visits every afternoon Beatrice was able to keep up with her teaching schedule. The faculty had also been discussing which new students should be admitted for the next year. Beatrice had overall supervision of those who were accepted.

One of the new post-doctoral fellows, Monica Tosi, came from Italy. Italian astronomer Alvio Renzini had interested himself in her career after they had met at a few conferences. Monica had a fellowship. She should try to be accepted by Yale, he had told her, as Beatrice Tinsley was the best person to learn astronomy from.

When the young Italian arrived at Yale at the beginning of 1980, she was dismayed by how quickly everyone seemed to speak. Her grasp of English was far from good, and the accents, too, were so different from those of the people who had taught her English in Italy. Beatrice's rapid way of speaking was particularly alarming:

> But, right from the beginning and quite apart from her obvious intellectual capabilities, what made her so special was the enthusiasm she put into any action, personal or professional. It was absolutely clear that she was fascinated by all the research fields she was talking about. Nobody could resist her enthusiasm. I would never have guessed that she had already been fighting her cancer for two years.[38]

Tosi's particular interest was galactic chemical evolution. Beatrice made her feel that her work was truly important, and took every opportunity to extend her student's grasp of the subject and point out further lines of attack. She also helped the newcomer settle into the Yale way of life.

'Thoughtful' is the adjective Monica Tosi chooses to describe Beatrice as mother-mentor to her students, considerate and perceptive of their needs beyond the usual boundaries of 'good' teachers:

> She definitely cared a lot about the young people around her. She didn't look affectionate, but she always did her best to help all of us. Despite her family and personal burdens, she was always 'present', both psychologically and physically, as long as she could be.[39]

In all this time, Richard was Beatrice's rock and mainstay. Terry's, too. Beatrice had been anxious, with some reason, about how well she and Terry would get on together in the new and inevitably heightened atmosphere of her illness, and in the small apartment. But there was very little friction. 'For every reason I'm glad that she's here,' she wrote to her family.[40]

She worried considerably that Alan would feel left out of this increasingly close relationship, but to look after two children, now, was obviously too much for her. She told herself that Alan seemed to be settled into the Dallas household. In temperament and in his attitude to life he seemed so very different from needy Terry. He was a survivor, she believed, and in letters and phone calls she sought to keep him close, and to give him news about her illness of the kind she hoped he would assimilate.

A new opportunity came her way, something which was close to her heart but which would involve her in considerably more work. Beatrice became the primary organiser of a 'non-optical' astronomy course which the department was able to sponsor when some funds became available. This was a novel idea, designed to give graduate students wider exposure to the broad field of astronomy, and to enable them to meet outstanding and inspiring astronomers from many different areas.

Yale's department was relatively small, and concentrated on optical and theoretical astronomy. Great things were happening in other areas – radio, infra-red and high energy astronomy, for instance. This spring course was taught almost completely by distinguished guest lecturers. The intention was that each would come to Yale for a week or two, give several lectures, and be available at times for student discussions. It was all very much in line with Beatrice's drive to expose the students to the broader community of astronomers and scientists generally.

News of her cancer had spread rapidly throughout the world community of astronomers. Such was her apparent calm about her illness, and her continuing presence in the heart of her particular area, however, that it was not hard for fellow professionals – those at a distance – to shrug off that dread word, cancer. It had been a shock, it was too bad, but here she was, all right again, working as usual ... the relief was palpable, and letters of invitation arrived in even greater numbers than before.

'After all, even many of those close to her were in a state of denial,' Curt Struck has said.[41] It seemed impossible to believe. Some, accepting that time could be running out, tried to make the best of whatever time remained. For many other astronomy centres, what became known as her 'brush with cancer' acted as a spur to their trying to get her to visit them. What she owed to her own people came first, but opportunities to broadcast the latest ideas and discoveries were hard to resist. Besides, Beatrice found most 'outside' meetings stimulating, even life-enhancing.

At the end of February 1980 she went off to speak at a public meeting in Massachusetts, and the next month flew to Austin for another meeting, taking Terry with her as far as Dallas. The child could visit and catch up with the rest of the family before Beatrice picked her up again on the way back to New Haven, where Jim was due for more joint work.

Edward Hill, too, liked to keep up with meetings and conferences in his sphere, both the Anglican Church and the Oxford Group. A congress of the Anglican Fellowship of Prayer was held in New Haven at the end of April, an excellent excuse for a reunion. Father and daughter had good memories of their earlier time on East Rock, so Beatrice drove him up to the top once more. There they had what turned out to be their last long conversation. She talked openly of what was most on their minds, that she was very well aware that 10 years were the most she could hope to live. She very much wanted to see five years, anyway, to make sure that Terry got to university. Terry was very intelligent but needed guidance and the stimulus to concentrate, and she was the best person to give this to Terry. Edward, thinking this over later, felt that Terry's need of help was concerning her more than her own research, or even her own life.

Edward's visit and her inability to have him to stay with her had brought home to Beatrice just how crowded her small apartment was, so they began looking for 'a proper two-bedroom place' in the same part of town. Accommodation was hard to find and seemed extremely expensive. After all their searching they at last found a comparatively roomy apartment – and under their noses, right next door and across a driveway. Instead of 433 Whitney Avenue, their new address was number 423, or would be when the tenants were ready to move out. The rent was considerably more but obviously the extra space was essential.

Visiting astronomers continued to interest Beatrice. She quickly made friends with Enn Saar, an astronomer from Estonia on a visiting appointment, a well-travelled man whose English was excellent and who was also fluent in Russian and Spanish. This would be helpful when it came to translating papers from those countries, where work she was interested in was being done. He was to be at Yale for six months, which Beatrice thought long enough to do useful work.

Every year made a considerable difference to the children, making them much easier to look after, she decided during Alan's visit. Finding childcare nevertheless took quite a lot of her energy. She took them both with her to Maryland where she was attending the American Astronomical Society's meeting, and managed to find the right people to take them touring around Washington so they were able to see all the famous places they had heard of. On their way home she took them to visit the historic battlefield at Gettysburg and then on through the beautiful Adirondacks.

But by 24 June 1980 she was writing that she had returned with an extremely painful left thigh, bad enough for her to be admitted to the Yale Health Services building, the infirmary. The problem was diagnosed as a muscular inflammation in her leg. Her doctors could not be sure of the cause but thought it could be a reaction to the radiation of several months previously, together with the shortage of lymph nodes in that area. The condition had responded to anti-inflammatory medication, reducing the danger of phlebitis, although the doctors did not know what kind of 'itis' it was. Beatrice was given physical therapy and was able to work and read in bed, but was most impatient to be discharged. By this time she was clearly a very special and greatly liked patient at the infirmary. The doctor in charge, Moreson Kaplan, said she could stay in her private room as long as she liked and come and go as her strength improved. Thus Richard was able to take her out for dinner on several evenings.

Linda Stryker had returned from Kitt Peak with a pile of data. She collected Curt Struck and went to visit her. They found Richard there with Alan and Terry, and a game of dominoes under way. In trying to move the game on to a tray it fell apart and had to be started again, this time with the students playing with the children. They managed to have a good time but the adults were only too aware of how very hard it all was, with Beatrice unable to talk easily with them as they had hoped.

Terry had earlier been booked into her favourite Girl Scout camp in Texas, where she hoped to meet up with some old friends for a large part of the summer before returning to New Haven. Both children hated seeing their mother in a hospital bed, the distinction between infirmary and hospital meaning nothing to them, so, with the Dallas Tinsleys' agreement, they all decided the children should cut short their time with their mother. Beatrice was to tell her family: 'The poor kids have just set off to Texas four days early, very unhappy at leaving me. Richard is taking them to the airport.'

Linda, trying to make sense of her new data and to resolve problems, found Beatrice more helpful than ever. 'She was always an inspiration. Just to observe her example of working, thinking, talking, solving, probing, creating ideas – all this helped me finish my thesis.' [42] They talked music, too, with Beatrice keen to take up chamber music again just as soon as possible.

A special visitor arrived, her old friend from her Christchurch student days, Guff France. He was working in New Zealand where he had become assistant registrar at the University of Waikato, and was able to detour to New Haven during a business trip. She found him the same as ever, and had the greatest pleasure in talking to him and catching up on so much news of other old friends from that time, years which now glowed in her memory as carefree and independent, when she could work without constraint.

The new apartment became vacant. Richard organised volunteers from faculty and students as well as professional packers for the move. Bill van Altena and his two sons were among the volunteers, and cleaned her old apartment for her.

In her closet were ball gowns from the 1960s. 'Gee, they were pretty!' Bill exclaimed, talking to Beatrice about what they had found. Her reply lodged in his mind.

'Those are from a former life of mine.' [43]

CHAPTER TWENTY ONE

ACCEPTANCE

'Friends can be counted on.'
BEATRICE HILL TINSLEY

With the children in summer camps, and Terry's hamster housed in the Astronomy Department, Beatrice flew to England in early August, to rooms in ancient Trinity Hall by the river Cam. She knew this could be for the last time. To her family she said that the only way in which this visit to the astronomers assembled at Cambridge would be different would be that she would not have the energy for her usual visits to her English family, even though the astronomers had been given one spare weekend. She spent most of those two days sitting in a beautiful garden at Trinity Hall, reading. The English could 'do' flower gardens so much better than the Americans, she wrote. On the other hand 'a very saggy bed and ill-kept bathroom facilities down the hall' made England not the easiest country in which to be lazy and comfortable. These were minor considerations, however, compared with the many interesting people, some worthwhile papers and many really good friends at the conference.[1]

But to the assembled astronomers this was an entirely different Cambridge summer as far as Beatrice was concerned. When it was her time to speak she walked slowly and stiffly to the front and gave a thoughtful, slow summary. Gone were her habitual break-neck speed and her extraordinary energy. Her sharp intellect was still evident but there was no escaping the realisation that the old Beatrice not there any more. People saw that it was terminal cancer.

At one of the receptions she and Ken Freeman headed for a corner where they could talk. Ken Freeman recalled:

We were intent on our discussion when up came a very large,

loud and intrusive individual whom I shall call X. He wanted to tell Beatrice about his latest discovery. Beatrice was recovering from some surgery and was a bit uncomfortable and irritable, and certainly not in the mood to be harangued by X.

"Piss off, X," she said.

X took the cue and away he went. He is usually a very difficult person to shake off. For some reason this incident has stuck in my mind.[2]

Possibly this was because Beatrice almost never used such language. Some of her students doubted she even knew the words.

She and Richard hired a car after the conference and explored the countryside around Essex, beautiful Constable country that she drank in. Altogether, though, they were away from Yale for a much shorter time than usual, less than three weeks.

As soon as she returned – 'a firm bed and a clean bathroom' – and had dealt with the usual pile of mail and backlog of student problems, Beatrice planned to use the rest of August for her own research and writing, including turning her attention to a new project which had been suggested by the Yale Press. She was to be one of three editors of a book, together with Richard and Gus Oemler. It was to contain about six chapters by various authors, adding up to a much-needed textbook on the galaxies and stellar populations. She told Rowena they had had great success at co-opting their favourite authors at the Cambridge meeting, so the project was under way. She was co-authoring a chapter with Gus Oemler, who had been at Yale for many years. 'It would be great to work with him again.'[3]

Then Terry arrived, unexpectedly early, from Texas. There had been more problems after she returned to her Dallas family from her summer camp. As Beatrice wrote to Rowena, 'Terry's wellbeing is a very top priority in my life – her good qualities are so tremendous, and she needs so much help.' Fortunately the child, now 12, was delighted to be back in New Haven, and her new term would soon begin. Again everyone who could help Beatrice did so; it was plain that her formidable energies were dwindling. Whenever they could, people undertook supervising or entertaining Terry. One of the instructors, Carol Christian, had a large empty space in her office, ideal for the child to use for gymnastics while Carol prepared her lectures. Others encouraged her to work on her designs and drawings.

Then Beatrice was hastily taken back to the infirmary. As she wrote to Rowena on 14 September, she had suddenly became extremely anaemic. Her pulse was 130 instead of the normal adult rate of 60–80, and she felt frighteningly weak. Six hours of transfusions on each of three days –

'incredible quantities of blood' – had helped, but the cause of her condition was still unknown. She hoped it was just more side-effects:

> I'm about fed up with the constant problems due to preventive treatments, and ready to have a long talk with the doctors about the relative advantages of stopping them.
>
> Richard is completely taking care of Terry, who is upset but good enough to spend hours playing cards on my bed. (She wants to board at the school if I don't get much better.) I think Richard suffers more because he's more aware of the possible implications. He's doing just everything possible for me, and I wish so much I could cheer him up.

All she could do was wait and see, she said. The trouble was that doctors could never be sure: tests took 10 days to get results. Weak though she was, Beatrice wrote on about her delight that Rowena was thinking about spending two years studying in India – 'You really are incredibly adventurous.' It was so good to hear that her sister was planning purposefully, and to gather that she was happier. In fact Rowena's main preoccupation was Beatrice, but she tried to entertain her with other matters.

As Richard saw it, things were rapidly going from bad to worse. Then came the inevitable day, the day in which nevertheless she had not quite believed, not quite accepted as truly inevitable. She had been feeling increasingly tired, increasingly sick. It was more than ordinary lethargy. She felt as if all her energies were leaving her, draining away, and she knew in her heart what was happening.

On 26 September 1980 they told her: the cancer had spread to her liver and lungs. This was the final sentence. Letting Richard and Terry know was the hardest thing she had to do. Telling everyone, in fact, and seeing or imagining the shock and pain on their faces, was something she had to steel herself to do, she was to say later. But she must not put it off.

The next day she wrote to her family the letter everyone had been dreading. She began casually enough:

> I haven't written for a while because of more medical problems, namely continued anaemia and time-consuming blood transfusions, plus a series of other tests. The anaemia is levelling off, that is I retain the bloodcount better after a transfusion, and I won't be getting any more of the medicine that caused it (thank goodness). However, the other tests have brought bad news.
>
> I can't think of anything better than to tell you it all at once. Yesterday I was told that I have tumours in the lungs and liver. The

liver disease is more serious. Its course is quite unpredictable in an individual case, but typically patients succumb in six-twelve months. So far, I feel okay except for a nagging pain in the side.

This isn't really surprising, and in fact it has been the most probable course of events since the original tumour on my leg two and a half years ago, but still – it is a shock.

My friends can be counted on to do everything possible, especially Richard, who has totally committed himself to caring for me and Terry. She will probably start boarding at the school when I get disabled, but so far she and I support each other well, most of the time.

Please, I don't want you to come rushing out to see me. I would much rather you remember all the good times we've had together, and not me as a terminal cancer patient. When I came to New Zealand in 1978, it was partly knowing that I might be paying such a last visit to you. Well, I'm sorry to lay all this on you. I'll write more next time.

She handwrote a similar long letter to Rowena the next day, but did not feel she had to be quite as restrained in giving details. So far only two people knew, Richard and Terry, but she realised the news would soon spread so had to try to tell people herself, rather than their hearing it from others. On the whole she felt all right apart from a pain in her side, aggravated by a persistent cough:

No accurate prognosis can be made yet, but the typical survival time is 6–12 months. How much of that is 'good time' varies a lot.

I heard all this just two days ago. It shouldn't really be such a shock, because it has been the most probable course of events ever since I got the disease two and a half years ago – but I've been clinging to the threads of hope, and I'm sure you and the rest of the family have too.

Sandra Olenik, a printmaker and a friend of Bill van Altena's, was helping enormously, she said, by teaching her self-hypnosis techniques of relaxation to relieve pain, techniques she had learned from a doctor. 'The relief is fantastic!' And again she stressed how much she wanted everyone in her family to remember the good times they had had together and not come rushing to see her, when the meeting would be full of grief and illness.

On 30 September she got out her Living Will, which she had signed two years previously, and signed it again. She made sure everyone knew about it. Her special friends needed to receive special letters or phone calls. Jill

Knapp, to whom she spoke matter of factly, has said the earth shook when they heard she was dying.

To Sandy Faber, further away at Lick Observatory, Beatrice settled herself to write on 5 October. Sandy, like virtually all Beatrice's friends, had been lulled into optimism. It seemed that years had gone past since that first diagnosis, and Beatrice always seemed so full of energy, so quick – still – to respond to the work of others while all the time producing her own stream of papers and continuing to push out the frontiers. Now, suddenly, came this thunderbolt of news

Dear Sandy
This will get to you on the grapevine soon enough but I would rather tell you myself first. In fact it is very hard to write what's coming and to imagine you reading it, because it is bad news. My health has taken quite a turn for the worse. Last week I learned that the cancer has spread to the liver and lungs, and there just isn't any effective treatment. (At least, the available treatments are so unpleasant, and so unsure of buying good time in return, that I have decided not to take them.) This has really been the most probable course all along, but I've been clinging to the threads of hope and pretending it wasn't.

The good news is that my friends, as usual, are marvellous. More offers of help for me and Terry than I can yet accept. Richard is doing an incredible amount, and is moving in to take care of us. (Did you know I moved to a bigger apartment, next door, in June?) I've been told to do exactly what I want to at work, and no more, which is a happy situation – no committees! And Gus is picking up the cosmology course I should have taught next term. Luckily I have this term off. Terry is my main concern; she very much wants to stay here rather than returning to Texas, and I'm sure this is where she can get the most emotional support. Richard and I both think she's the greatest asset in our lives, easily making up for the effort and responsibility by her bubbly cheerful nature.

It's hard to say any more. There isn't any accurate prognosis yet, but the typical survival time at this stage is 6–12 months. You probably know how little those numbers mean.

I'm sorry to write such a dismal letter, and I hope all is well with you, and your back is leaving you in peace.
Love, Beatrice.

Sandy, the resolute, dazzling woman of science that she also was, could not cope with the news. She tried to block it from her mind. She had to go to the observatory on Mt Hamilton in the mountains past Los Gatos, some 75 miles from Lick Observatory, for six days of observation. There was no time to write. Then a night storm lashed the area. Everyone else left. Sandy was left alone at the dome, in a read-out room. She could not delay her reply any longer. First she played some Beethoven, screwing up her courage to face what Beatrice had told her. What could she say? What would she want to read if she were Beatrice?

It was 8pm. The room was unheated and in poor light, and the dome was creaking. Sandy wrote on, finishing her first draft at 2am. Then she worked on it again until 6am. Later that day she edited it, and wrote it out again, dating her letter 15 October 1980:

Dear Beatrice

I received your letter the evening before I left for this six-day observing run. It was shameful of me not to pick up the phone that instant to let you know I had gotten it. But I couldn't bring myself to talk to you right then. I told myself that I had to cook dinner, had to finish my finding charts, had to get some sleep since I'd be staying up the next night. Ever since, I've been scurrying around, busying myself with the myriad small details of observing, grateful for the lack of any quiet moments when thoughts of you might come crashing down.

But now the weather has changed. I am sitting here alone in the 120-inch dome, a winter storm howling outside. The night assistant gave up and went home hours ago. I can't hide from your letter any more.

I am writing this to say what's in my heart, the thoughts I'd be too shy to say face to face. It's been hard to get started – I keep breaking down in utter despair. I ask myself, what can I write that might be helpful to you? What words might comfort me were I in your place?

Beatrice, the most important thing I can say is that you have mattered a lot to me. You have been a faithful and a generous friend. I am reminded of the two letters I got from you this month – so typical of you. The first, your conscientious, excellent comments on the Working Group Report. You took time out to read it, even though time is now precious. None of the regular committee members did as much. And then, your second letter, in which your main worry was how I might take the news. I treasure also the memory of the cheery plant you sent when my silly back first acted up, never mind the fact that your own health was in infinitely greater danger. Then there was our M-dwarf paper on which you generously insisted I be senior author even though you provided the original idea, the brains, and all of the hard calculations. Through all, I have valued your loyal support as a colleague. You've always been generous with words of approval and encouragement, both out in front and, I know, behind the scenes as well. For all of this I have been deeply grateful.

Even beyond our lasting personal friendship, though, there has been a vital tie between your scientific work and mine. Your ideas have been the rock upon which most of my own research is based. To pull myself together this evening, I thought it might somehow help if I re-read some of your papers. I eventually wound up reading all of your entries in *Astronomy* and *Astrophysics Abstracts* back to 1969, plus a few of your very best papers. Seeing it all at once, Beatrice, I was staggered by what you have accomplished. Your oeuvre is prodigious. You have played a critical role in developing so many of the new concepts in galactic evolution that are now part of our everyday vocabulary. Evolutionary corrections, primeval galaxies, the extragalactic background light, chemical evolution, infall, stripping, dynamical friction, mergers, the slow formation of disks – they're all there.

When you burst upon the scene with your revolutionary thesis over a decade ago, galactic evolution was a sleepy little field where no one ever attempted a quantitative analysis because the problem seemed too hard. You showed the world that, provided one was willing to master the details of stellar evolutionary theory, stellar dynamics and hydro dynamics, comb endless catalogues of data for every conceivable clue to the origin of nearby stars in our own Milky Way, and write fiendishly complex computer programs, the subject could indeed be put on a firm foundation. In my view, the science of galactic evolution began with Beatrice Tinsley.

There's only one problem. No one but you has ever managed to do all these things, to put all the pieces together the way you have. In the past, we could all afford to be lazy, because you were there to answer the really tough questions. Your elegant and lucid reviews lulled us into a false sense of security – they made the subject seem so deceptively simple. I personally feel suddenly paralysed by the thought that I'll have to think on my own. Who will make sense of our element abundances when they are finished two years from now? Someone will have to pick up the pieces, but I frankly don't know who can.[4]

It's now 2am and I feel slightly better having written these words. I hope I've not been too candid in pouring my heart out this way. But as I began to write, the thought of anything less than the full truth of my grief was abhorrent. I hope, too, that the effect of these words is as positive as I intended. But if I have instead achieved this measure of personal catharsis by jeopardising your own peace of mind, I apologise deeply.

I very much appreciate your writing to me personally, Beatrice. The courage in your letter was awe-inspiring.

Love, Sandy.

Then she added a postscript.

> I wanted to add that I would very much like to visit you in the not too distant future if you would find it convenient. I will be in Baltimore in December, and the extra distance up to New Haven would be no trouble. If on the other hand you would prefer not to be bothered by visitors, I understand.
>
> If there is any way at all in which I could be of help, do not hesitate to call on me, even if a trip East is involved. Clerical work, editing, helping out with Terry; nothing would be too much to ask. Please let Richard know he can call on me at any time. Sandy.

It is impossible to think of a letter, from anyone, that could have given Beatrice more comfort. Letter-writing did not come as easily to Sandy as to Beatrice. Sandy had given intense thought to what could mean the most to her friend, and then had written three drafts before she felt this was the very best she could do.[5]

Beatrice, who always wrote as rapidly as she spoke, apparently without need to take stock of what she wanted to say, replied at once:

> Your letter arrived yesterday and it has moved me very deeply. You really are the most extraordinary person, astronomer and friend! I can't imagine anyone else spending hours reading someone's abstracts and papers in these circumstances, all alone in a storm on a mountain-top. Somehow it is wonderfully gratifying to think of my career being reviewed in that way, and even gratifying in a horribly selfish way to think that doing so would make you grieve. I most sincerely hope that you worked all those emotions well out of your system and can return unhindered to productive thoughts of your own again. Anyway, by spending all those sad hours you have given me something great.
>
> Of course, you exaggerate my efforts and contributions. Most of the ideas you mention as appearing in my papers were not my own – I've mostly developed and synthesised other people's ideas. For

example, evolutionary corrections started with Sandage (HMS), chemical evolution with Schmidt, infall with Oort and Larson, mergers with Ostriker, and slow disk formation with Richard – as far as I know. And I can't do hydrodynamics and I don't write difficult computer codes!

If I have anything to offer all the people who do burst with originality and do the complex computations, it's enjoying the process of linking different aspects of a subject together. (My most original idea was for my thesis, to do numerical galaxy models, and that ironically seems to have been buried by a paper five years later that used much of my work without referring to it.)

I've often felt sorry for people who get marvellous eulogies after they're dead, wondering if they ever knew how much they were appreciated. Your letter spares me that fate! You should also know that I often feel quite inferior to you, since observations are really the cutting edge, and I so admire your powers of concentration and the way you manage all the facets of your life.

It will be wonderful to see you in a few weeks. Richard is really grateful for all your offers of help, and I am enormously grateful for your offer to give Terry a break.

Love, Beatrice.

After she had written carefully to Alan, to Brian and to close friends, Beatrice's next major concern was her students. Everyone in the department prepared to do their utmost. As chairman, Bill van Altena had at once called a meeting. Gus Oemler would take over her graduate students.

Her old friend from Dallas, Bea Wolf, arrived at the end of October to stay for a few days, to see what she could do to help. She had hoped to come earlier, thinking she could at least drive Beatrice out to see the glorious New England autumn leaves, but she was too late. Most had turned drab and dreary on the rain-soaked ground. On the journey up from Texas she had tried to prepare herself for what she would find, as Beatrice had been quite explicit when she phoned her about the cancer's recurrence. Nevertheless, 'We can never be quite prepared to see pain, and the abandonment of hope, in those we love.'[6]

To her first shocked gaze, Beatrice, very thin, looked to be pregnant. She explained this matter of factly: her belly was swollen because the cancer had reached her liver. This stopped her from bending over to fasten her shoes so she wore slippers when there was nobody to help her put her shoes on. The swelling also stopped her from sitting behind the wheel of her car, but Bea took her wherever she wanted or needed to go.

One of their first outings was to see Sandra Olenik, the printmaker who had made her home in an abandoned church. Beatrice was drawn to the lively Sandra, who had helped with pain-relief techniques, but Sandra's views on alternative medicine were not for her. Sandra thought she could help Beatrice psychologically, such as getting her to adopt various rituals to develop her mental power over the disease. Beatrice resisted this. Bea Wolf saw gleams of hope. She and Sandra later had telephone conversations, and Bea stayed with Sandra for a night, when they talked about ways in which they might be able to help.

Looking back on her stay with her friend, Bea said, 'Beatrice was not self-pitying, nor did she appear depressed. She seemed matter-of-fact, and quite clinical in discussing her condition and situation. She was, actually, a marvel in that way.' But one sequence of behaviour, repeated, began to disturb her:

While I was there it was sometimes rainy. Her apartment was rather chill and dank all the time, but especially on rainy days. I had been with her only a short time when I began to notice a curious thing – at least it was curious to me. It was a little like a play.

Scene 1. Rainy day. Beatrice is wearing a loose-fitting dress, slippers and a cardigan-sweater.

Me: The house is pretty chilly. Is the heat turned on?

Beatrice: No, it's not cold enough for that yet.

Me: But you're shivering! How about a warmer sweater, or a blanket to wrap around you?

Beatrice: No, no, I'm fine.

Scene 2. Same setting, later on.

Me: You're shivering, and I'm getting really chilly. I'm going to make some hot tea for us, or maybe some hot cocoa.

Beatrice: None for me. Just some ice water, please.

So I had hot tea, and Beatrice, shivering, drank ice water. We played these scenes, with variations, several times, and I found it all bewildering. Why not turn on the heat? Why not coddle and comfort yourself as much as possible? The expense would hardly be a factor if you knew you were dying. Most ill people want to be as cosy and comfortable as possible. It seemed to me that Beatrice almost wished to be in discomfort.[7]

Sandra and I had several phone conversations and she too was concerned about this. Sandra was a kind of New Age person before the term was invented. She was bright, energetic, totally dedicated to the power of positive thought. There was a book by a cancer specialist she really wanted Beatrice to read. One of its themes is that positive mental and emotional processes can help alter physical processes. You see, Beatrice's seemingly calm acceptance that she was dying, her very resignation, were in fact horrifying to Sandra and me. We would have done anything to help.[8]

Bea Wolf loved Beatrice dearly, and could not bear to stay silent when she thought she knew about something that might help. But Beatrice the scientist, priding herself as a matter of course on having an open mind and looking squarely at possibilities, believed she had evaluated all of them without prejudice. After all her study and reading about cancer, all her discussions with her doctors, those fuzzy globs of hope from Bea and her ally, the artist Sandra, were too much to bear.

By now Beatrice's nerves were raw. She had to carry the burdens of helping her friends come to terms with her illness, her anxieties about Terry, the pain and the very real nausea eating away at her, together with her overwhelming knowledge of the imminence of her death.

The women were at cross-purposes. Beatrice exploded to Richard – 'with all the energy she could muster' – after the two friends had tried to talk to her about her attitude. 'My best friends have concluded I've given up!' She who had nursed hope for so long and done everything in her power to help herself, now believed she was being realistic in the face of inescapable facts, and behaving accordingly, whereas her two friends were persisting in looking for any crumbs of hope, perhaps for themselves as well as for Beatrice.

She was sharp with Bea, and hurt her. Bea Wolf understood. Her greater pain was not being able to help Beatrice.

Looking back, Bea Wolf has said she can see that there was much subtext in the situation:

> For instance, the great terror that the impending death of a beloved friend can inspire in the healthy ones – that creates a powerful subtext, especially if that death is occurring in a relatively young person with so much to offer the world. I know now that the world of the sick is vastly different from the world of the well, but at that time there was no way I could begin to understand where Beatrice was.
>
> I so regretted that the situation between us had been strained, but by the time came for me to depart we were better. We hugged. But driving away I looked out at the bare trees and the faded fall leaves along the sides of the street and felt indescribably desolate because I knew I would never see her again.[9]

They continued to write to each other, and telephoned often, talking mostly about the music Beatrice was listening to.

Bea Wolf has long wondered whether there was a subconscious part of Beatrice, a saboteur which turned against her for rebelling against her parents' religious beliefs and for leaving her marriage and children, and whether that deeply unconscious part of her made her do penance, endure the cold when she could have made herself warm, even judge her as undeserving of a fulfilling life, unfit to live. 'We will never know the answers, of course, but I think the questions will haunt me for the rest of my life.'[10]

If nobody yet knows how to measure the extent to which the mind and emotions can influence the body, and in what circumstances, Beatrice herself continued to believe it was the trauma of leaving her children which had triggered her dormant melanoma.

Rowena believes the stresses of her childhood had made her sister more susceptible. She herself had been in Jungian analysis and found it enlightening and transforming, a stimulus to the creative processes. Remembering that Beatrice had been responsive to the psychologist in Pasadena, she had earlier tried to persuade her sister to undergo analysis herself, but Beatrice would have none of it. That pathway was not for her. In any case, by the end of October 1980 when Bea Wolf visited her, any help of that kind would have come far too late.

The main decision Beatrice and Richard now made was for him to give up his apartment and move into hers. Very soon she would need even more help. He was already doing the shopping, laundry and housework, as Beatrice could no longer handle daily chores. The move meant he no longer had to go back and forth between the two apartments. It also meant that Terry would have one parent able to care for her, to put in place loving but firm boundaries for her behaviour. That is what Richard had already become, a father, in all but the narrow legal sense.

When the likely course of the melanoma had first been apparent, they had investigated the legal position concerning Terry. Beatrice did not want Terry to go back to Brian. This time she herself had raised the question of marriage. If they married, would not Richard become Terry's legal guardian? No, she was told. It would take a special court order, by no means certain to be forthcoming and particularly not if, as they expected, Brian opposed it. The notion was quietly dropped.

Richard, the reserved, long-time bachelor, had given up his own space and moved his record collection and such other of his possessions as he could fit in, to Beatrice's small apartment, a space redolent with sickness and exhaustion. He faced an indefinite and uncertain future as father to a distraught teenager and as live-in housekeeper and giver of care to a woman preparing to die. Their friends and colleagues looked on him with the awe one reserves for a saint.

One of Richard's sisters and her husband, Carole and Roch Vaillancourt, visited from Quebec and went on a few excursions with both Terry and Richard while Beatrice rested at home, joining them for meals. 'Rest' meant that she sat in her special reclining chair while working hard on research. Terry, suffering from the strain of the overall situation, was receiving help from her school adviser and once a week from Carol Morrison, her doctor at the Yale Health Services, a young woman who got on particularly well with her and who had offered to be a supportive counsellor. She had earlier referred them to the Yale Child Study Center, where the psychologist put on the case was Marie Cohen.

To her father, Beatrice was matter of fact about her new living arrangements:

> It's kind of funny and ironic that we should be living together now in these circumstances, after deciding for so long that we prefer to live separately (a simple reason for not getting married, in case you wondered), but it will work out very well. The new apartment is big enough and his 600 or so records are a feast![11]

Before Edward received this letter, however, he phoned Beatrice. He was so shocked and disoriented by what she told him, including the fact that Richard had moved in to look after them both, that she decided she should try to make him understand, even if this meant fudging the exact situation. Tired and affected by medication she certainly was, but her next letter in her usual careful handwriting was densely packed:

> My health is at present reasonably good. Going off chemotherapy has restored my blood counts, so I have more energy. However, I'm on 24-hour rather strong pain relievers, which make me alternately sleepy and lively over intervals of a few hours. So I'm spending only four or five hours a day at work, and some of that time dozing at my desk! But the waking hours are reasonably comfortable.
>
> I suppose you never knew how close a friend Richard has been to me for nearly five years now. We've often thought about getting married, but always decided that we both really liked the solitude of living alone, and that it didn't really matter since no-one asked questions or bothered us the way we were.[12] He's also become very fond of Alan and Terry, and has been an incredible asset to me since she came to stay; it's obvious now how tremendously committed he is to her!

> Of course Richard was the first person to know when I got cancer, and we've been through countless emotions over it all together since 1978. I don't know how I would have come through the whole scene without him. Certainly now it's wonderful to have someone else in the house – a feeling of security, for Terry too, as well as practical help.
>
> The doctor is as keen as I am that I should be at home just as long as possible. If ever I do have to move out because I need more nursing care than friends could reasonably provide, I would go to the familiar Yale Infirmary nearby, not the hospital. My doctor is its medical director, and says they have terminal patients there, among the cheerful students with athletic injuries and so on, and pride themselves on a 'hospice' atmosphere. It's good to know that's all available if necessary.[13]

Beatrice now took a big step. She decided that the right thing to do was to destroy her diaries. It would be cruel to leave them for others to read. She said no more to Richard other than that the diaries were now no more. She did not have to tell him why. He knew they had been her escape, her private place where she had screamed her frustration and loneliness and anger at what she called her Texas imprisonment. Beatrice had also written some poems in her multi-year diaries, mostly poems which cried out about what she called her wasted years. If she did refer to her diary-keeping, it was to say to Richard that on this particular day in such and such a year they had done something specific. She liked to reread them, observing anniversaries of significant events.

Linda, visiting Beatrice in her office, saw the torn-up diaries and other shredded papers before they were taken away and burnt. Poems on separate sheets of paper she did not shred. Linda could keep these, she said, although she did not necessarily want anyone else to see them. They were about some of the people who had been important in her life, and her states of mind and feelings at the time.[14]

Now it was as if she had begun to come to terms with herself, a stage which Sandy Faber later called the gaining of wisdom. She saw that everyone would have to be helped, in their turn, to deal with her death, and that she was the one who would have to help them.

Terry having agreed to the arrangement, Richard took her to become a boarder at Cheshire Academy at the beginning of November. Many weekend visits home were planned. The strains of her living at home were becoming too great for both mother and daughter – particularly for Terry.

Beatrice had said she wanted colleagues, especially Sandy Faber, to have any relevant material of hers: as Sandy said, 'Beatrice decided she had stuff in her files I could use.'[15] Sandy had a busy professional life, her own students to care for, two still-young children and a schedule of meetings. But in her diary was a long-standing commitment to fly from the west to the east coast for a colloquium. Now she could detour to New Haven to see Beatrice, as long as she caught a particular train back to fulfil her commitment to give a talk.

When Sandy arrived, Beatrice was noticeably weak but full of plans for their time together. They talked for most of one day and were talking again the next day, with Richard, when a call came from Cheshire Academy. Terry had broken her arm while doing gymnastics. Could Richard come for her at once? He hurried away from the two women, collected Terry and took her to the emergency room at New Haven Hospital. She was in pain but very brave, Richard thought, because it was several hours before she had a large cast put on her arm, and he could take her back to the apartment.

Terry, whom she had met at Santa Cruz, concerned Sandy but her attention was centred on Beatrice. There was still so much to talk about. On the Saturday, Sandy's last day, they talked again. Eventually Beatrice decided they needed papers from her office. At her insistence and with Sandy assisting, Terry tagging along too, the two women made their way to the Gibbs Lab and the Astronomy Department, seemingly deserted. Among other papers they particularly wanted summaries of stellar population models that Beatrice had made with Linda Stryker's computer output. They rifled Beatrice's files and then decided to copy a pile of papers. Still with Terry in tow, they hurried to the Xerox room, only to find that the machine was not working.

What to do? The clock was ticking for Sandy's train. They rushed to a classroom, spread out armfuls of files and began sorting papers. With all her heart Beatrice wanted to help her friend and her other colleagues. Sandy wanted to do what Beatrice wanted, but kept thinking of her other commitment. Both scooped up material and tried to sort it in piles on tables. They paid no attention to Terry. In spite of her broken arm she was busy, they thought, drawing pretty pictures on the blackboard.

Sandy looked up. There, on the big three-panel blackboard, Terry had written right across the three, in big sprawling capitals:

I LOVE YOU MOM ! ! !

She had drawn a circle right around the words.

Beatrice had not noticed, and Sandy, too, had not really absorbed what the child was saying. They went on desperately sorting and gathering up the papers. Terry was still at the blackboard. Now she had underlined each word, and the exclamation marks, in yellow chalk.

When they were ready to go, Terry was still busy. She had signed her name, in big flowing letters, to her message. 'Beatrice was frantic and weak and wanting to help me, and I was trying to help her, and thinking of my train. Beatrice should have grabbed Terry and even cried with her, but she was thinking of my train too. I felt so guilty because all the rush was for me.'[16]

Terry was able to stay with her mother until the next day. Beatrice was obviously very ill and weak. It was out of the question for her to look after a partly immobilised Terry for long. 'Terry understood why she couldn't stay at home all the time,' Richard observed. 'She was very sweet and grateful for the six hours I had spent with her that first day.'[17] He took her back to boarding school.

Sandy could not get her visit to Beatrice, and Terry's blackboard message, out of her head. So much had happened in so short a time. She knew she must face up to Beatrice's death or she could not live with herself. This might be the last time she would ever see her friend. It was a watershed experience, and Sandy knew she must not leave unfinished business. On 8 November 1980 she sat down and drafted a six-page letter, then typed it, making sure that Beatrice was in no doubt about what her example of courage and unselfishness had meant. Their opportunity to talk and reminisce quietly together had meant so very much, as had her generosity with her gift of research material. But Sandy wanted to talk about Terry:

> I do want to make an observation which I feel is important in helping Terry come to grips with your death. Perhaps you are already doing this, and if so you can be further encouraged. If not, you might give these thoughts some consideration.
>
> Beatrice, it has been very difficult for me, an adult, to reveal to you even a fraction of my true feelings about loving you as a friend and as a scientist. Without the detached and slightly impersonal medium of written letters, most of my thoughts would probably have gone unsaid. Your own incredible outward calm and desire to avoid unpleasant scenes seem incompatible with the display of emotion in those around you. On a day-to-day basis, this is a good thing, as it is better that your household and apartment not be kept in a continual state of upheaval. For my part, however, I have found writing to you an invaluable way of confronting my own grief and, I sincerely hope, as a way of adding to your own peace of mind. Your letters to me have held great meaning for me, too.
>
> What do these things have to do with Terry? A lot, I think. Terry

has even stronger emotions but cannot resort to letters.

Sandy wrote on about what was, to her, the single most moving event of her visit, Terry's work at the blackboard, dramatically conceived to capture attention, yet, in the pressure of the situation, ignored. She, Sandy, should have abandoned any thoughts of her train and given Terry the support she needed right then.

It was not until she was sitting in this train that the close parallel between her behaviour and Terry's became apparent to her:

> Like me, Terry was communicating in writing. If that's all she can do, it isn't enough. Beatrice, if you haven't already, make sure that Terry has the opportunity to express her love, anger, and loneliness fully to you, face to face. She needs to pour out her emotions to you and to have you reciprocate fully. She needs to see directly into your heart and to experience the full intensity of your admiration and passionate devotion to her.
>
> With your help now, she can come to grips with her feelings, gaining from your love and faith the courage and emotional energy she needs to go on. These experiences will be painful – they must be painful. To deny her pain is to ask her to deny her own capacity to love deeply and unsparingly, and that is perhaps the single quality she needs most to go through life as a healthy human being. Help her to close this important chapter in her young life with the all the dignity and solemnity it deserves.

Sandy apologised for touching on such intensely personal matters, and even daring to offer advice but, in the circumstances, 'such expressions somehow seem permissible'.

On that same day that Sandy was writing to her, Beatrice was writing to Rowena, one of the few people she might have told about Terry's reaching out with her blackboard writing. She said nothing about that, but wrote that the child's doctor was 'tremendously positive about Terry's future, which is a marvellous diagnosis for me to hear!' Many people at Yale had offered to help in totally generous ways; what was shocking was the lack of support from Dallas. They had made it clear they would have her back if absolutely necessary but obviously preferred not to be involved, especially as Richard would pay a huge fraction of her living and educational expenses.

Terry's needs were obvious, but Beatrice almost certainly did not understand the extent of Alan's, although she had realised that each child should have time alone with her. Because Alan seemed so quietly accepting, even – as the older brother – showing that he too could see how desperately Terry needed her mother, Beatrice was inclined to let him stay in the background, quite apart from the fact that Brian and Yvon had sent her Terry but had kept Alan themselves. It was not a matter of unequal love. Her feelings for Alan were much more straightforward, comparatively uncomplicated by guilt. Alan would be all right. This was almost an article of faith with her, something she often said. The dangers she could see ahead were centred on Terry.

She did not repeat these thoughts to Rowena in this letter, but finished strongly:

> Richard is absolutely the strength of my life – morale, love, practical help in every possible way. The illness has opened all sorts of new depths in our relationship, which I feel as a lasting good.
> When I die, he will stay here, keeping Terry's room for her to come home to. But death feels far away; life is so full, fulfilling and busy. Take care of yourself. Much love. B.

With Terry back at school, on 11 November Richard went off to work in his office as usual. Some time earlier, he and Beatrice had got tickets for a Bizet concert at Yale that night, but he had given the tickets to Linda Stryker and Nelson Caldwell. That morning Beatrice had had some muscle weakness, which they both knew was a bad sign, but nothing had happened during the day and there were no phone calls. Richard went home at the usual time. He brought with him Linda's small tape-recorder on which they had recorded a visiting speaker whom Beatrice wanted to hear. He put it down by Beatrice's armchair:

> She was sitting there, the phone beside her. Suddenly she had violent convulsions in her right leg and arm. I jumped up and tripped over the tape recorder. I dialled 911 – and dialled it again. Clearly it was a catastrophe. Through her convulsions she couldn't speak, and there was a look of horror on her face.
> She indicated I should put a pill in her mouth. I did this and got ready to go with her to the hospital. It seemed that everyone in the building came to watch her stretcher taken outside and put in the ambulance. I told the ambulance attendants that most likely it was a brain tumour – we had some idea what to expect – and that she was going to get some treatment.[18]

While this was taking place, Linda was at the concert with Nelson, using the tickets they had been given. Suddenly, just as the orchestra was playing

Symphonie Fantastique, Linda burst out with uncontrollable tears and muffled sobs. 'It was so unlike me. I didn't know why I was doing it.' She knew that something dreadful must be happening.[19]

Richard, for the second time in four days, found himself in the hospital's emergency room. He waited. Beatrice was unconscious. Machines were hooked up to her and a doctor told Richard what the procedure would be.

By then it was 10pm. The doctor explained to Richard that it was indeed a brain tumour and that they would treat it with radiation. Beatrice had told Richard she did not want this to happen. At crisis time there is a compulsion to carry out the wishes of the person one loves, so Richard tried:[20]

> I say Beatrice wouldn't want this radiation. She wouldn't want extreme measures to keep her alive. The doctor says the treatment will shrink the tumour. I'm trying to give him Beatrice's wishes, make him understand. We try to call her doctors. We can't reach them. I say Beatrice must be allowed to have her say. He says we'll discuss it tomorrow.
>
> And the next day she's lucid. She grills the doctors and agrees to the first radiation treatment. She's taken control again.
>
> The doctors say she will live another few months. I don't believe them.
>
> I ask Beatrice and she says yes, she'd like to see Terry. So I go back to Terry's school and drive her to the hospital's cancer ward but Beatrice is beginning to be nauseated and throws up in front of us after her radiation. She can't say much to Terry, and looks like death.
>
> After 10 minutes we go. Then Terry and I talk. Terry seems to be without any emotion at all.

CHAPTER TWENTY TWO

LETTING GO

'The task of philosophy is teaching how to die.'
SOCRATES

Beatrice rallied, as her doctors thought she would. Sooner than anyone had counted on, she was able to leave the hospital and be taken to a room in the Yale Infirmary. There Terry could visit her. She could see all her dear people again, work with her students and advise them, and carry on with her research. One of the first things she did was apologise to Linda for any damage that might have been done when the tape-recorder had been knocked over as she was convulsing.

She felt great relief to be back and to be made welcome, and to be able to sit, supported, in her special chair so she could work and dictate replies to at least some of the letters and cards which were pouring in. So shocked were their writers that many spoke about where they were and what they were doing when they heard the terrible news, the scenes indelible in their memories. Vera Rubin, for instance, was observing at Kitt Peak, in the console room of the 4-metre telescope, when she got the call telling her that Beatrice was now seriously ill. Years later she was to say,' "I still remember where I was sitting when I heard this.'[1]

A man who liked to call himself one of her greatest fans, Nobel Prize-winner Arno Penzias, had earlier offered Beatrice a research fellowship at Bell Labs in New Jersey but she had turned this down in favour of Yale. Now, when he heard the news, he said 'Oh my god' and swept up to the Yale infirmary in a limousine, causing quite a stir among the staff.[2]

Beatrice was told that for some weeks she was to return to the hospital for more radiation, despite its side-effects. The brain tumour had left her part-paralysed in her right leg, side and arm, and the right-hand side of her face. Her speech was slurred, but her energies had come back in part and her spirits had rebounded. She was now again cheering up her friends, dictating notes to them, and considering her students' needs.

They in turn were keen to take dictation and type for her, although she wanted to be as independent as possible. The most important thing she had to do was teach herself to write with her left hand. There were papers she was determined to finish. At first she dictated notes of thanks – messages and gifts were still pouring in – and then wrote her name, but soon she was able to write short notes herself.

Room 506 in the Yale Infirmary looks out on some of the university buildings. The mathematics department is closest, with the engineering buildings and Strathcona Tower, and West Rock in the distance. The window also gives a glimpse of the cemetery.

Her room soon became decorated with pictures and plants. The wall by her bed was covered with photos, many of them of her children and Richard taken on summer vacations. Her student David Guenther took photographs of almost everyone in the department for her so that she could always feel in touch. He took a special one of Terry with her cast, which staff and students had autographed, to cheer her up. So many visitors came with potted plants that extra stands had to be brought in for them.

It was not only Yale people who sent notes and gifts. Another wall carried messages from well-wishers around the world – individuals, observatories and astronomy departments. One particularly striking image of an interacting galaxy system came with a note from François Schweizer, then in Chile. He often sent her pre-prints of his papers on galaxies, sometimes including prints of his most interesting photographs. Schweizer was using the 4-metre 'Blanco' telescope at Cerro Tololo, then brand-new and the world's second largest telescope after the Palomar 200-inch 'Hale' telescope. In the days before the internet, distributing extra photographic prints was a gesture Schweizer could make towards colleagues interested in the same subject.

Beatrice had earlier faced the fact that a number of students and colleagues were likely to miss out on receiving references from her unless she hurried to do something before she died. She made herself concentrate on dictating a number of carefully considered individual references. The Astronomy Department's business manager, Mary S. Albee, typed these together with covering notes explaining how Beatrice had written them to be used as needed, at any time and after her death.

A typical testimonial, for a woman she had not worked closely with but whom she had observed, Beatrice signed and dated:

> I knew Dr Carol Christian quite well for the year that she was employed at Yale as an instructor. She made an extremely good

impression on me, with her obvious dedication to, and abilities in, all aspects of her job. She was undoubtedly the most successful teacher we have had for our large introductory course on astronomy.

Teaching necessarily took up most of her time, but she energetically pursued research wherever possible and was co-organiser of a highly successful 'neighbourhood' astronomy meeting.

I have not been in a position to follow her research closely in the past year, but I believe that she has been working with characteristic energy and enthusiasm; her list of research projects speaks for itself. Carol Christian is a promising young astronomer with unusual qualities of leadership.

In the Yale Infirmary Beatrice continued dictating careful individual letters, particularly for her graduate students, and carried on with papers for publication. She also doggedly practised writing with her left hand and managed to write at some length to Rowena again, in what was becoming a distinctive printed scrawl. She was particularly happy that Alan had been able to fly up from Texas the previous month, even if for only two days. This had been arranged instead of the usual visit for Thanksgiving, as her doctors could not guarantee that, by then, she would not be much worse. She mentioned visits by Sandy and by Jim, without saying what they had talked about.

In fact, both the Fabers and the Gunns had come, separately, to the same remarkable decision. The very best they could do for Beatrice and her peace of mind was to offer to adopt Terry. They were all fond of the child and saw something of her possibilities. Above all they wanted to show their love for Beatrice. Richard was central to Terry, they all saw that, and in the imaginings of both families Richard would continue to play this central part.

They had all realised what Beatrice apparently had not taken into account. Everyone respected Richard and held him in deep affection. What he had already done was selfless in the extreme. But Richard was a bachelor. Would the authorities permit him to continue being a father to Terry? And would Brian agree to another man's becoming her father, taking his place?

Beatrice was grateful for her friends' concern. There was no need for discussion, however. She and Richard and Brian had arranged matters, she said. Terry herself wanted what they wanted for her: to stay on as a boarder at her school, with Richard as her substitute father, and with all her New Haven friends around her. She would, of course, keep in contact with Alan and Brian, and all Beatrice's close friends would maintain a special bond with her.

There was an understanding, supported by a notarised statement from Brian, that Richard was authorised to act on his behalf in arranging schooling and care for Terry while she was living in Connecticut. The statement had never been needed, and nobody had ever raised any questions about Richard's authority or legal status.

All would be well, Beatrice told her friends. And that was that.

When continuing her letter to Rowena, Beatrice made no mention of this although she was forthright about her condition. Nevertheless she wrote:

> But I don't feel I'm about to die! In fact, one can live for a long time in my present state, but I could deteriorate rapidly if a tumour got to a vital spot. So I'm preparing for the worst, to the extent necessary, but generally hoping and living as tho' there is much time ahead.

She added that it was not easy having visitors. She still hoped her family would not come to see her – fond and happy memories were so much better than a sense of crisis, and tension – and ended by saying two students were about to appear with their thesis problems.

For her part Rowena still longed to come to her but Beatrice remained adamant. Rowena had to tell herself that it was Beatrice's feelings, not her own, which she should consider. Later Rowena was to say this:

> I cannot help seeing her cancer as the deep-seated reaction to those stresses set up in her childhood. I find the image of her in a hospital bed – with her illness and her latest set of important equations developing side by side with equal force – terrifying, as well as admirable, and infinitely sad.[3]

So many wanted to see Beatrice that visits had to be rationed. One day a bunch of friends, including Jim and Jill from Princeton, were all gathered in her room. A party got under way. Jim was the target of the laughter. Because of New Haven's one-way streets, he and Jill had lost their bearings and had driven in circles for 30 minutes. The great astronomer could not navigate.

Beatrice wanted to talk. Altogether it was the best everyone could hope for. Furthermore, they found they did not have to leach any sensitive topics from their talk. They need not be constrained. Follow my lead, she seemed to be saying. Here was the old Beatrice, this in spite of her speech difficulties. She was jovial, making fun of her slurring. 'I'm not as dumb as I sound.' Staff were attracted to the room by all the laughter.

Fighting pain and debility each day, she was always open to ideas and possibilities. It is sometimes said of people, usually disparagingly, that they become 'more so' as they age or come close to death: more self-centred, or cynical, perhaps. Beatrice's characteristics became even more pronounced. It seemed as if she dedicated every minute of her now very constrained days to working on new papers, helping her students and concerning herself with her friends and with Terry and her needs. By now this last was mostly left to Richard, including driving Terry back and forth from her school.

As ever, Beatrice did not let anything slide by. Once when Jim and Jill came to visit, she asked Jim if he had got some data she had asked for. As Jill saw it, 'When Jim mumbled that he hadn't done so as yet, she fixed Jim with a beady eye and said, "I'm dying, you know." And Jim did get what she wanted.'[4]

As Beatrice particularly enjoyed the comic strip *Doonesbury*, by Gary Trudeau, Richard brought it to her from each day's paper. Trudeau lived in New Haven, and Linda had the idea of telling him about Beatrice and asking if he could visit. By then he had moved to New York but he sent a book of his cartoons with an inscription to her. She showed visitors her current favourites.

Her former students Barbara and Bruce Twarog kept in touch, sometimes phoning her. The last paper she wrote, *Chemical Evolution in the Solar Neighbourhood IV, some revised general equations*, particularly impressed Bruce because it typified the role, virtually unique at the time, which she played in the studies of galaxy evolution:

> Beatrice's greatest talent came in synthesising the results from a variety of sub-disciplines in a way that collectively shed light on a problem or question that individual investigators failed to recognise because they were too focused on one small piece of the puzzle. She was constantly looking for new ways to update and revise her models, or view, to incorporate new observations and theoretical constraints because, as was apparent to everyone in the field at the time, there were far too few constraints and a large degree of uncertainty in those that existed.
>
> My thesis was just one example of her desire to improve on the observational constraints. She gladly worked with observers and theoreticians because they were the lifeblood of the research she did. They supplied her with the pieces she needed to make sense of the picture.[5]

Beatrice regained partial use of her leg, and she was able to walk a little with a stick. She was happy to accept help with her wheelchair, and to be assisted from her bed and back again, calling Linda and Curt her 'two elves'. But her right hand remained useless and it was a painful struggle to learn to speak again almost normally, to take herself to the bathroom, and to look after herself at least in part.

One day she wanted to see her apartment again. Richard and some of the students helped with the transport. Then Richard lifted her from her wheelchair and carried her in his arms up the steep and narrow steps. Nobody forgot that scene.

To some it was clear that she could not live long. Others moved in a blur of hope and shock, tuned in to the smallest indication of what she might want. For Beatrice this meant being able to do as much work as possible. Curt in particular helped her with this. She wrote notes so Curt could do computations, and discussed her ideas with him. Other students went back and forth like errand boys.

Visitors came from distant centres. Jill and Jim usually drove up from Princeton but on one occasion came by train. They had seen for themselves what Richard was doing, for Terry as well as Beatrice, and all around them people were saying Richard was a saint. This was not the sort of thing one said to Richard, but when he drove them back to the railway station they thanked him warmly. Jim was to say, 'That small smile came on Richard's face. We both realised what a really small thing this was, driving us, compared with everything else he was doing. He was not a saint but a hero.'[6]

Vera Rubin visited, bringing a set of pillows with astronomical themes, and found Beatrice listening to music on a system which Jim had rigged up, and writing poetry: 'It was clear the visit wasn't easy for her, but she reported in some detail all that the doctors had told her.'[7] Other friends brought gifts of a bed jacket, books and always cassette tapes.

Linda found herself the person in charge of visits, telephoning people, taking their calls, arranging a roster, and trying to keep most visits to 10 minutes at most. To conserve her strength and her time, Beatrice decided not to see some people, either because she knew them only slightly or because the meeting would be too emotional. Linda was deputed to make explanations. Her students of course were different. They could visit almost any time. As they called in to help her each day, Beatrice in turn questioned them about their theses, tweaked a fact there and an assumption here, and expected to be involved in day-to-day progress as well as being consulted on any problems.

She still had things to get in order: what was to be done with her possessions, the payment of bills and matters of that sort. Here Richard helped her most. Beatrice was not a sweet and docile invalid. She was in pain, and felt she had much to do before she died. Sometimes her tongue was sharp. Linda did most of the shopping around for cassette tapes which more than anything helped Beatrice ride over the pain and nausea. Linda said:

Once she misinterpreted something I said about one of the tapes. It was Dvorak's violin piece in A minor, and I said I'd always wanted that one. She thought I meant that, since she was going to die, the tape should pass on to me, and scolded me for my supposed lack of feeling. I'd only meant that since I'd always wanted that tape, I thought she would like it too – nothing more. I can't remember speaking up for myself. I just hoped she'd see another interpretation later.

On the other hand she co-operated fully with a graduate student from another department who was writing a PhD thesis on a topic connected with death and dying, and the attitudes and concerns of a dying person. If it were a search for the truth, she was wholehearted in co-operation.

Ministers of religion would come to the infirmary and ask if she wanted to talk with them 'to make your peace with God'. She always said that she was not at all religious and had no interest in such things.[8]

Her musical boundaries were strictly drawn. Compromising in this, as in other areas, was not in her nature, even at the risk of hurting someone's feelings. Jill Knapp brought her a number of records when she was in hospital. One was of Luciano Pavarotti singing Neapolitan songs. Jill noted that Beatrice quietly refused to listen to it.

Because of the effects of radiation, she had been wearing a small knitted cap to cover her balding head. Some people were unhappy about this. Beatrice thought she should perhaps get a wig so that her visitors would feel more comfortable. She called up a wig shop in North Haven, then dispatched Richard to choose one for her. It had to be something as close as possible to her own natural hair, long and black. Richard found it was not an easy job, particularly with the shop's proprietor saying, 'They send a man to do a woman's job!' He selected a wig which Beatrice called Liz Taylor, and her visitors decided she looked gorgeous in it. Her hair was loose around her face again.

The next time Bill van Altena visited her in the infirmary he thought she looked much better, and commented on how pretty her hair was. Beatrice would allow him no illusions. She told him flatly that it was a wig. 'I felt bad,' he said. But softening the reality of her cancer was not for her.

Among the many roles which Beatrice had been filling was that of model for the female students. They would never be able to find another Beatrice, Bill knew that, but as department chairman he continued to look out for a woman they could recruit.[9]

The whole department, staff and students, continued to be concerned although people reacted differently. Some shrank back, not knowing what to say or do. Others stepped forward. Richard believes he could not have got through this time without the help of Terry's psychologist, Marie Cohen. 'Most of the time I felt in a daze with things falling on me. I saw Marie Cohen every week to talk about Terry and what was happening, and I could call on her whenever some new and awful situation arose.'[10] He came to think that he benefited almost as much as Terry did from her counselling, and from the calm observations of Terry's pediatrician, Carol Morrison. It had earlier been plain to Dr Morrison that Terry needed professional help, and Richard, too, if he were to get more understanding of what was driving Terry, and not buckle under the load he was carrying.

Beatrice herself had no counselling. In the early 1980s, psychological help for people under stress had not become the commonplace of a decade or so later. She was suspicious of those she called 'the touchy-feelies'. She had seen and heard something of possibly ill-trained counsellors whom she thought 'over the top', and had no patience with them. Certainly she had been surprised and impressed by the Pasadena psychologist, and what she had come to see as his informed insights. But this was different. She would face death in her own way, relying on herself as a fully rational human being.

Marie Cohen came to the infirmary to talk with Beatrice about Terry. She decided Marie was not a touchy-feely, and came to value her input greatly.

Beatrice asked for an information package from the Hemlock Society, including lists of what people should know and do to prepare for their deaths. She chose a funeral director, someone used by Yale people, and told Richard she did not want to be buried with a gravestone, 'which would just take up space'. Better to be cremated and her ashes scattered, she said, so there would be no waste of land.

Not wasted but preserved, Richard thought, and he argued for a gravestone. It would be something positive, and a record for generations.

'I'll go with that,' Beatrice said. Providing her body was cremated, there could be a headstone, the cheapest possible. He managed to persuade her that something one step up from this would be better. She agreed that Richard could find a modest but not cheap-looking gravestone.

The next item on Beatrice's list was a site for the grave, and she asked Richard to look at cemeteries to find a plot. She had been thinking again about using her maiden name, Hill, and when he returned she asked him if any of the women's graves had 'née' and their maiden names inscribed. He had not noticed any. She did not speak to him about this again.

Beatrice had been matter of fact when she asked Richard to go off to look at cemeteries, just as the poet John Keats, dying of consumption at 25 in a small room by the Spanish Steps in Rome, had sent his friend Joseph Severn out to select a plot and arrange for a gravestone. The end was coming soon and everything should be ready. For Richard, 'It was very heavy. But then I found walking in cemeteries was strangely soothing. It's what we all come to. It gave me a different perspective.' They decided on a plot in the same cemetery that Beatrice could glimpse from her bed in the infirmary.[11]

Staff who helped nurse her retained vivid memories. Nurse-aid May Guthrie said Beatrice was someone who could never be forgotten:

> Such a pleasant woman – she knew no self-pity. She was so interested in everyone and everything, and crammed her life full to the utmost. I felt she did so much more in her 40 years than others do in a whole lifetime.
>
> I never saw her despondent.
>
> Four or five of her students seemed to come in every day. She would smile a greeting and throw questions at them – a really lovely, vibrant person. There was one problem. When she didn't have visitors she read all the time. You could get a book away from her only when she had to have a bath. I bathed and showered her, and it made me feel good just to know her.[12]

May Guthrie also had vivid memories of Terry, 'bouncing in, so affectionate that she'd hug you.'

Even when Beatrice's students were not immediately under her eye she continued to think about them. A few weeks earlier Monica Tosi had had to return to Italy for a professional commitment which kept her away longer than expected. Beatrice had been as active as usual when she left. But in mid-November a letter came to Italy with the shocking news of her relapse. Beatrice had dictated it, explaining that she was unlikely to be around to greet Monica on her return, as her illness had reached a critical stage. A detailed reference was enclosed. Monica Tosi has written:

> I still feel the shock of that letter, but even more I'm moved that she thought of me in her situation. The vast majority of the male astronomers I have known over the years seem to care more about their own careers, or professional or personal or bureaucratic problems, than they do about their students. Not Beatrice.[13]

Monica hurried back to Yale as soon as she could, and went to the infirmary. They were delighted to see each other again, but Beatrice was obviously so ill that Monica decided she must not 'talk science' with her. That did not suit Beatrice. When she learned that Monica had written the first draft of her thesis on galactic chemical evolution, she asked for the draft and within a few days had written detailed comments on it, in pencil. She printed these comments herself, as by then she had taught herself to use her left hand reliably. 'I keep that sheet of paper, as well as her farewell letter, to this day, as relics.'[14]

Early on the morning of 9 December, Nelson Caldwell and Linda Stryker walked together to visit Beatrice, and report to her on their work. She had been listening to her radio. Nelson Caldwell remembers:

> In her characteristic clipped words she repeated the radio report. The previous night a man had crept up behind Beatle John Lennon on a New York street, and shot him dead.
>
> The way she expressed her disgust at the lack of decency in the world was peculiarly affecting. It was further affecting because we knew she wouldn't live much longer herself.[15]

One card for Christmas and three letters, all written with her left hand, were the last she sent to her family. Her father wrote to her with a full heart, telling her of his delight and pride in her. On 16 December, using a pencil and obviously carefully guiding her left hand, she wrote a packed aerogramme to 'Dear Daddy', ending with a seemingly light-hearted, 'How's this for four weeks' practice with the left hand? Very slow, but I hope legible.' Signing it 'with much love, Beetle,' she wrote:

> I do want to reply to your letter of 3 December. It will be very slow, but it's hard to dictate personal things.
>
> I've been one of the luckiest people – as I said to Theo – really to realise my lifelong ambitions, and far more so than I could have hoped.
>
> I have vivid memories in my early high school years, of studying astronomy in the *Oxford Junior Encyclopaedia*, and wanting to contribute knowledge in cosmology in particular. Must admit I always wanted to be famous, as you and Mummy tell me from a young age! I have to thank you and Mummy for the total encouragement you always gave me; I always thought you believed that I could do practically anything I tried, and the difference that made in my life must be incalculable. And unlike so many 'ambitious' parents, you made it clear that the choice of what to do was my own. (I have since learnt how many parents discouraged daughters from

science.) I owe you a vast amount in life!

You would be very moved, as I am, by the outpouring of wishes that have arrived from astronomers around the US and the world. Not only individuals, but groups from big places. The latest was a card signed by about 20 people from Cambridge, England. Astronomers are in general exceptionally nice people, and there is a genuine sense of 'community' which is now supporting me in a marvellous way.

I so wish that you didn't feel dissatisfied with your life's achievements. When we were children, we were often embarrassed because our father was the important person – vicar, mayor! The greatest thing is how you broke free of English society and carved your own life in New Zealand. I had no idea as a child how hard that must have been for you and Mummy – we kids, trying usually to be like 'the other kids', must have made everything so much harder! And you have had a real impact on New Zealand. I am proud of my father.

The medical and nursing staff had thought Beatrice would be able to be taken home to her apartment for Christmas Day. Richard and Linda planned and shopped for this, with special gifts for Terry. Jill and Jim came up from Princeton in the morning.

Then, early in the day, it was plain that she could not be moved, even in a wheelchair. It would be too hard on her. Linda recalls Jill cooking a magnificent feast in Beatrice's apartment, and afterwards she and Jill drove to the infirmary:

> Beetle was so happy to see Jill – she always was. Maybe she and Jim talked on the phone ... We had to walk the mile or so back to the apartment as my car door lock froze up. It was minus five degrees Fahrenheit. The extreme cold would have made it even harder for Beatrice to come home.[16]

This Christmas Day in 1980 as ever stirred memories of the day, six years earlier, when Beatrice had left her children opening their Christmas presents as she drove away from Dallas to a new life. If Rowena had been there, she would have turned to her to talk, but she had expressly forbidden her sister to come. It seems most likely that she was protecting herself as well as Rowena. The emotions of such a reunion could well have been too much for her. Instead she turned to Linda to be a kind of substitute sister, Linda who was so close to her in age and interests and in many attitudes:

She used me as a sounding-board for some of the things she had to clear up before leaving the world. She talked about the men in her life, and sometimes laughed, and she told me about things that one might have told to a minister or counsellor.

One of her most pressing concerns was having left the family on Christmas Day, when the children were small. She asked several times, 'Did I do the right thing?' 'Was it wrong to leave on Christmas Day?' 'Did I do right to leave?'

My responses were always those of confirmation – yes of course it was the right thing to do. And I would back this up along the lines of her having a gift which she owed the world, and opportunities to pursue a great career which brought benefits to the world. I told her what was so true, that she was a role model for other women, showing what they could accomplish.

I said things such as if a woman was in a situation where she wasn't being encouraged and supported and helped, then she did the right thing if she left that situation. If there would be a court battle if she took the children with her, then it was better to leave them. It had to be done.

I didn't say that actually Christmas Day would probably ring in the children's minds all their lives as the day Mom left them … but I truly felt that she had to leave her family and go about her destiny in science.[17]

Because Beatrice had so much to do before she died, she was protective of her health. One day Linda had a cold but came to see her as usual. Beatrice scolded her, so Linda, contrite, said she would sit further away, on the couch. That was not good enough for Beatrice who said that she was very vulnerable in her condition and that Linda should leave. 'Which I did at once, of course. I felt such a dolt, and apologised later.'

On the other hand Beatrice sometimes had Linda actually lie on the bed with her so that they could better look at graphs together. 'Oh how she used time! She got the marrow out of time.' Linda was her last dissertation student, and when Beatrice decided she was going downhill she passed Linda over to Gus Oemler to supervise.[18]

Jill and Jim observed that Beatrice was always forthright with her friends, with everyone. On one occasion she said to Jill over something in a paper, 'You made a real howler here.' Jim remarked that it was uncharacteristic of her not to confront a one-time collaborator about a mistake he had made. 'She could deal with people, and she did. Besides mistakes, she made short work of pomposity, of sloppiness. Respect for science, for scientific truth, was her life.'[19]

Sandy had been 'sitting it out' in California, fretting about what she should do. Finally she decided she would regret it for the rest of her life if she did not see Beatrice one more time. But what could she say? Beatrice took the initiative when she arrived, however, and put her and everyone else completely at their ease, as if the cancer were just a broken leg:

> We talked about how she felt when she was about to die. You had to talk about it, and she made it possible. She was exiting just as the curtain was about to go up in her field. We talked about this a lot.
>
> Beatrice knew that not many of her facts would survive because papers decay or go out of date and are superseded. But she knew she had shown us a whole new way of looking at things. I said we could compare her to a talented chef in a supermarket, the supermarket being analogous to the astronomical library, both with so many ingredients or facts. What to choose, and what to combine? In astronomy a huge amount of understanding is involved, which is where the supermarket analogy ends.[20]

Back in New Zealand, Edward had taken Mattie to visit her sister in her hometown of Invercargill in the far south. He had driven Mattie through some of the most beautiful places in the country, which he described when he telephoned Beatrice in February. Her letter to him, written on 22 February 1981 as she entered the last four weeks of her life, began by referring to his description of this drive, and to New Zealand's controversial right-wing Prime Minister, Robert Muldoon.

> I've got pretty fast writing with my left hand by now. Thank you so much for your latest call on return from the south; I'd intended to call you myself. I'm glad the drive was so beautiful – New Zealand really is unbeatable.
>
> Well, if you think you have problems with Muldoon, look what Reagan's going to be like! Budget cuts into programmes like child nutrition and heating fuel for the poor (who literally die of cold without it), and tax cuts that mostly benefit the rich. There's little enough social justice in this country already! Maybe Congress won't let the worst of it pass. And his statements about foreign policy seem so dangerously naive.
>
> The freeze has ended here and it's unusually mild for a change.
>
> Terry is very loving and nice when she visits and we play dominoes and Scrabble. But she's very nervous about my state. It's been a long strain on her (and others!) knowing that I was expected to die before Christmas, and not having any idea what to expect now.

Richard is a pillar of strength and he's wonderfully understanding of Terry. After a bit of mutual bargaining, they get on very smoothly at home. Richard deserves a lot of credit for learning how to be an effective parent to a troubled 12-year-old. What an age for all this to happen to her. She's also getting much help from her psychologist who specialises in helping kids with terminally ill parents – a warm, very impressive woman. And both Terry and Richard talk every other week to her pediatrician, also an excellent person. So we do have a lot of professional help.

Rowena has been calling me every week or two. She found a crazy phone that gives you three minutes of international talk for one coin, then gives back the coin after cutting off. But we only talk for three minutes because there's always a long, impatient queue for the phone. Her university's on strike, as usual! Otherwise, things seem to be going well for all of them.

Theodora wrote me a nice happy letter about their holiday at the beach. Please will you give her my thanks, and share this letter with her and David – and Mattie of course. I don't have enough news to write more letters.

Alvio Renzini in Italy wanted Beatrice's comments on a paper he was writing, 'Energetics of Stellar Populations'. He sent the manuscript to her at the end of February, not knowing how ill she was. A week or so later he received a reply, 'written with insecure handwriting', for which she apologised. She praised Alvio's approach, but said she was concerned about its limited usefulness on a short time-scale.[21]

On 11 March she printed out for her old friend Pip Jackson in Sydney the kind of letter that is cherished – a remarkably philosophical and loving letter of their joint memories, 'all those happy times'. She wrote about how very young she was when she married, but said nothing more on that subject. Their mutual friend Guff France from student days had visited her:

> Guff goes back nearly as far as you! He hadn't written to me at all since the divorce, making the New Zealander's usual assumption that I was simply the wicked deserter. Well, he passed through New Haven and we really caught up on each other, which was most gratifying.

In a reference to Pip's tragedy, the death of her young son, she wrote that the two of them knew only too well how vulnerable everyone was. But Pip had written her such a joyful account of her other children and all their activities that she was filled with delight:

The kids will remember all their lives how you gave them a start. I suppose my nearest equivalent is the huge, in fact often overwhelming, response I've had from young astronomers, middle-aged (yes, we're 40 and more!) astronomers and entire observatories from Australia to England.

These people are a tremendous support to me. Doing astronomy has been just the most fun, and apparently not in vain.'

Rowena in Venezuela wrote a poem for her, writing in Spanish which had become the language of her poetry. She made a translation for Beatrice, saying in part:

> The solstice has fallen
> behind, and I don't know
> if you've died or are staying
> on time's defile
>
> My hand punishes itself
> with cuts for the paralysis of your ending.
>
> It seems our footprints
> diverge since we dug
> in the garden to get to
> Australia or Hell.
>
> You followed the precision
> of baroque music to the limit
> of the measurable cosmos
> and spoke of the beginning
>
> and of the end – I drill
> through the floors of memory
> groping after origins
> and laws of growth.
>
> The betrayal of blackened
> cells is taking you –
> Sometimes I stand at the door
> of the reflecting chamber
>
> The light dilates.
> Is it you seeing

> through my eyes in passing
> out of time?
>
> Where there are no longer
> eyes or memory
> our forces will flow in the same universe.

Now friends observed that Beatrice was calm, almost relaxed. She had done everything she could to help her family, friends and students through this time, and to tidy up her affairs. When they visited her, they tended to find her with her hands folded, quietly waiting.

One day Curt Struck found himself alone with Beatrice. Usually he visited her with other students. When they were alone, their conversations mostly revolved around such topics as her health, his thesis and his post-doctoral research, although they shared interests in music, history and philosophy.

On this occasion, a clear, cold, sunny afternoon, and without any particular lead-in, Beatrice suddenly questioned him: 'Do you believe that we go on after death, or that, as Bertrand Russell says, consciousness simply ceases?'

> She knew that I was like her in my religious beliefs. And of course she knew she didn't have long to live. Probably she had such conversations with all her friends. But I was very surprised, and felt very much on the spot. I didn't want to hurt this truly beloved adviser of mine.
>
> I think I equivocated – 'Who can say, how can we know?' But she pursued me, something like, 'But what do you think?' And we went around it a bit. In the end I said I guessed I agreed with Russell. She seemed satisfied, and said something like, 'I thought so. Me too.'
>
> I didn't leave this conversation satisfied. It has haunted me ever since. I've alternated between cursing myself for not finding a way to give her a little more hope, and thinking that Beatrice was not one to spare anyone her truth, and she expected the same in return.
>
> Nowadays I am more agnostic than atheistic. I am quite open to Eastern beliefs such as that while the ego personality vanishes at death, some part of the life essence continues. I also like parts of the theory of memes, that is, intellectual-cultural fragments that we can pass on through our students as easily as genes to our offspring. If that conversation had happened in this part of my life, I'd like to think I could contribute a little more. But it is unforgettable because it happened with Beatrice.[22]

She would surely have responded to what Edwin Hubble said to the poet Edith Sitwell. As told by Gale E. Christianson in her biography, *Edwin Hubble, Mariner of the Nebulae*, Hubble showed Edith Sitwell some plates of what she called 'universes in the heavens', millions of light years away. 'How terrifying,' she said.

'Only at first,' Hubble replied, 'when you are not used to them. Afterwards, they give one comfort. For then you know there is nothing to worry about – nothing at all!'

As the days passed, everyone agreed on the need to give Terry a holiday during her spring break from school. Richard needed time off, too. The strain on the adults around Beatrice was at times almost insurmountable; on the child it was acute. What would give Terry most pleasure, and help her the most? A friend in Chicago had invited her to stay, and she liked the idea. The psychologist suggested involving Brian, too. Phone calls went back and forth, with Richard co-ordinating.

Beatrice wrote again on 14 March to her father, saying she expected him to share the letter with Mattie and the Lee-Smiths.

> Terry leaves today for her spring break, and I'll miss her. But she can always come in a day if I get really worse, which is a comfort to both of us. She's spending a week in Chicago, at the invitation of a girlfriend.
>
> And Brian is going up just to meet Terry on the last weekend; staying together in a hotel. We suggested that as a way of their getting together without involving the rest of the family (Terry gets on so badly with Yvon), at no more cost than her going to Texas, and thank goodness, it has worked out. The second week she is staying with one of Richard's sisters with whom she gets on well, near Ottawa.[23] And I hope Richard gets a real rest!
>
> How beautiful the South Island of New Zealand sounds! I think it's better for you now than having a bach, to be able to drive around. Lots of love, B.

Two days later she wrote her last letter to Rowena, giving her much the same news. She added:

> Richard has learned amazingly fast how to be father to a troubled teenage girl! You never know how much is in someone until it has to be there.
>
> The main thing I want to say is that I love you very much, as always. And love to your children, too. Alan is in my head more than I write. Love. Beatrice.

Alan undoubtedly was. In the almost all-consuming, sheer busy-ness of how best to look after Terry and help secure her future, Beatrice was comforted by her belief that, somehow, Alan would stand tall. He would be all right. She said so repeatedly to those close to her.

On 13 March 1981 a paper which she had written in the infirmary was received by the *Astrophysical Journal*. Called 'Chemical Evolution in the Solar Neighbourhood.1V. Some Revised General Equations and a Specific Model', it was published in November 1981. It ends with an unusual note: Beatrice's thanks to Curt Struck for editorial help, and to 'innumerable friends', particularly Richard Larson and Linda Stryker, 'for personal support while it was being written'.

Word began to spread about her feat, writing such a paper in such circumstances. It was indeed a memorable accomplishment since the process, as Richard has said, was very laborious. People spoke of her in awe.

It is not always easy to be precise about when she wrote a particular paper. The usual time-lags – between completing a paper and having it assessed and then published – sometimes seemed particularly long in Beatrice's case. But in 1980 she must have put up a near-record for publications: 10 papers in all, as she fought to use each good hour she could. The culmination was her swan song, 'Evolution of the Stars and Gas in Galaxies'. This review article summarises her special subfield, and is a large-scale sustained analysis of the subject, one then at the frontier of research. It was published in 1980 in *Fundamentals of Cosmic Physics*, and up to April 2006 has had more than 620 citations in refereed journals. It is still often referred to. With her death so imminent, Beatrice tried to include in this paper everything most important in her life's work. In 1981, four more of her papers were to appear.[24]

Beatrice wrote home one last time, on a card that showed all the signs of the zodiac. The note was carefully printed:

> Dear Daddy and Mattie, This is HAPPY BIRTHDAY to both of you. I know D's is on the 30th so the card should be (!) too early, but I have no record of M's birthday except early March. And I don't have dramatic episodes to recall as D. did for me! Anyway I think of you a whole lot, not only on birthdays, and wish you strength and happiness in the coming days. I honestly don't think the length of life is important.
> With very much love from Beatrice.'

The note was signed off, for the last time, with a drawing of the symbolic beetle, the mark of her childhood.

Beatrice did not speak of euthanasia as her body continued to break down. She had every confidence in Moreson Kaplan and her nurses to keep pain at bay if it were too severe. But she was still firm that no unusual steps must be taken to prolong her life.

She had spelled this out very clearly in her Living Will. In no way did she want Richard to have to be responsible for any decisions. He had gone through the agony of trying to fulfil her wishes after the convulsions caused by her brain tumour. Now Doctor Kaplan was well aware of what she wanted, and respected her.

Then, on 21 March 1981, she had a brain haemorrhage. It was certain to be fatal very soon. She was given morphine to treat her pain.

The infirmary called Richard in the late morning, telling him that something catastrophic had happened. The end was now very close. He went there at once, at noon, and sat with her until late afternoon. By then she was almost comatose. She talked a little in small, unconnected bursts. It was as if she were in a dream-like state. Her children were in her mind and she spoke of them.

Linda came to spell Richard, and put cold facecloths on her brow to ease her pain. Nurses watched over her, and gave morphine.

Then Beatrice said Rowena was out in the hall. She had heard her voice – she was sure she was there. Linda gently said that she did not think so, or Rowena would have come into the room. Beatrice insisted, so Linda went out to check for her. Nobody was there except the nurses. 'She would have come if she could,' Linda said. Beatrice still seemed to think that Rowena had come to her, but did not insist on further checks.[25]

That morning, as 21 March was Bach's birthday, Jill had phoned Beatrice to wish her a happy Bach-birthday. She had answered the call but something, clearly, was very wrong.

Jill had immediately called Linda, and explained that Jim was at Palomar, observing, and that she herself was at Bell Labs in Holmdale, New Jersey, working on an antenna there. They agreed that she would arrive by train. Linda met her the next day while Richard again stayed with Beatrice who was comatose, and they went straight to her room.

Richard and the two women sat beside her. By now Beatrice was on the edge of death. Richard told them of the prognosis. No more morphine was needed. She was beyond it. The women wanted to stay with her, talking to each other and to her.

Beatrice's brain had ceased to function and her breathing came in gasps, Cheyne-Stokes breathing. They all knew this was the end. Richard and Beatrice had long since said their goodbyes. He left her with these two close friends.

Back at Beatrice's apartment he called up Brian Tinsley who was staying in a hotel in Chicago with Terry, as they had arranged. Then Richard fell asleep.

Although Beatrice was in a coma and could not talk, Jill and Linda thought she might be able to hear them. They talked to her, and about her, sitting together on the small couch in her infirmary room. They talked, cried, laughed, and went over so many of the things in Beatrice's life, reminding her, and themselves.

At last they too said their farewells. Linda took Jill back to her apartment to spend the rest of the night.

Linda dreamed, a good dream about Beatrice taking her on a flying journey high over the city. 'I feel the wind, see the stars and city lights. It is pretty exhilarating ... no airplane or parachutes – just the two of us. I awoke then and looked at the clock. It was 4.05am on March 23.'[26]

Richard was wakened at 8am on March 23, by a phone call from Brian. Were there any developments? Richard said no, but 10 minutes later the funeral home called him to say they had Beatrice's body. They and the infirmary had let him sleep.

The nurse on duty, making her regular check of her room, had found her dead at 4.15am.

Richard called Brian who broke the news to Terry. She was devastated because she was away on holiday when Beatrice had died.

A group of students had left Yale on Friday for the weekend, still talking about how incredible it was that Beatrice could work and complete so many important papers in hospital when she was so ill. And then, as Monica Tosi said, 'When we came back on Monday she was gone, as if she had waited for everything to be accomplished before leaving. This is something else that makes her unforgettable for those of us who were there.'[27]

Beatrice had sent Curt Struck away that same Friday, in pursuit of a job. She had insisted he arrange an interview with Len Cowie, at MIT, as he had a possible post-doctoral position coming through. Len Cowie was one of three brilliant young men whom MIT had hired all at the same time, the others being Scott Tremaine and Charles Alcock.[28] Beatrice had rehearsed Curt in the way he should present himself, and waved him off:

> I didn't want to go away. It was a very hard thing for me to do, but I did it. You couldn't say no to Beatrice in this sort of situation. I knew the end was very close. She was nearly incapacitated and in much pain but she was still my adviser, and not about to abdicate.
>
> Cowie was very pleasant. As it happened, the money for the post-doc position didn't come through, and I've no idea if I would have

had a chance at it anyway. But the point is that Beatrice was slipping away just as I was doing what she had planned for me.

As I came back to Yale there was a great deal I was looking forward to talking to her about, things about the interview I knew she'd be interested in. But when I got back to my apartment I didn't want to call Linda right away, to see how Beatrice was. And I wouldn't call her directly in the infirmary. She was too weak, and didn't like phones ringing.

I was thinking all this when my phone rang. It was Linda.

Oh my god. I knew. I knew before she said. It was one of those 'Kennedy assassination' moments that I will never ever forget.

I talked to Linda while standing and looking out the window of the front room of the apartment. We exchanged some conventional sentiments: it was for the best, she had been in so much pain, for us it was such a great loss ...

Nobody could know, then, just how enormous a loss it was.' [29]

CHAPTER TWENTY THREE

AND AFTERWARDS

*'Even when all the questions of science have been answered,
the meaning of life will remain untouched.'*
LUDWIG WITTGENSTEIN

As happens when such a death occurs, nothing in that world where Beatrice had left the people closest to her seemed quite real to them. Everything looked skewed, different. The vital link had gone.

It fell to Jill and Linda to break the news to so many people, to say what did not seem possible. As the responses of distress mounted, each call seemed harder than the last. Jill first called Jim, then Jerry. She needed to save Sandy till last, and it was as she had thought and hoped: 'Sandy propped me up.'[1]

Richard and Linda worked their way through what had to be done. Beatrice's children were very much on everyone's minds. Brian had put Terry on a plane for Canada, to Richard's parents, telling the stewardesses what had happened. She was staying on in Canada for a week, as arranged, then Richard would keep to their plans by driving to New York and meeting her at La Guardia. She would be back in time for the memorial service. Alan, in Dallas with Brian and his new family, tried to be self-contained, as usual. He was told he would be brought to New Haven in the summer.

There was one thing that Beatrice had not organised, although she had agreed to a memorial service after Richard had pointed out that her friends would want one. 'Anything is fine by me,' she had said. 'It won't be my business any more.' But she had not said where it should be held.

Richard and Linda were her executors. They arranged for Beatrice's cremation and the placing of her ashes in the ground in that part of the cemetery she had approved. Then, for a memorial service featuring music, they went together to look at the Divinity School's chapel, and then Dwight Chapel on campus. Dwight seemed to be what Beatrice would have preferred, so they wrote a brief note to be sent to 'all friends'. There would be no formal burial service, but friends were invited to come to the Grove Street cemetery at 11am on the next Saturday, 28 March, when the ashes would be buried. Anyone wishing to bring flowers or make brief remarks would be free to do so. Similarly, suggestions for the service, to be held at Dwight on the afternoon of Friday, 3 April, would be welcome.

Beatrice, they knew, would have wanted all her students to know they were especially invited. Their note also said a fellowship fund in Beatrice's memory was planned, and a scientific symposium in her honour had been suggested. 'Again, any suggestions would be welcomed.'

Nobody said much at the gathering in the cemetery. Grief was too raw. The silence spoke for everyone. Something more formal should be arranged for the memorial service, they thought, and agreed that eulogies should be given by Sandy Faber, Jim Gunn, Bill van Altena and Linda herself.

Too late they discovered that Linda's tape-recorder was too small to cope with the acoustics in Dwight Chapel. It had been brought along at the last minute so that the children and family in New Zealand and Venezuela could feel they were part of it all. From where it was positioned it could record the music well enough for the musically minded family members to recognise what was being played, but spoken words were to sound muffled and obscure.

The music and musicians naturally had been chosen with particular care. The organist was a Yale pre-medical student, Charles Brown, who began the service with preludes by Bach and Couperin, and featured Bach's Prelude in G. He and a piccolo trumpeter, Mark Richards, played Albinoni's Sonata in A, breathtakingly beautiful. Friends Leona Francombe, Heather Kurzbauer and Kim Cook – not Linda, who had decided she could not trust herself – played Telemann's Partita in E Flat.

Sandy Faber had been asked to speak at some length. She reminded all who knew, and explained simply to those from the university who did not, what Beatrice had given to astronomy. First she outlined Beatrice's life and how she had earlier encountered 'several of the difficulties which have often beset women scientists'. But it was her work that Sandy wanted to bring home to everyone present. She had drafted and re-drafted her eulogy, striving to put into the simplest possible words what Beatrice had done, and would continue to do, for astronomy, beginning with her 1967 PhD degree that she had completed in record-setting time:

> Her thesis broke wholly new scientific ground. Its breadth and boldness foreshadowed her future research, and changed the focus of many other workers' researches. In this work she tackled the evolution of stellar populations in galaxies, and modelled the

complex interplay between star formation, stellar evolution, and the recycling of the interstellar medium with a degree of realism never achieved before.

She was able to demonstrate that evolutionary changes in the properties of galaxies are large enough to be clearly observable over the range of look-back times of five to ten billion years accessible with present telescopes. As Beatrice Tinsley emphasised, the expected evolutionary trends were of such a magnitude as to completely overshadow the small differences in galaxy brightness predicted in open and closed cosmological models of the universe.

This work, together with the many increasingly detailed papers she wrote on this and related subjects over the next 14 years, had a profound impact on the course of cosmological studies. Formerly a favoured cosmological tool, the redshift magnitude test and other similar observations of distant galaxies now came to be viewed as even more fertile avenues for studying galactic evolution. Her work was thus instrumental in opening up an active and ongoing new field of research.

She, more than any other single individual over the last decade, succeeded in illuminating and unifying the complex processes which together constitute galactic evolution.

Concepts which she either originated, or played a significant role in developing, include the chemical and isotopic evolution of matter via nucleosynthetic processes, the importance of the stellar luminosity function and the physics of highly evolved stars in understanding the integrated properties of galaxies, and several new methods for studying the way in which stars die and return mass to the interstellar medium. Any one of these diverse topics would provide ample material for years of study, but Beatrice Tinsley was highly regarded for her intimate familiarity with all of them.

To increase further her scientific breadth, she collaborated actively with dozens of fellow astronomers from a wide variety of backgrounds. In one of the most fruitful of these associations, her work with Richard Larson of Yale produced the first realistic estimates of spectral and composition gradients within elliptical and spiral galaxies. Also appreciated were her many lucid and widely-read reviews of galactic evolution, which helped greatly to consolidate knowledge in this complex area and make the subject accessible to astronomers in other fields.

Beatrice Tinsley's activities at Yale were central to that institution's present eminence in the fields of stellar and galactic evolution. Her broad range of interests unified the Yale department scientifically, and helped to create an atmosphere where a wide variety of research projects flourished.

In her role as director of graduate studies, she was warmly appreciated by Yale students, for whose needs she showed special concern. Women received much encouragement, and, perhaps as the result of her attentions, Yale produced several talented women PhDs during her tenure.

No tribute to Beatrice Tinsley would be complete without emphasising her strongly positive, even inspirational, impact on colleagues and students. Those around her were enlivened by Beatrice's obvious zest for her own scientific endeavours, a joy she never relinquished even during her long illness. Equally stimulating was her enthusiasm for research going on around her. She was invariably interested in new results and plied one with astute questions, all the while radiating appreciation and encouragement. In relation to my own work, for example, Beatrice's critique was the one I could always count on, and the one I valued most highly.

Her unflagging cheerfulness in the face of fate, even though she had to give up attending her beloved chamber music recitals and became more and more confined to her hospital room, was a remarkable and uplifting experience for all. She kept up a voluminous correspondence and wrote and edited papers, either dictating or writing with her left hand …

It is distressing at any time to see a scientific career cut short in mid-stride, before the full flowering of scientific thought has run its course. In Beatrice Tinsley's case, the loss to astronomical science is exceptionally keen. Her role as catalyst and synthesiser was truly unique.'

Bill van Altena spoke of the huge contribution Beatrice had made to the Astronomy Department, to the professorial staff and to the students. The resounding success of Yale's great galaxy conference of 1977, he said, was almost entirely due to her. Many groups were now undertaking the extensive observations needed to discover evolutionary effects in galaxies. Exciting results had already been reported.

Linda Stryker, speaking particularly for the students, used anecdotes of Beatrice in unforgettable action, drawing threads together to show her character and what everyone felt to be her continuing presence. Encouragement and enthusiasm were among her greatest gifts, she said:

You would be taken to an astronomical meeting in another city, find yourself standing bewildered among the giants of astronomy, then

Beatrice would carry you off and start introducing you to everyone, urging you to explain to them what you had been working on, and to ask questions. This was done partly because she knew we would need jobs later on, but also because she felt it was important for us to talk to other scientists.

She encouraged us in our professional writing by painstakingly editing our outlines and drafts, telling us where we had done something particularly well but also pointing out our weaknesses, and making thoughtful suggestions.

She had a unique knack for expressing difficult scientific ideas in clear language, and it was one of her main goals to help us to write clearly. Whenever we became discouraged, she always found the right words to set us on our feet again.

Beatrice's enthusiasm was marvellous. In class, her teaching always included the classic as well as current literature on her beloved topic, galaxies, and, in order that we cover this vast field, her lectures were very fast-paced. So much so that it forced us to ask questions every 10 minutes or so – if for no other reason than so we could finish writing down our notes before she took off again.

Perhaps her greatest gift of encouragement and enthusiasm occurred when you worked with her on a research project. She treated you as a collaborator and a colleague – not only by asking your opinion in all matters great and small but by working long hours at the computer centre, discussing the astrophysics in detail, interpreting results and comparing with other workers, and often writing a paper with you, in a joint contribution to astronomy.

Linda brought some needed lightness into the chapel when she spoke of how the students lovingly parodied Beatrice in their annual Christmas revues featuring their teachers, how the student playing Beatrice would speak extra-excitedly about galaxies, at shot-gun speed, and how greatly Beatrice herself enjoyed these send-ups. Her passion for music and for playing chamber music meant that others, including some visiting astronomers, were drawn from time to time into expanding their regular trio. 'We always thought it was curious how many musically-inclined graduate students came to Yale while she was director of graduate studies.'

Jim Gunn wrote a eulogy but then saw that he would not be able to speak. Nor, even, could he bring himself to come to New Haven for the service. He folded the pages he had written, wrote 'I love you' on the back for Jill, and gave them to her to read in his place. Then he went off, alone, into the mountains. Jill, too, found that she could not stand up and read Jim's words, so asked Jerry Ostriker to do this. He read:

Beatrice is gone.

In the last months, seeing her often in her hospital room, talking as we always talked about astronomy and astronomers, it was easy to believe that her imminent doom was not really coming, that this was simply a new status quo, that when we came next she would still be there, alive and smiling in the face of waiting death. We finished the most ambitious paper of our long collaboration there, and there seemed no reason to believe there would not be more.

But there will be no more. I, her many other friends, and perhaps most of all this peculiar discipline we pursue – which she loved, I think, more than anything else in her life – will be infinitely poorer for her absence. For me, she was the colleague whose sense and industry and intelligence I valued more than any other, and a friend whose love and friendship withstood all difficulties and sustained me more than once in disaster in my own life.

She incurred a debt in me I could never have repaid, and certainly cannot now.

For astronomy she offered both a unique love and a unique talent. She was a synthesiser in a small but fragmented discipline. There are many who have attempted what she has done, but no one really came close to doing as well.

We will understand the universe more slowly, and very much less well, without her, but have understood it very much better because she was here. I think she would have wanted no higher tribute.

The silence which followed these words had a special quality.

The minister, the man who regularly officiated in the chapel, had been told that Beatrice had not wanted any prayers or religious invocations but he ignored this, and also read some celebratory psalms. After the eulogies he asked if Richard wanted to say or read anything. But it was plain that this was beyond Richard, so the minister read a poem that Beatrice had written not long before her death – reading it perhaps not in the way her friends would have liked, but then he had not known Beatrice, and no doubt took it as a prayer to his God:

> Let me be like Bach, creating fugues,
> Till suddenly the pen will move no more.
>
> Let all my themes within – of ancient light,
> Of origins, and change and human worth –
> Let all their melodies still intertwine,

> Evolve and merge with ever growing unity,
> Ever without fading,
> Ever without a final chord ...
> Till suddenly my mind can hear no more.

Edward Hill, looking for the right words, much later, to finish his book, *My Daughter Beatrice*, chose this same poem, and ended with words both from his heart and from what he thought he knew of his daughter: 'That poem, to whomsoever it was addressed, was surely answered.'

Afterwards, many of those who had been at the service walked down to the campus cemetery. Sachiko Tsuruta found it the most moving of funerals. Beatrice had arranged for her to come to New Haven to visit her in the infirmary; the date had been set for the following week. Her sorrow grew with the weight of goodbyes never said.

Jill and Scott Tremaine had carried armsful of forsythia to the chapel, and now brought the flowers with them. Everyone stood where the memorial stone was placed. Richard saw to the simple inscription:

> *Beatrice M. Tinsley*
> *January 28, 1941*
> *March 23, 1981*

As Richard had foreseen, family and friends and colleagues, and even disciples who had never met her, were to need this place as a focus for their feelings. He was himself helped by having to focus on Terry. His parents in Canada had warmly taken her into their family. She was to say later, 'They are my grandparents of choice. They treated me as if I was Richard's daughter. So did Anne and Carol and Mary, his sisters.'[2] When she returned to New Haven, to Beatrice's apartment, both Terry and Richard saw Beatrice's absence everywhere. Terry was particularly upset that the New Haven newspaper had printed an obituary of Beatrice. It was an intrusion. 'Why is it in the paper?' Perhaps returning to school would help.

They chose a print by Sandra Olenik and presented it to the infirmary in memory of Beatrice. Then Richard drove Terry back to the academy, having earlier explained the situation to the staff as best he could. The head teacher told the whole school at assembly that Terry's mother had died. Terry was outraged. This was a private matter.

'I skipped school.'[3] It heralded even more turbulent times. Psychologist Marie Cohen continued to make herself available, and mediated at the academy when there were more troubles.

Richard began to believe that they would come through and that Terry would be able to settle to the education that Beatrice had so dearly wanted and planned for her. She still had many conflicts but was beginning to accept the shape and pattern of her new life. Then, three months after Beatrice's death and just after Terry's 13th birthday, Brian and Yvon with Alan and their younger children arrived in New Haven on a summer excursion. They would take Terry back to Dallas with them for the rest of the vacation.

Richard had made arrangements for the fall semester at Cheshire Academy, and he and all his friends and colleagues expected Terry back in weeks. Everyone talked over what they could do to help when Terry returned. It was something they could continue to do for Beatrice, as well as for Terry.

Instead, it was four years before Terry came back to New Haven. Brian had decided her place was with him, not Richard. There was nothing that Richard could do.

Terry has said that she found the only way to get attention in Dallas was to raise hell. In the seventh grade when she was 13 years old she ran away seven times; 'bad things happened.' She managed to come out the other side of those most difficult years and said of herself, 'I'm a survivor.'[4]

As an adult, Terry reverted to her real name of Teresa. She had come to believe she had been wrongly diagnosed as hyperactive, and was on the wrong medication. A later diagnosis was manic depression, or bi-polar illness. She may have suffered from both conditions. Her customary mood swings and need for reassurance and affection had intensified in the last months of her mother's life. The shadow of cancer hung over the child. She would visit, but it was often just a 'Hi Mommy!' and in and out of Beatrice's room, and talking to the nurses. Then time ran out, the time Beatrice desperately wanted so she could help her daughter, and the pain of Terry's loss was all the sharper. Her mother's death had not been the time for another uprooting.

In June 1985, four years after Brian had taken Terry away to Dallas, he suddenly sent her off in a plane for New Haven again. Yvon had issued another ultimatum. When this happened, Richard was away in Toronto on vacation. As soon as he heard, he hurried back, to another crisis situation.

He became her guardian again, caring for her until her school re-opened. Terry talked frankly to him. 'Richard always listened,' she said, and made it plain that she needed and wanted strict and consistent boundaries. As Richard was to say, she demanded of him that he be a firm disciplinarian. 'She needed to be able to say to her friends, "Richard says I must be home by 11pm." If by any chance she was going to be late, she had to call me.' She had so many talents and possibilities, as he and others recognised, but her problems seemed to include an inability to stay focused on any one thing. They worked out a mutually acceptable way of living for the next two years, until Terry was 19 and decided she wanted to strike out on her own, in Texas.[5]

The editor of the *Quarterly Journal of the Royal Astronomical Society* asked Richard to write Beatrice's obituary. He found this difficult but eventually wrote it in a spurt, then asked Linda to work on it too: 'Linda added, polished, amended and edited.' They strove to contain themselves to her work as a scientist, but her character and personality shone through.

Groups and individuals everywhere suggested memorials to her. A number of people including Alar Toomre, Jim Gunn and Jerry Ostriker had thought another galaxy conference could be held in her honour, a memorial meeting. This did not happen because, as everyone soon realised, the only person who could have organised and run it was Beatrice herself.

Members of her department at Yale decided at once that their best way of commemorating her was to establish the Beatrice Tinsley Fund to help outstanding students. The annual award was to be used for any legitimate research purpose, such as paying for travel to an important conference. Preference was to be given to activities most important for the completion of a PhD.[6]

Sandra Faber was the driving force behind lobbying key members of the American Astronomical Society to endow a special award to commemorate her unique achievements. It became the Beatrice M. Tinsley Prize, and in accordance with Beatrice's principles both men and women are eligible for it. As she would have wished, too, no restrictions are placed on a candidate's citizenship or country of residence. The Beatrice M. Tinsley Prize is normally awarded every two years, and recognises 'an outstanding research contribution to astronomy or astrophysics, of an exceptionally creative or innovative character', and one that has played a seminal role in furthering understanding of the universe.

The prize carries with it a medal and an endowment, and was first awarded to the outstanding English astronomer, S. Jocelyn Bell Burnell, in 1986.[7]

With Frank Bash of the Astronomy Department at the University of Texas at Austin as chairman, a committee from the Macdonald Observatory and the Department of Astronomy Board of Visitors raised $100,000, matched by another $100,000 from the university, to endow the Beatrice Tinsley Visiting Professorship. The income was to be used to bring distinguished astronomers or physicists to the university. It was decided that the first holder would be someone who had known Beatrice well. Holders of the professorship would be asked, typically, to conduct a graduate seminar in their research specialties, to cite the professorship in any publications resulting from their visits, and to talk briefly to the Board of Visitors. Because the organisers did not want in any way to interfere with the American Astronomical Society's memorial fundraising nationwide, they confined their efforts to Texas. This meant, as Frank Bash wrote to Linda Stryker, that they had 'approached people who otherwise would not be aware of Beatrice Tinsley's eminence'. The single most important influence in encouraging donations to the endowment, he said, was being able to reprint and use the Larson-Stryker obituary.

The first holder of the Beatrice Tinsley Professorship was Gillian (Jill) Knapp, in September 1985. Others have included Vera Rubin, Richard Larson, Virginia Trimble, Robert Kennicutt and Ken Freeman.[8]

Beatrice had made it known that she wanted the graduate students to divide her books among them. After her memorial service, Richard and Linda found that organising this gave them something practical they could do for her. They decided that Curt Struck and Nelson Caldwell should have the first choice. About 10 of these books sit on Curt's shelves today. Some, such as Allen's *Astrophysical Quantities*, he uses often. Others, including Rindler's *Essential Relativity*, and Chandrasekhar's *Stellar Structure*, (with her signature and the date, 1964, written in,) are treasured heirlooms. He also keeps the proceedings of the Eighth Texas Symposium on Relativistic Astronomy, the meeting to which Beatrice drove him all those years ago.

Nelson Caldwell has given most of the books he received to his own students. He particularly refers to Beatrice's own article, 'Evolution of the Stars and Gas in Galaxies', in *Fundamentals of Cosmic Physics* 1980, which appeared not long before her death. He considers it has a clarity seldom reached in more recent papers.

Not that anyone who knew her needed her books to have Beatrice still very much in their lives. It was plain that so much of what she had accomplished, and the directions she had pointed to, were living on in the work of astronomers she taught or collaborated with.

Her first PhD students were Bruce Twarog and Barbara Anthony-Twarog, who are now both professors of astronomy at the University of Kansas. Barbara says of her:

> I have been conscious of her example as an astronomer and as a woman scientist (she would insist on that order!) every day of my professional career.
>
> Like many women who went before us, she had conflicts that she could not reconcile. But bitter? Never. A bitter person does not write letters of recommendation for her students to be used in job searches after her death.
>
> At the time I was her student, and ever since, I have felt that

Beatrice set a standard of responsibility and concern for students that I have measured other professionals against.[9]

Monica Tosi is a professor at the Bologna Astronomical Observatory in Italy, working mainly on stellar populations and galaxy chemical evolution. She has published many papers in refereed journals, and review papers in conference proceedings. All, she says, are in some way connected to Beatrice, although only the first 'CNO Isotopes and Galactic Chemical Evolution', published in the *Astrophysical Journal* in 1982, was conceived while Beatrice was alive, and actually inspired by her. Monica Tosi says:

> There is no doubt whatever that Beatrice has been a great teacher for me from many points of view, what in Latin we would call *Magistra vitae et studiorum*.
>
> Professionally my career is certainly based on what I learned from her on the evolution of galaxies; she gave me both the cultural bases and the technical tools to work in this field.
>
> Today I still use her notes and her review papers, both to check on specific issues and to provide my own students with the most accurate and extensive texts on galaxy evolution.
>
> Today, still, most of her findings on fundamental aspects – like the evolution of the light elements synthesised during the Big Bang, the effect of gas infall on the chemical abundance gradients, and so on – are not only valuable but much more reliable and robust than more recent results by other renowned people.
>
> When I am at an international meeting today and someone quotes Beatrice's still valid predictions, I look at the audience reaction. I have the feeling that the astronomical community is divided into two parts: those who met her (and they all still miss her), and those who didn't have this opportunity. The younger generations seem to consider her a legend. Once in a while I am asked if it's really true that I worked with her for a year – as if they were talking of a myth rather than a real person.
>
> The best compliment I ever received at an astronomical meeting was a few years ago from an old friend of hers, who told me that my talk somehow reminded him of her. Presumably he just wanted to be nice, but the compliment may have originated from a certain 'imprinting' she may have left on us.
>
> According to colleagues who knew and appreciated her, there is another imprinting of Beatrice which characterises the astronomers involved in studies of galaxy chemical evolution: it's a community

that some consider dominated by women! Actually there are several men doing excellent researches in this field, but I guess it's because of Beatrice that we appear more 'visible' than they do.

I may be biased, but my feeling is that she was possibly the first to introduce 'a female way to astronomy'. This includes caring for students and collaborators.[10]

Jerry Ostriker has another way of looking at women astronomers. He said:

Many of the most important ideas in cosmology have come from women – and in Beatrice we had a truly brilliant scientist.

When I was a student, galaxies were considered fixed. Beatrice pointed out that they were not, and caused a whole change of direction in everyone's thinking. How did she do it? Well, she was smart! And to attack the big problems she had intelligence, independence and courage – young people should know that science is easy if you have these three qualities.

Women have an advantage – they don't have a whole lot of mental baggage to carry round. Men are so fixed on solving the problems of the universe. Women look at things from scratch. Vera for instance discovered dark matter. Her observations were the most critical. Beatrice discovered the evolution of galaxies – nobody had systematically investigated it before she did.

The essence of science is asking the right questions.

The greats in science are the first who ask the questions.

Beatrice asked – and she answered.[11]

If women have an advantage in the way Jerry Ostriker claims, most agree that it does little to balance the scales. Women are still not encouraged in the upper ranks in astronomy, although universities are generally now more welcoming. At some, too, women undergraduates in astronomy outnumber men. Jill Knapp considers that women have had a hard time in the field. 'Intellectually we know we can do things, but to have this denied is so discouraging.' Women scientists, she says, have little professional security, so it is hard for them to decide to have children. Many of them have quite dismal personal lives. In Beatrice's case she thought she should have children. 'She had it laid on her. She did her best and tried to talk herself into it.'[12]

Curt Struck is now professor of astrophysics at Iowa State University. His work is mainly in the field of galaxy evolution, and he has published many papers in the journals and in conference proceedings. He decided that his main duty to the memory of the master-student relationship he had with Beatrice was to move on, try to find a place in the world of astronomy, and to help others as she had helped him.

Nelson Caldwell is an astronomer at the Smithsonian Astrophysical Observatory in Cambridge, Massachusetts. He too has published many papers, many on the topics he began working on, as a student, with Beatrice. He is constantly reminded of her 'large legacy' by the many references to her work that appear in publications and lectures, and that confirm her ideas.

Jean Audouze has said:

> Beatrice still has a great impact on me after some 30 years. She was a great person, not only in herself but great for so many of the friends with whom she worked, the many students she inspired, all those who had the chance to attend her lectures or to read her papers.
> One of the great chance happenings in my life has been meeting this woman from New Zealand – small, fragile with a very bizarre English accent, not especially pretty but with a deep intelligence, and astonishing enthusiasm and determination, and a charismatic personality which, as I remember it, makes me still enjoy life and humanity.[13]

Vera Rubin remembers her like this:

> Even in her all too brief a life, she was one of the giants of astronomy in the twentieth century.
> There is no telling how much more we might have learned if she had had a longer time to teach us ... I miss her very much.[14]

Harlan Smith at the University of Texas in Austin said, 'Had Beatrice lived another 10 or 15 years, I think she would have been the pre-eminent astronomer in the world.' Nobody can tell, of course. And in any case, there are many pre-eminent astronomers, depending on their special fields. But she was special, even then.[15]

Sandy Faber has said, simply:

> We were so fortunate to have known Beatrice. Everyone she loved, and who loved her, was fortunate beyond compare.

We still are, for she lives on, waiting to be discovered by scientists to come.[16]

In 1985, four years after Beatrice's death, the Astronomy Department of the University of Texas in Austin held a meeting in her honour. It was also to welcome the first holder of the Beatrice Tinsley Visiting Professorship, Jill Knapp, married to Jim Gunn and working with him at Princeton. Few things could have pleased Beatrice more.

Most of the people who had figured largely in her working life managed to be present at this meeting in Austin. Edward and Rowena Hill were there, too. Both were taken aback by the numbers and calibre of those attending, with Rowena in particular fascinated to put faces to the names long familiar to her from her sister's letters and conversations. Besides many of Austin's good and great, New Zealand's ambassador to the United States, former Prime Minister Bill Rowling, came to the sessions which focused on what Beatrice had contributed, her legacy to cosmology. With New Zealand's population a mere three million people, Rowling said,[17] the country had given the world three great scientists: Ernest Rutherford, William Pickering and Beatrice Tinsley.[18]

Edward spoke proudly of his multi-talented daughter, but understandably showed little knowledge of her work. He revealed an abyss of misunderstanding of what drove her when he said, in his address, 'When Beatrice left New Plymouth Girls' High School, it was a toss-up whether she would pursue music or science.' And when round-the-table reminiscing about Beatrice painted her as unstoppable in her pursuit of some goal, Edward picked up one word in particular, and said in genuine surprise, 'Not my little daughter formidable?'

'You didn't know the half – not the quarter!' came, *sotto voce*, from the other end of the table.

New Zealanders are only just beginning to realise what sort of person grew up and was educated among them. Scientists were the first to see this. In the South Island, two observers and members of Canterbury's Physics Department, Alan Gilmore and Pamela Kilmartin, discovered at the Mt John Observatory, Lake Tekapo, on 30 August 1981, Asteroid 3087, which they named Beatrice Tinsley. Edward Hill called it 'a kind of moving tombstone'.

A street has been named for her in the North Shore City district of Albany. Wellington East Girls' College has named a laboratory after her.

A play, also called *Bright Star*, by Canterbury playwright Stuart Hoar was performed at Wellington's Circa Theatre in 2005. It was, however, based on some of the letters selected by Edward Hill for his book *My*

Daughter Beatrice, and so could scarcely give a picture in the round.

Her old school and her old university as yet have named nothing in her honour.

When astronomers get together and play the game of 'what if', they sometimes speculate on what Beatrice might be involved in if she were still alive in the 21st century.

Monica Tosi believes: 'She would likely be dealing with the controversy between hierarchical and monolithic galaxy formation, and with the applications of galactic chemical evolution to light elements and primordial nucleosynthesis, a subject she pioneered.'[19]

Jerry Ostriker had a quick response to the question. 'I'd be doing whatever Beatrice was doing, because she would have been up in front.'[20]

Richard Larson thinks such speculation is pointless, and that Beatrice's whole story is a lesson in the unpredictability and capriciousness of life. 'We can only take what comes, and try to make the best of it.'[21]

Sandy Faber has said: 'With Beatrice the door opened one inch. The first important inch. She laid the foundations for how ordinary or baryonic matter behaves. Now it's dark matter. If she were still alive she would of course have followed up dark matter because all theories of galaxy formation today are based on this. Edwin Hubble showed that galaxies are moving outwards. Beatrice showed that galaxies are also changing, and interacting with one another. She knew there was so much more work she was on the verge of doing.'[22]

Jim Gunn says that Beatrice would have moved well beyond her own early work. It would have been superseded by a torrent of new ideas:

> Cosmology is now much more of a science than it was then, with much better data. It is still incredibly exciting and attracting the best minds. The same old questions are flying around. They may have been answered if she had lived, so she would have moved on to the next questions.
>
> The formation of galaxies is still an imprecise subject. Her areas – the ones she invented – are galaxy evolution and stellar populations, and the importance of interactions in galaxies. She was the architect of all this, and the whole area took a nosedive after her death.[23] There have been many talented young people in the field since then, but no Beatrice.
>
> She was the synthesiser. She brought synthesis to the field. There's a discipline in physics called phenomenology. She was the first phenomenologist. Others were specialists in a fairly narrow range,

a narrow window – even the theorists were. Beatrice was so bright and interested that, reading avidly, she would spot a connection that nobody else would see. She saw that this tied up with that, and so on. It would have been a real tragedy if she had specialised. She would have been brilliant – but her incredible synthesising ability would have been wasted.

With Beatrice it would still have been the big picture, the promise of being able to understand. Most astronomers, if they're not derailed, at least get diverted to another path. She didn't.

I do believe Beatrice was a genius. Breadth, not depth, was her forte and she didn't get hung up in details. She had piercing clarity.[24]

Beatrice was both blessed, and ill-starred. Luck, with its bearing on every life from the moment of conception, gave her exceptional genes from each side of her family, and an exceptional temperament. If it is usually possible to make at least some of one's own luck, Beatrice did come up against some intractable circumstances. In other instances she made mistakes, startling mistakes, mostly arising from early imprintings, and her interpretation of her upbringing.

If it is the grit in the oyster which makes the pearl, the challenges and adversities Beatrice had to face almost certainly spurred her on to achieving greatness.

When one looks back over her life, it is striking how many 'ifs' there are, with such far-reaching consequences. Her life may well have been different if she had not been born prematurely, after an air raid, with her mother at first ill and weak, then lacking in the usual maternal feelings and attitudes, largely because of her own upbringing. Beatrice herself firmly believed it would have been different if she had not been separated – twice – from her nanny, resulting in what she repeatedly described as the 'black hole' in her life. It does seem possible, even probable, that she unconsciously focused on this loss rather than face the even more painful fact of her distant relationship with her mother, and her near-obsessive need to placate her parents.

There was the fact that the Hills had been taken over by the religion of the Oxford Group and its Absolutes, including their version of Absolute Purity. The mother's hysteria at her oldest daughter's necessary rebellion so affected Beatrice that she became the almost overly dutiful daughter: that way she would not risk more loss of love. If she had been able to have had boyfriends normally instead of feeling that she must become engaged – at 16 – and then have married, so very young and inexperienced at 20; if she had known more of life and been more worldly wise; and if in her reaction against her parents' religious beliefs she had not had such a need for someone who 'talked science instead of religion', then her life would almost certainly have taken a different turn.

Chance had it that the new husband and wife went to a centre where she could not flourish, and where she could not use her scholarship as she had intended. After his original contract period and then its renewal, her husband found it hard to take active steps to find a university with work for them both, and move on, as they had told each other they would do. Had she been able to conceive a baby when she planned to, if she had had more idea of the realities of motherhood and indeed of what is involved in warm, close mothering, if their marriage relationship had been more compatible – all these things militated against her and her family.

Her final tragedy might have been averted if Beatrice had consulted a doctor years earlier, instead of summarily rejecting her husband's suggestion that she 'prettify' herself, as she called it, by having the mole on her leg removed.

There could very well have been other, different, problems if he had not refused to let her take the children away with her, as she had planned to do. As it was, she believed that her pain and guilt at leaving the children was the trigger for her melanoma. Even then she was told she had a 50 percent chance of surviving her cancer. One chance in two ... and again she lost.

One could go further, and speculate. What if she had been born into a different family in a different place at a different time, even 20 years later when the women's movement was beginning to spread its ideas and convictions deep into most crevices of Western society?

And what if she had been born a man?

But then she would not have been Beatrice, the golden girl who grew into a woman whom nobody, anywhere, could forget.

The mistakes she made in her personal life – errors of judgment, and omissions as well as commissions – were at least partly because she was subtly and inexorably nudged into a web of conventional expectations. In her later Dallas years these hardened into the bars of a cage. Her personality changed. New Zealand friends and colleagues would scarcely have recognised her. For a time the golden girl became cynical, suspicious and inward-turning. Yale helped her, just as she helped Yale. In her later years, and in the years when she was dying, she became a mature version of her earlier self, her real self. Sandy Faber said of her, 'Beatrice acquired wisdom at the end of her life, but it was so, so sad that she and the children had to wait till then.'[25]

Rowena Hill, poet and student of philosophy, has said this:

> Without blame now, (knowing something of how they themselves were brought up), just trying to understand – I would say that both our parents were responsible for making us grow up unable to see what is because of the pressure to see – and be – what should be ...
>
> Beatrice and I, in our very different ways, became possessed by the search for other answers to the big questions, since we couldn't accept the family ones. We were always being made to think about God, ultimate destiny and universal values, and this, obviously, is part of what set us off on our search ...
>
> Probably what harmed us most was the lack of basic emotional security and support, which we never got from our mother – her inability to give it, plus the feeling that approval and even love depended on our obeying the rules, left us with a shaky base on which to build our personal universes.
>
> An insecure personal base together with an obsession with answering the deepest questions (and a passionate nature!) made us into tightrope-walkers. Beatrice's tightrope led her out into the spaces of the physical universe (mine went into inner spaces).
>
> I am not overlooking her immense talent, but trying to see her career from a more personal point of view – and while intellectually she was perfectly at home out there, the kind of strains her obsession set up at other levels of her life – whether her marriage or relationships with others involved in the same obsession – were, in themselves, critical.
>
> And, at the same time, she did not have the emotional resources to cope with them without damage to herself. There were some far-down falls ...[26]

Far-down falls, yes. And also creativity, immense joys and fulfilment which are the lot of few people, anywhere.

Before she died, Beatrice had time to lie back on her pillows, her hands folded, and think over her life. She did not have her sister's insights into her own nature. She was not a philosopher. But, like her father, she sought truth as she conceived it. Just before her health finally deteriorated in her last three days, Jean Audouze managed to speak to her on the phone from Paris. He was so moved by his memories of her, and by what she then said to him, that he at once wrote down her words as he remembered them.[27]

Beatrice had known that this was almost certainly the last time she would ever be able to put into words, to a trusted friend and colleague, how she really felt, what she believed to be true, and what she wanted to

be remembered of herself:
  She said:

I have no regrets and no resentments. I have been spoilt by life.

  Although I was a woman and a foreigner, I have been able to make my way in science in the United States.

  I have enjoyed working and interacting with so many friends here in the States and throughout the world, especially in England, in France, in Italy and in Europe in general.

  My children have been a blessing to me. I feel I have been lucky in many ways, and I am happy for all these reasons.'

Epilogue

Edward Hill lived on in New Zealand another 20 years after Beatrice's death, dying in 2001 at the age of 94. There is an Edward Hill Collection at the Alexander Turnbull Library, Wellington.

Rowena Hill retired from the University of the Andes in Merida in 1998 and continues to be based in Venezuela. Her interests remain India, Buddhism and poetry.

Theodora Lee-Smith teaches the piano in Upper Hutt, where her librarian husband, David, has retired.

Alan Tinsley graduated from the University of Texas in Austin in electrical engineering, then went on to a career designing microchips for computers and cell phones. Married to Michelle, he lives in Phoenix, Arizona.

Terry Tinsley, who has reverted to her given name of Teresa, has a son, Jacob. She lives in Texas and retains an affectionate relationship with Richard Larson. She currently manages a small business.

Brian Tinsley is still at the University of Texas in Dallas. His work in spectroscopy takes him to many parts of the world, and he has been involved for more than 40 years in observational and theoretical research in upper atmosphere processes. He has served on many national and international organisations in this field.

Richard Larson remains on the faculty of the Department of Astronomy at Yale. He continues to be interested in star formation, and the formation and evolution of galaxies, with recent interests in star formation in the early universe, the growth of black holes in galactic nuclei, and the origin of stellar masses.

Sandra Faber has continued to study observational cosmology with an emphasis on galaxy formation and high redshift galaxies. Soon after Beatrice's death, she and colleagues George Blumenthal and Joel Primack published the Cold Dark Matter theory of galaxy formation, which has since become standard. She participated in the building of the 10-metre Keck Telescopes and was one of a small team of astronomers who diagnosed the optical flaw in the Hubble Space Telescope. After Hubble, she led the contruction of the state-of-the-art DEIMOS spectrograph for Keck and is now using it with colleagues to conduct the DEEP survey of the distant universe, a redshift survey of 50,000 galaxies that is unravelling the secrets of galaxy formation over the last half of cosmic time. Currently department chair at UC Santa Cruz and a staff member of UCO/Lick Observatories, she feels it is one of the saddest things in her scientific career that Beatrice never got a chance to see how the story of galaxy formation evolved.

Jim Gunn and Jill Knapp married in 1982, adopted two children from Peru in 1990, are now grandparents, and are professors of astronomy at Princeton. Since 1990 they have worked with a large multi-institutional team to build and operate the Sloan Digital Sky Survey, for which Jim built the multipixel camera, and is project scientist. The aims of the SDSS include mapping the locations of galaxies and quasars on small to large scales and measuring the structure and evolution of the universe. The SDSS's successes include the discovery of the epoch of the initial ionisation of the universe, the measurement of the total masses of galaxies, detailed and accurate measurements of the cosmic parameters, detailed information about the star formation and chemical history of galaxies, and a major shift in the sociology of astronomy. This has made an enormous archive of forefront cosmological data available to anyone, not just those connected to large NASA projects or at institutions with large telescopes.

Jerry Ostriker, who succeeded Martin Rees as Plumian Professor of Astronomy at Cambridge, returned to Princeton where he heads PICSciE, the Princeton Institute for Computational Science and Engineering. He works on theoretical cosmology and on numerical hydrodynamic simulations of cosmic structure formation. In his role as department chairman, and later as Provost of Princeton University, he was largely responsible for the political acceptance and financial support of the Sloan Digital Sky Survey.

Jean Audouze is director of research at the Institute of Astrophysics in Paris, having been IAP's director 1978-89 and scientific adviser to President Mitterand. He has written about 200 research papers in nuclear astrophysics and cosmology, and some 10 popular books on astrophysics.

Linda Stryker is associate professor of astronomy at Arizona State West University and now holds the chair of the integrative studies department. She changed her focus to a broader view of science and its connections with other areas of study such as art, astronomy and philosophy, with emphasis on Beatrice Hill Tinsley's 'big questions'.

Said Jill Gunn at the beginning of 2006, 'We often wonder what Beatrice would have made of all this ...'

References

The following abbreviations are used:

B.H., then B.H.T., for Beatrice Hill Tinsley; E.H. for Edward Hill; R.H. for Rowena Hill; T.L.-S. for Theodora Lee-Smith; D.L.-S. for David Lee-Smith; B.T. for Brian Tinsley; R.L. for Richard Larson; S.F. for Sandra Faber; J.G. for James Gunn; J.K. for Jill Knapp; L.S. for Linda Stryker; J.A. for Jean Audouze.

Author's Notes & Acknowledgements
1. *Springboard for Women*, Cape Catley, 1985.
2. See Appendix II.

Introduction
1. Evelyn Lind to author, 19 December 1989.

Chapter One
Beginnings
1. The manuscript of 'Half a Life' is in the Edward Hill Collection in the Alexander Turnbull Library, Wellington.
2. E.H. to author, 9 December 1994.
3. Mattie Hill (whom Edward married after his first wife died) to author, 13 July 1994.
4. 'Half a Life'.
5. E. H. to author, 10 December 1994.
6. Ibid.
7. Ibid.
8. Formerly called Monmouthshire.
9. E.H. to author, 10 December 1994.
10. Ibid.
11. A later Gavin Morton, great-great-nephew of the first Gavin, was killed in World War II when he was a pilot of a Royal Air Force bomber. He was an outstanding mathematician, gaining the major scholarship to Trinity College, Cambridge, of his year.
12. *Three Generations in a Family Textile Firm*, Jocelyn Morton; Routledge and Kegan Paul, 1971.
13. E.H. to author, 9 December 1994.
14. 5 November 1938.
15. 'Half a Life'. Edward Hill says: 'Whatever may be thought now of the public school system of education for the upper class, it was a fact then that it taught the young men involved a sense of obligation to people for whom they became responsible.'
16. 'Half a Life'.
17. Ibid.
18. *My Daughter Beatrice*, published by the American Physical Society in 1986, is a selection from Beatrice's letters home over the years, with a framework narrative by Edward Hill.
19. Ibid.
20. E.H. to author, 7 April 1992.
21. R.H. to author, 10 January 1992.
22. E.H. to author, 4 December 1994.
23. Kate Morton to author, 30 December 1991.

Chapter Two
A Sea-Change
1. R.H. to author, 17 January 1997.
2. Kate Morton to author, 30 December 1991.
3. Teddy Fardell to author, 4 October 1990.
4. E.H. to author, 4 December 1994.

5. Teddy Fardell to author, 7 May 1994.
6. All Beatrice's writings in this book are reproduced exactly as she wrote them.
7. Teddy Fardell to author, 7 May 1994.
8. Mattie Hill to author, 13 July 1990.
9. R.H. to author, 31 August 1994.
10. T. L.-S. to author, 16 April 1994.
11. 'Half a Life'.
12. Teddy Fardell to author, 4 October 1990.
14. R.H. to author, 12 December 1985.
15. T. L.-S. to author, 16 April 1994.
16. E.H. to author, 12 November 1994.
17. Vinnie Ross to author, 19 June 1990.
18. E.H. to author, 3 December 1994.
19. Jean Ross to author, 18 May 1994.
20. This diary was inherited by her daughter, who became Dr Ruth Flashoff.

Chapter Three
Towards a Life's Work

1. R.L. to author, 2 September 1991.
2. Sylvia Philips to author, 18 February 2001.
3. Hilda Veale to author, 19 June 1990.
4. Ibid.
5. Unless otherwise stated, all schoolgirls and teachers quoted in this chapter were originally interviewed for the school's centennial history, *Springboard for Women*.
6. Teddy Fardell to author, 7 May 1994.
7. Kate Morton to author, 30 December 1991.
8. The mountain is now called Mt Taranaki.
9. New Zealand's population in 1953 was 2,037,000.
10. E.H. to author, 12 November 1994. Edward Hill therefore remained licensed to conduct marriages, and did so as late as the early 1990s.
11. Teddy Fardell to author, 4 October 1990.
12. Kate Morton to author, 30 December 1991.
13. Orders were given for the manufacture of hundreds of thousands of small Union Jacks for schoolchildren and many adults so they could wave to the Queen. By the tour's end an estimated seven out of every 10 New Zealanders had managed to see the Queen.
14. E.H. to author, 3 December 1994.
15. *Springboard for Women*.
16. Ibid.
17. Ibid
18. Crowds gathered everywhere, as in the small North Island rural settlement of Tirau. Its normal population of 600 swelled to 10,000 on the day the Queen and Duke of Edinburgh slowly drove through, without stopping.
19. E.H. to author, 12 November 1994.
20. Virtually only one cheese, cheddar, and very few wines, indifferent at that, were produced in New Zealand up to the 1970s when foundations were laid for the production of the range of gourmet cheeses and wines now rated as among the world's best.
21. R.H. to author, 17 January 1997.
22. Mary Heward to author, 18 June 1980.
23. Teddy Fardell to author, 4 October 1990.
24. *Springboard for Women*.
25. E.H. to author, 1 December 1994.
26. In lieu of its own university, the province of Taranaki enjoyed an income from especially endowed lands. For many years this income was applied to Taranaki Scholarships, which were available to all locally educated young people who passed with credit the annual university scholarship examinations.
27. It is popularly claimed that Hoyle intended 'the Big Bang' to be a derisive term. Hoyle refuted this notion. In Len Crosswell's *The Alchemy of the Heavens*, Oxford University

Press, 1995, Hoyle explained that he wanted a 'striking' phrase that would arrest and hold the attention of his radio audience. 'There was no way in which I coined the phrase to be derogatory.'
28 Joyce Jarrold to author, 2 March 1984.
29 Lynne Hall to author , 3 January 2001.
30 Hilda Veale to author, 19 June 1990.
31 Janice Griffin to author, 24 June 1994.
32 *Springboard for Women.*
33 E.H. to author, 12 November 1994.
34 *Springboard for Women.*
35 Alma Davies to author, 20 June 1984.
36 *Springboard for Women.*
37 Ibid.

Chapter Four
Freedom to Learn
1 Colin Broadley to author, 20 February 2001.
2 The two men stayed in touch over the years, to the extent that Edward Hill asked Colin Broadley to be master of ceremonies at his 80th birthday celebrations.
3 Colin Broadley to author, 20 February 2001.
4 Pip Jackson to author, 8 February 1991.
5 When the university left its Christchurch city site for Ilam, a new hall of residence on Ilam Road was named Helen Connon.
6 This room, Rutherford's Den, is to the right in the Clock Tower building of the old Canterbury University in Worcester Boulevard in what is now the Arts Centre of Christchurch. It was restored and opened to the public in 1978.
7 It is unclear why Beatrice said that atomic science was what she wanted to do as it is only one area of the physics important in astronomy. Had she tried to spell out her meaning for her parents, she might have said that the analysis of star light shows what kinds of atoms are present, leading to the question of how they got there.
8 Glen Metcalf continued her research into methods for analysing steroid levels in women with endocrine disorders. She gained her PhD in 1972 and has published some 70 papers.
9 R.H. to author, 16 January 1991.
10 Nita Hanna to author, 18 October 1990.
11 Ibid.
12 Jan Heine to author, 5 September 1994.
13 John Ritchie became professor and head of the Music Department. A clarinetist and composer, he founded the John Ritchie Orchestra, out of which grew today's Christchurch Symphony Orchestra.

Chapter Five
What If
1 Nita Nitschke (Hanna) was later to do research for the New Zealand Wool Board on the chemistry of the microbes on wool fibres.
2 Nita Hanna to author, 18 October 1990.
3 David Hanna to author 18 October 1990.
4 Nita Hanna to author, 25 May 1994.
5 Jan Heine to author, 5 September 1994.
6 Penny Saunders to author, 24 November 1991.
7 Rae Julian to author, 14 July 1990.
8 Undated letter to the Hills.
9 Gordon Ogilvie to author, 20 June 1998.
10 2 October 1959.
11 This ms, 'Chronicles of an Unquiet Land, A Study of Taranaki, 1816-1881', is in a private family collection. It was a considerable undertaking for the man who now became 'a retired gentleman of means' in Wellington. Had Edward Hill consulted appropriate authorities as he researched, his work could have been more fruitful. Publishers have

indicated that, although well researched and written, it has no clear readership, being neither an academic text nor accessible enough for general readers.
12. Don Locke at that time was a temporary lecturer in philosophy. He later became professor of philosophy at Reading University.
13. Undated letter to the Hills.
14. Father and son shared a Nobel Prize for this work.
15. The letter is illegibly dated, probably 3 July 1960.
16. Richard Easther to author, 29 September 2005: Bragg was describing the mechanics of DNA, still big news then, and Beatrice's letter shows that her interests were broad enough for her to appreciate the importance of this work.

Chapter Six
A Different Language
1. Nita Hanna to author, 18 October 1990.
2. Guff France to author, 27 June 1994.
3. B.T. to author, 30 July 1990.
4. Ibid.
5. Now Mt Taranaki.
6. Brian Tinsley dropped from 8 1/2 stone to 8 stone during the day.
7. B.T. to author, 30 July 1990.
8. Harvey McQueen became a poet and an education consultant, including a term as education adviser to Prime Minister David Lange. His book, *The Ninth Floor*, is about this period in his life.
9. Harvey McQueen to author, 14 July 1990.
10. E.H. to author, 8 December 1989: Arnold Nordmeyer and Edward Hill became friends. Towards the end of Nordmeyer's life he told Hill that the only reason he had included in the Budget those 'savage taxes' – as he and almost everyone else called them – on liquor, and particularly the country's main drink, beer, was because Treasury had insisted they were essential.
11. Harvey McQueen to author, 14 July 1990.
12. Guff France to author, 27 June 1994.
13. Teddy Fardell to author, 4 October 1990.
14. E.H. to author, 12 November 1994.
15. *Springboard for Women*.
16. Archie Ross to author, 5 October 1990.
17. Ibid.
18. B.T. to author, 7 October 1990.
19. Phyl Wardell to author, 4 October 1990.
20. Teddy Fardell to author, 4 October 1990.
21. Harvey McQueen to author, 14 July 1990.
22. B.T. to author, 7 October 1990.

Chapter Seven
Mr and Mrs Tinsley
1. This book uses the name of Beatrice Hill Tinsley, in accordance with her wish.
2. 26 May 1961.
3. William Pickering was responsible for the launching of America's first satellite Explorer 1 in 1958, the Ranger and Surveyor missions which photographed the moon to prepare for the epochal manned lunar landing, and the early deep space robotic missions to the planets.
4. Marilyn Head, 'Star Man', *NZ Listener*, 5 April 2003.
5. 'Gerry Gilmore', by Marilyn Head, *NZ Listener*, 17 January 2004.
6. *Letters Home: Correspondence 1950-63*, edited with a commentary by Aurelia Schober Plath, London: Faber 1975; New York, Harper & Row, 1975.
7. Phyl Wardell to author, 10 October 1990.
8. Guff France to author, 27 June 1994.
9. Ann Ballin to author, 8 May 1994.

10. B.T. to author, 7 October 1990.
11. Mark Sadler to author, 20 May 1990.
12. These included a 1962 *Canta* feature article in which Brian Tinsley called himself 'Second Thoughts', and a feature under his own name in *Hemlock 4*, February 1963.
13. Interviews with Bruce Moon and Joan Williman, 11 September 1995.
14. New Zealand's first computer, an IBM 650, had been imported from the United States by Treasury in late 1960, and was used at first for public servants' salaries.
15. *Exponential Growth: Lincoln University's Computing Story to 1990*; Neil Mountier, Centre for Computing and Biometrics, Lincoln University.
16. Alan Williman to author, 11 September 1995.
17. A speaker at the wedding of Joan Lester to Alan Williman said, 'I don't know if Alan has married Joan for love or to get more time on the computer.'
18. Bruce Moon to author, 11 September 1995.
19. An ionised particle, or ion, is an electrically charged particle. When an atom or molecule is ionised, an electron is split off it.
20. Rowena's and José's son Andres, usually called Sito, was born on 12 September 1962. Fourteen months later they were to have a daughter, Cecelia, usually called Lila.
21. Jean Hanlin to author, 8 May 1994.
22. This was also observed by New Zealand amateur astronomer Frank Bateson, who was responsible for establishing Mt John University Observatory.
23. Guff France to author, 27 June 1994.
24. 4 September 1962.
25. 3 November 1962.
26. *My Daughter Beatrice*.
27. Phyl Wardell to author, 10 October 1990.
28. Beatrice's coaching did not enable her Ugandan student to pass his exams.
29. Edward Janus to author, 6 November 2000.
30. 6 July 1963.

Chapter Eight
To A New World

1. 30 October 1963.
2. He was Welsh.
3. 30 October 1963.
4. Ivor Robinson to author, 28 July, 1991.
5. Lyndon Johnson.
6. Lee Harvey Oswald, later shot by Jack Ruby.
7. Some years later, Ivor Robinson was to come across a paper whose Russian authors gravely said they were unable to find any use of this term, relativistic astrophysics, before 1963.
8. They were quasars.
9. 15 February 1964.
10. S. Chandrasekhar, 'Shakespeare, Newton and Beethoven', Ryerson Lecture, University of Chicago, 1975; reprinted in S. Chandrasekhar, *Truth and Beauty*, University of Chicago Press, 1987.
11. In 1915 Karl Schwarzschild gave an exact solution to Einstein's equation which described the geometry of space around *non-rotating* stars and predicted the extreme conditions of what is now called a black hole, but which he called a singularity. But since it was known that stars do rotate, Einstein declared, 'Schwarzschild singularities (black holes) do not exist in physical reality!'
12. Kip Thorne is professor of theoretical physics at Caltech.
13. *Black Holes and Time Warps: Einstein's Outrageous Legacy*, by Kip Thorne, W.W.Norton & Co, 1994.
14. Writing about Roy Kerr in 'Man of Mystery', *NZ Listener*, 25 September 2004, Marilyn Head commented that most of those present at the Texas Symposium still failed to grasp that it was this 'esoteric' solution which would ultimately describe the mysterious source of energy of the very objects they were interested in.

15 José Fajardo did not speak English, and altogether there were too many stresses and strains for them to settle in New Zealand. After some months they returned to Italy.
16 3 April 1964.
17 1 May 1964.
18 Ibid.
19 Ibid.

Chapter Nine
The Doctorate
1 Alan McDiarmid, Nobel Prize-winner in Chemistry, was New Zealand's third scientist laureate, after Ernest Rutherford and Maurice Wilkins.
2 When faculty wives were asked to submit recipes for a book, Beatrice offered one as from Beatrice Tinsley. It was presented as from Mrs Brian Tinsley. Friends years later remembered her annoyance at this.
3 R.L. to author, 3 September 1991.
4 Ibid.
5 Harlan Smith to author, 10 December 1985.
6 18 July 1964.
7 Ibid.
8 R.L. to author, 16 January 2000.
9 Ron Angione to author, 27 September 2001.
10 From de Vaucouleurs's testimonial when supporting Beatrice's job application in 1973.
11 *My Daughter Beatrice*.
12 Graham Hill to author, 22 January 2001.
13 John Williams to author, 9 December 1985.
14 Terry Deeming to author, 13 December 1985.
15 Ron Angione to author, 27 September 2001.
16 de Vaucouleurs to author, 14 December 1985.
17 Beatrice did not tease her father by sending him one of her favourite pieces of history based on the Bible, although he later discovered it as an aside in a clipping she had sent him – and laughed. It was to the effect that Bishop Ussher of Armagh, Ireland, sat down in 1648 to count the number of 'begats' in the Book of Genesis. He was thus able to conclude that the universe began in 4004 BC. Within a few years the Vice-Chancellor of Cambridge University, Dr John Lightfoot, fixed the day and hour as Sunday, 23 October in the year 4004 BC, at 9am.
18 Interview with Samuel E Bleeker, *Who's Who in Astronomy, Beatrice M Tinsley's contribution to cosmology*, 1979.
19 Nick Woolf to author, 14 December 1985.
20 4 April 1965.

Chapter Ten
The Biggest Change
1 New Zealand was a world pioneer in 1952 with its Parents Centre movement, which offered classes for pregnant women and their husbands. They aimed at preparing them for childbirth, and stressed their babies' emotional as well as physical needs. The Parents Centre movement spread throughout New Zealand, and, under various names, throughout much of the Western world, but in 1966 its work was still little known. See *The Trouble with Women*, by Mary Dobbie, Cape Catley Ltd, 1990, a history of the New Zealand Parents Centre movement.
2 B.T. to author, 20 January 2002.
3 *The Trouble with Women*.
4 1 June 1966.
5 Even the eminent Marie Curie, already a Nobel Prize-winner, had such an effect when she first taught physics at a girls' lycée in Sèvres, France. Susan Quinn in *Marie Curie, A Life*, Heineman, London, 1995, writes of how bored her students were, and how they made up a sarcastic ditty:

> 'Wouldn't she be better off
> Cooking for her husband-prof
> Instead of talking in a stream
> To a class that's bored enough to scream?'

'The students did not understand Marie Curie's presentations, largely because the formulas and equations which accompanied them went beyond their mathematical knowledge.' After this first year of teaching, she figured out how to reach her class, and became much loved and valued.

6   Bob Tull to author, 10 December 1985.
7   29 January 1967.
8   Sachiko Tsuruta to author, 8 November 2003.
9   James Gunn is sometimes said to be the leading astronomer of the 20-21st centuries. He and his team at Princeton are working on one of the most ambitious projects in astronomy, the Sloan Digital Sky Survey. They are constructing a huge colour scanner to give an electronic three-dimensional map of the universe. It is expected to show a million quasars and a hundred million galaxies.
10 J.G. to author, 31 August 2001.
11 29 January 1967.
12 It was published in November 1967. Gérard de Vaucouleurs's introduction to this paper was: 'The present report by Mrs Tinsley arises from her remarkable dissertation work on the evolution of galaxies, 1967.' Beatrice was not pleased by that demeaning use of the word *Mrs,* which she underlined in her copy.
13 Penny Saunders to author, 24 November 1991.
14 Published by Harper Collins, 1991.
15 Gérard de Vaucouleurs to author, 13 December 1985.
16 R.L. to author, 30 August 1991.
17 *Status* January 2005, based on Robert C. Kennicutt Jr's address to the Garching conference in 2003, the conference dedicated to the memory of Beatrice Tinsley.
18 4 January 1968.
19 Ibid.
20 Bea Wolf to author, 16 March 2001.

Chapter Eleven
The Family Of Four
1   17 May 1968.
2   12 July 1968.
3   Guff France to author, 27 June 1994.
4   27 December 1968.
5   B.T. to author, 29 July 1991.
6   Ibid.
7   J.G. was the first of many to say to author, 2 September 2001, that Beatrice would have bombarded colleagues nearly into the ground with a constant barrage of ideas if she could have had email as a tool.
8   It was published in September 1969 in *Astrophysical Letters.*
9   Published in *Astrophysics and Space Science* in 1970. Vol 6.
10 Attributed, in various wordings, to Edmund Burke, but not found in his writings: *Oxford Dictionary of Quotations,* 1992.
11 19 January 1970.
12 15 April 1970.

Chapter Twelve
Family Years
1   Alan Tinsley to author, 10 January 1992.
2   10 May 1970.
3   13 July 1970.
4   2 August 1970.

5   5 October 1970.
6   E.H. to author, 8 April 1992.
7   12 November 1970.
8   Mattie Hill to author, 13 July 1990.
9   16 January 1971.
10 7 February 1971.
11 12 August 1971.
12 6 September 1971.
13 Margaret Burbidge said, 'It is high time that discrimination in favour of, as well as against, women in professional life be removed.'
14 26 September 1971.
15 4 December 1971.
16 The world's population did not quite double in that time. In 1970 world population was officially 3,707,610,112. In 2000 it was 6,080,141,638.

Chapter Thirteen
Rocky Years
1   John Beverley Oke.
2   J.G. to author, 31 August 2001.
3   Ibid.
4   The 4-Shooter went fully into operation in 1980 and the first data was received after Beatrice's death. The data is known as the PDCS or Palomar Deep Cluster Survey.
5   J.G. to author, 31 August 2001.
6   Ibid.
7   Ibid.
8   Willie Fowler was to win a Nobel Prize in Physics in 1983 for his work on stellar nucleosynthesis.
9   Richard Easther, note to author, 2 April 2006: The expressions 'chemical evolution' and 'nucleosynthesis' are crucial to understanding Beatrice Tinsley's work. Immediately after the Big Bang, the universe was very hot and very dense. While high school science teaches that atoms are the 'fundamental building blocks of nature', not even atoms existed in the first moments of the universe.

   Instead, all the atoms in the universe today – in the distant galaxies, in the earth and solar system, or in our own bodies, were created some time after the Big Bang. Atoms all have a heavy central nucleus made of protons and neutrons, and the nuclei form first – hence the term 'nucleosynthesis'. The outer cloud of electrons is then added to make a stable and electrically neutral atom. Some three minutes after the Big Bang, the conditions were just right for an initial period of 'nuclear cooking', and most of the helium in the universe, along with traces of other light elements such as lithium and beryllium, were created in this epoch.

   The universe cooled as it expanded and, once the temperature dropped far enough, nuclear reactions became impossible. At this point the initial mixture of elements in the universe was fixed, and it did not change until the universe became big enough and cool enough for stars to form – some hundreds of millions of years later.

   Stars shine because of nuclear reactions in their cores, and the ash from these nuclear fires consists of more helium, plus heavier elements such as carbon and oxygen. Consequently, as a galaxy's stars burn, the chemical composition of that galaxy will slowly change – and this is the 'chemical evolution' Beatrice Tinsley was concerned with.

   Beatrice's work was chasing answers to several very different questions. On the face it, she and her colleagues were simply trying to understand the life history of the galaxies. By knowing the rate at which different chemical elements are created in stars, however, (and rare processes involving cosmic rays), they wanted to work their way backwards to find the composition of the primordial gas from which the first stars formed. This gives a window through which we can view the universe as it was a few minutes after the Big Bang.

   The scientific payoff does not stop here, however. It turns out that the detailed mixture of light elements produced during the phase of nuclear burning in the first few minutes after the Big Bang is exquisitely sensitive to the exact conditions that applied during this epoch, so by measuring the primordial abundances of rare, light elements like lithium and beryllium, rival theories of the Big Bang can be tested.

10. R.L. to author 16 January 2000.
11. Ibid.
12. J.G. to author, 2 September 2001.
13. S.F. to author, 14 December 1991.
14. 6 April 1972.
15. S.F. to author, 9 January 1992.
16. Introduction, Edward Hill's *My Daughter Beatrice*.
17. E.H. to Teddy Fardell, 15 March 1972.
18. Mattie Hill to author, 13 July 1990.
19. *My Daughter Beatrice*.
20. Jeno Barnothy and his wife Madeleine, astronomers then based in Chicago, had been corresponding with Beatrice. She was one of the few who seriously evaluated their many, often unorthodox, theories.
21. The Steady-State theory. Beatrice disagreed with it, and evidence was beginning to destroy it. Hoyle did not admit defeat.
22. 30 October 1972.
23. Glen Metcalf to author,
24. Teddy Fardell to author, 4 October 1990.

Chapter Fourteen
The Unravelling Family
1. 21 September 1974.
2. 28 January 1973.
3. Ibid.
4. Ibid.
5. 19 February 1973.
6. Sandy Faber to author, 9 January, 1992.
7. Virginia Trimble to author, 19 November, 2003.
8. J.K. to author, 9 December 1985.
9. Vera Rubin, Senior Fellow at the Department of Terrestrial Magnetism, Carnegie Institution of Washington DC. She is credited with work that led to the discovery of dark matter.
10. Vera Rubin to author, 26 June 2001.
11. J.K. to author, 2 September 2001.
12. J.A. to author, 17 July 2001.
13. Their joint paper, 'Galactic Evolution of the Light Elements', eventually appeared in the *Astrophysical Journal* in 1974.
14. 15 August 1973.
15. 10 September 1973.
16. Roger Cayrel built up the Canada-France-Hawaii observatory, and was its director for some 10 years; Giusa Cayrel is noted for work in stellar evolution.
17. Gérard de Vaucouleurs died in 1995.
18. Gregory Shields is now the Jane and Roland Blumberg Centennial Professor in Astronomy at the University of Texas in Austin.
19. Greg Shields to author 10 December 1985.
20. Alan Tinsley to author, 10 January 1992.
21. 18 December 1973.

Chapter Fifteen
Decision
1. 25 January 1974.
2. Ibid.
3. Craig Wheeler to author, 10 December 1985.
4. 7 February 1974.

[5] J.K. to author, 3 September 2001.
[6] J.G. to author, 3 September 2001.
[7] Alan Tinsley to author, 10 January 1992.
[8] 'Synthesis of Stellar Population' was published in 1975 in the *Memorie della Societa Astronomica Italiana*.
[9] R.H. to author, 24 June 2001.
[10] 14 July 1974.
[11] Penny Saunders to author, 24 November 1991.
[12] Alan Tinsley to author, 10 January 1992.
[13] 21 February 1975.
[14] To Theodora, Beatrice said it had taken her six weeks to realise this.

Chapter Sixteen
To A New Life
[1] Pierre Demarque to author, 31 August 1991.
[2] R.L. to author, 31 August 1991.
[3] Ibid.
[4] Pierre Demarque to author, 22 May 2003.
[5] 12 December 1974.
[6] The paper had been sent first, in 1973, to *Nature* which had rejected it as 'inappropriate'.
[7] David Schramm was to die when a plane, which he was piloting alone, crashed in 2000. Richard Gott is a professor of astronomy at Princeton.
[8] Richard Easther to author, November 2005.
[9] Ibid. 'To answer the question one needs to know two numbers: how much matter the universe contains (usually expressed in terms of its density), and how fast the universe is expanding – which is quantified by Hubble's constant. The mutual gravitational attraction of all the matter in the universe slows its expansion, and if there is enough matter (that is, if the density is high enough) this deceleration will be strong enough to bring the expansion to a halt, after which the universe recollapses towards a "Big Crunch". One has to make a couple of assumptions about the nature of the matter inside the universe, but the critical density, below which the universe is destined to expand forever, can be expressed in terms of the measured value of Hubble's constant.

The paper itself surveys a large number of arguments for establishing the current density of the universe, and concludes that they all tend to point towards a density that is well below the "critical" value, so that the universe will thus expand forever. With the passage of time, the paper's technical details are largely obsolete, but it caused a huge stir in cosmological circles.'
[10] *New York Times*, 23 December 1974.
[11] George Blumenthal to author, 7 January 1992.
[12] Bill Matthews to author, 8 January 1991.
[13] Rosemary Gunn to author, 13 August 2000.
[14] 7 February 1975.
[15] 21 February 1975.
[16] 15 March 1975.
[17] S.F. to author, 14 December 1991.
[18] Ibid.
[19] In this paper, which continues to be much quoted, the two authors present a compendium of their joint knowledge of the observed abundances of the elements in nature. They propose an analysis of the evolutionary processes which occur at a galactic scale, apply it to the solar neighbourhood, and propose some ideas regarding the evolution of more remote galaxies.
[20] The discussion revolved around the concept of the cosmological constant. Einstein invoked, then repudiated it, saying it was the worst mistake he had ever made. Beatrice said Einstein should not have repudiated it. Currently cosmologists believe that it may be vindicated, in ways not then foreseen. The consensus is that the universe contains a measurable cosmological constant, which works to increase the rate at which it expands. Einstein originally proposed the cosmological constant to allow him to construct a static

cosmological model, which neither expands nor contracts. Once it was realised (thanks to Hubble) that the universe was expanding, there was no necessity for a cosmological constant. It turns out, however, that the cosmological constant is a realistic possibility. Beatrice saw the theoretical possibility, but the first observational evidence was not produced until the late 1990s.

[21] 4 May 1975.
[22] S.F. to author, 8 January 1992.
[23] 4 May 1978.
[24] J.G. to author, 31 August 2001.
[25] Ibid.
[26] Psychologist William Domhoff.
[27] 2 June 1975.
[28] David Burstein to author, 10 January 1992.
[29] Gail Burstein to author, 7 January 1992.
[30] Alan Tinsley to author, 9 January 1992.
[31] S.F. to author, 23 January 2000.

Chapter Seventeen
Going Somewhere

[1] 27 June 1975.
[2] J.G. to author, 31 August 2001.
[3] Virginia Trimble to author, 4 November 2000. 'Beatrice had a feeling for fringe people. She could see the wheat in the straw.'
[4] J.G. to author, 2 September 2001.
[5] Virginia Trimble to author, 16 August 2000.
[6] Ibid.
[7] This article was published in *Scientific American* in March 1976.
[8] 29 July 1975.
[9] She was referring to Gus Oemler and Chris Wilson.
[10] Beatrice soon learned that he preferred to be called Richard.
[11] R.L. to author, 22 January 2000.
[12] Bill van Altena to author, 26 August 1991.
    The influence of the writer Betty Friedan, whose 1963 book *The Feminine Mystique* almost single-handedly revived feminism, and who in 1966 helped found NOW, the National Organisation for Women, spread rapidly through most areas of American culture.
[13] More than one observer has said that Beatrice would have been opposed to some of the ways in which affirmative action has been implemented in the US in the years since her death, in spite of the good intentions behind such moves.
[14] Pierre Demarque to author, 28 August 1991.
[15] Ibid.
[16] Dorrit Hoffleit to author, 29 August 1991.
[17] J.K. to author, 3 September 2000.
[18] Bill van Altena to author, 26 August 1991.
[19] R.L. to author, 1 September 1991.
[20] 24 September 1975.
[21] R.L. to author, 1 September 1991.
[22] 16 October 1975.
[23] Jim Rose to author, 29 October 1991.
[24] 12 November 1975.
[25] S.F. to author, 13 December 1991.
[26] Jim Gunn was to laugh at being included in this, saying, 'What, me, a poor boy from Texas, knowing Plato?' J.G. to author, 3 September 2000.
[27] 7 November 1975.
[28] Ibid.
[29] 7 December 1975.
[30] R.L. to author, 14 January 2000.

31 Nelson Caldwell to author, 26 June 2002.
32 R.L. to author, 1 September 1991.
33 Bruce Twarog to author, 5 July 2003. He is a professor of astronomy at the University of Kansas in Lawrence.
34 Barbara Anthony-Twarog to author, 7 March 1996.
35 Ibid.
36 Alan Tinsley to author, 10 January 1992.
37 Nelson Caldwell to author, 26 June 2002.
38 6 April 1976.
39 Alvio Renzini to author, 28 October 2003.
40 23 April 1976.
41 26 May 1976.
42 R.L. to author, 2 September 1991.
43 J.K. to author, 10 December 1985.
44 Ibid.
45 Proceedings of this session were published by Rydel.
46 Scott Tremaine to author, 10 October 2003.
47 Curt Struck to author, 4 April 2001.
48 Ibid.
49 Nelson Caldwell to author, 26 June 2002.
50 23 September 1976.
51 R.L. to author, 15 January 2000.
52 Martin Rees became Astronomer Royal.
53 Bob McClure to author, 10 August 2003.
54 R.L. to author, 31 August 1991.
55 Alan Tinsley to author, 9 January 1992.
56 Gary Steigman to author, 22 June 2003.
57 Curt Struck to author, 4 April 2001.

Chapter Eighteen
Year Of The Galaxies
1 18 February 1977.
2 R.L. to author, 14 January 2000.
3 R.L. to author, 3 January 1992.
4 Curt Struck to author, 4 April 2001.
5 Gary Steigman to author, 22 June 2003.
6 15 March 1977.
7 Curt Struck to author, 4 April 2001.
8 Ivan King to author, 18 July 2003.
9 Nelson Caldwell to author, 26 June 2002.
10 Ibid.
11 R.L. to author, 14 January 2000.
12 Nelson Caldwell to author, 26 June 2002.
13 Curt Struck to author, 4 April 2001.
14 The Hubble constant measures the rate at which the universe is expanding.
15 Curt Struck to author, 26 June 2002.
16 Ivan King to author, 20 July 2003.
17 Ken Freeman to author , 20 October 2003.
18 R.L. to author, 14 January 2000.
19 J.G. to author, 3 September 2001.
20 Barbara Anthony-Twarog to author, 25 April 2001.
21 Laser printing was not yet available. The first laser-printed thesis from Yale's Astronomy Department was in 1981, Linda Stryker's 'Stellar Populations in the Large Magellanic Cloud.'
22 5 August 1977.
23 Dorrit Hoffleit to author, 29 August 1991.
24 R.L. to author, 15 January 2000.

[25] 'Half a Life' has never been published, nor did Edward Hill complete the second part of his memoirs although he wrote a number of vignettes of family members and notable people he had met.
[26] 12 August 1977.
[27] R.L. to author, 15 January 2000.
[28] 12 August 1977.
[29] L.S. to author, 25 June 2001.
[30] 30 October 1977.
[31] Terry Tinsley to author, 5 January 1992.
[32] 20 November 1977.
[33] Dorrit Hoffleit to author, 29 August 1991.
[34] 2 December 1977.
[35] L.S. to author, 24 June 2001.
[36] Ibid.
[37] Curt Struck to author, 4 April 2001.
[38] 15 January 1978.

Chapter Nineteen
Time Of Crisis

[1] B.T. to author, 22 June 1999. Brian had taken a photo in 1970 which clearly showed this mole, but he thought he had mentioned it to Beatrice well before that. He also thought it possible that it had existed when he first met her.
[2] R.L. to author, 15 January 2000.
[3] Ibid.
[4] Ibid.
[5] L.S. to author, 24 June 2001.
[6] J.K. to author, 7 December 1985.
[7] Curt Struck to author, 14 April 2001.
[8] L.S. to author, 11 January 1992.
[9] 15 April 1978.
[10] W.W. Morgan mapped the spiral arms of the Milky Way Galaxy.
[11] 28 May 1978.
[12] Gary Steigman to author, 22 July 2003.
[13] Vera Rubin to author, 26 June 2001.
[14] 16 July 1978.
[15] R.H. to author, 16 January 1997.
[16] Peter Stetson to author, 1 August 2003.
[17] Frank Andrews to author, 2 May 2003.

Chapter Twenty
Face To Face

[1] 21 April 1979.
[2] 6 May 1979.
[3] Nelson Caldwell to author, 26 June 2002.
[4] R.L. to author, 31 August 2001.
[5] Jim Schombert to author, 16 December 2000.
[6] Ibid.
[7] Edwin Hubble had found this intermediate type of galaxy and named it S0. Of the two basic kinds of galaxies, the kind which form stars are spirals and flat, and the non-star-forming kind are elliptical, and round, whereas the S0 type is flat, but is not star-forming.
[8] 'Evolution of Disk Galaxies and the Origin of S0 Galaxies', *Astrophysical Journal*, 1980.
[9] 10 May 1979.
[10] Alan Tinsley to author, 10 January 1992.
[11] 4 June 1979.
[12] Published in 1980 by North Holland in a book, *Cosmologie Physique – Physical Cosmology*, edited by Audouze, Balian and Schramm.

[13] J. A. to author, 17 July 2001.
[14] 27 July 1979.
[15] The main thrust of what Jim Gunn and Beatrice were trying to do was to take the observations of galaxies and of stars in our own galaxy, and find out what mix of stars made the spectrum of the distant galaxy. This used the evolutionary synthesis technique which Beatrice pioneered, but at that time there was no homogeneous data set on either galaxies or stars. While Beatrice was still at Caltech, Jim began a long programme to get observations of stars.
[16] 'Evolutionary Synthesis of the Stellar Population in Elliptical Galaxies. lll. Detailed Optical Spectra.' Published in *Astrophysical Journal*, 1981. In 1983 Linda Stryker went to Princeton to work with Jim Gunn so that this part of the project could be finished, and published as a stellar atlas.
[17] L.S. to author, 12 June 2001.
[18] 27 July 1979.
[19] 22 August 1979.
[20] Matthew Hill to author, 13 August 2001.
[21] Gary Da Costa to author, 3 August 2000.
[22] This thesis was published in three parts: in the *Astronomical Journal* in 1983, the *Astrophysical Journal* in 1983, and the *Publications of the Astronomical Society of the Pacific*.
[23] Nelson Caldwell to author, 26 June 2002.
[24] Ibid.
[25] 5 October 1979.
[26] 22 October 1979.
[27] 29 October 1979.
[28] Virginia Trimble to author, 3 November 2000.
[29] R.L. to author, 4 November 2000.
[30] Virginia Trimble to author, 3 November 2000.
[31] Glen Metcalf to author, 6 April 2003.
[32] 8 December 1979.
[33] 17 December 1979.
[34] Nelson Caldwell to author, 18 January 2002.
[35] Terry Tinsley to author, 5 January 1992.
[36] Yvon Tinsley was later to take her children and leave her husband in Dallas.
[37] 24 January 1980.
[38] Monica Tosi to author, 21 April 2001.
[39] Ibid.
[40] 23 February 1980.
[41] Curt Struck to author, 26 April 2001.
[42] L.S. to author, 12 June 2001. Linda Stryker dedicated her thesis 'to Beetle'.
[43] Bill van Altena to author, 31 August 1991.

Chapter Twenty-One
Acceptance
[1] 21 August 1980.
[2] Ken Freeman to author, 20 October 2003.
[3] This book did not eventuate because of Beatrice's illness.
[4] S.F. to author, 15 February 2005: 'No single individual emerged although the field did continue to move on and flourish without her.'
[5] S.F. to author, 23 January 2000.
[6] Bea Wolf to author, 9 March 2001.
[7] R.L. to author, 22 May 2003: the heating was turned on or off by the building manager, on set dates, and Beatrice disliked being coddled.
[8] Bea Wolf to author, 9 March 2001.
[9] Ibid.
[10] Ibid.
[11] 13 October 1980.

12. In Richard's recollection, Beatrice had made it clear that she would never marry again so the subject was not discussed, except once, in connection with Terry's future.
13. 17 October 1980.
14. Linda Stryker has these poems. Beatrice also gave Richard some, written at low points in her life.
15. S.F. to author, 23 January 2000.
16. S.F. to author, 7 January 1992.
17. R.L. to author, 22 May 2003.
18. R.L. to author, 14 January 2000.
19. L.S. to author, 12 June 2001.
20. R.L. to author, 14 January 2000.

Chapter Twenty-Two
Letting Go
1. Vera Rubin to author, 26 June 2001.
2. Arno Penzias, along with Robert Wilson, discovered the cosmic microwave background. They shared the Nobel Prize for Physics in 1978.
3. R.H. to author, 11 December 1985.
4. J.K to author, 3 September 2001.
5. Bruce Twarog to author, 5 July 2003.
6. J.G. to author, 3 September 2001.
7. Vera Rubin to author, 26 June 2001.
8. L.S. to author, 12 June 2001.
9. More than a year later they were able to appoint J. Patricia Vader as assistant professor. Her field was dwarf galaxies.
10. R. L. to author, 4 September 1991.
11. Ibid.
12. May Guthrie to author 30 August 1991.
13. Monica Tosi to author, 23 April 2001.
14. Ibid.
15. Nelson Caldwell to author, 26 June 2002.
16. L.S. to author, 12 June 2001.
17. Ibid.
18. Ibid.
19. J.G. to author, 1 September 2001.
20. S.F. to author, 7 January 1992.
21. Alvio Renzini to author, 23 October 2003. He pondered for two weeks before replying to Beatrice with a long telex. He was too late. Bill van Altena answered him, saying Beatrice had died the day before. Renzini told this story, and others, greatly moving his audience at the Garching Conference, which was dedicated to Beatrice, in September 2003.
22. Curt Struck to author, 4 April 2001.
23. This was Richard's sister Mary, in Oxford Mills near Ottawa. She also visited Richard's mother and sisters Anne and Margie in Toronto.
24. See Publications.
25. L.S. to author, 7 September 2001.
26. Ibid.
27. Monica Tosi to author, 23 April 2001.
28. The New Zealander Charles Alcock is currently director of the Harvard-Smithsonian Center for Astrophsics. He has been principal investigator in the search for massive compact halo objects and their contribution to the dark matter component of the Milky Way's halo.
29. Curt Struck to author, 6 May 2001.

Chapter Twenty-Three
And Afterwards
1. J.K. to author, 3 September 2001.

2   Terry Tinsley to author, 5 January 1992.
3   Ibid.
4   Ibid.
5   R.L. to author, 20 May 2003.
6   This Beatrice Tinsley Award is now combined with the Newton Fund.
7   Besides S. Jocelyn Bell Burnell, other recipients of the Beatrice M. Tinsley Prize have been Harold I. Ewen, Edward M. Purcell, 1988; Antoine Labeyrie, 1990; Robert H. Dicke, 1992; Raymond Davis, 1994; Aleksander Wolszczan, 1996; Robert E. Williams, 1998; Charles Alcock, 2000; Geoffrey W. Marcy, R. Paul Butler, Steven S. Vogt, 2002; Ronald J. Reynolds, 2004; John E. Carlstrom, 2006.
8   The Beatrice Tinsley Visiting Professorship at the University of Texas at Austin has been held by: Gillian Knapp, September 1985; Vera Rubin, October-December 1988; Richard Larson, October-December 1990; Robert Kraft, September 1991-May 1992; Virginia Trimble, April-June 1992; Bengt Gustafsson, January-May 1994; Robert Kennicutt, October-December 1994; Anneila Sargent, selected in 1996 but unable to take up the professorship; Ewine Fleur Van Dishoeck, January-March 1997; James Truran, April-June 1999; Keith Horne, October 2000-May 2001; Ken Freeman, March-June 2001; Ralf Bender, March-April 2004 and February-April 2005; Tom Geballe, November-December 2005.
9   Barbara Anthony-Twarog to author, 26 April 2001.
10 Monica Tosi to author, 23 April 2001.
11 Jerry Ostriker to author, 31 August 2001.
12 J. K. to author, 31 August 2001.
13 J.A. to author, 17 July 2001.
14 Vera Rubin to author, 26 June 2001.
15 Harlan Smith to author, 9 December 1985.
16 S.F. to author, 20 January 2000.
17 New Zealand's population reached four million in May 2003.
18 Rowling omitted Nobel Prize-winners Maurice Wilkins and Alan McDiarmid.
19 Monica Tosi to author, 23 April; 2001.
20 Jerry Ostriker to author, 2 September 2001.
21 R.L. to author, 20 May 2003.
22 S.F. to author 22 Jan 2000.
23 Galaxy evolution remains an important area of astronomy. Beatrice's line of work has been continued, resulting in the development of ever more refined models of galaxy evolution.
24 J.G. to author, 2 September 2001.
25 S.F. to author, 22 January 2000.
26 R.H. to author, 17 January 1997.
27 J.A. to author, 17 July 2001.

Publications of Beatrice M. Tinsley

'Analysis of the optical absorption spectrum of neodymium magnesium nitrate,' *J. Chem. Phys.* 39, 3503-3508, 1963.

'Equivalent widths of interest for studies of the composition and evolution of galaxies,' Publ. Dept. Astron. Univ. Texas, Ser. II, Vol I, No. 15, 1-47, 1967.

'Evolution of the stars and gas in galaxies,' *Astrophys. J.* 151, 547-565, 1968.

'Distribution of the redshifts of quasars,' with T.N.L. Patterson and B.A. Tinsley, *Astrophys. Letters* 4, 55-56, 1969.

'Possibility of a large evolutionary correction to the magnitude-redshift relation,' *Astrophys. Space Sci.* 6, 344-351, 1970.

'Evolution of the M31 disk population,' with H. Spinrad, *Astrophys. Space Sci.* 12, 118-136, 1971.

'The color-redshift relation for giant elliptical galaxies,' *Astrophys. Space Sci.* 12, 394-407, 1971.

'The luminosity function of old-disk red giants compared with theoretical rates of evolution,' *Astrophys. Letters* 9, 105-108, 1971.

'Possibility that the far ultraviolet excess in M31 is due to main sequence stars,' *Astron. Astrophys.* 15, 403-405, 1971.

'Analysis of the magnitude-redshift relation including possible effects of evolution,' with J.-E Solheim, in IAU Symp. No. 44, *External Galaxies and Quasi Stellar Objects*, ed. D.S. Evans (Reidel, Dordrecht), 397-400, 1972.

'The possible line feature in the X-ray background,' *Astrophys. Letters* 10, 31-35, 1972.

'A first approximation to the effect of evolution on q,' *Astrophys. J. (Letters)* 173, L93-L97, 1972.

'The magnitude-redshift relation in Brans-Dicke cosmology,' *Astrophys. J. (Letters)* 174, L119-L121, 1972.

'Galactic evolution: program and initial results,' *Astron. Astrophys.* 20, 383-396, 1972.

'Effects of evolution on the diameter-redshift relation,' *Astrophys. J. (Letters),* 178, L39-L42, 1972.

'Stellar evolution in elliptical galaxies,' *Astrophys. J.* 178, 319-336, 1972.

'Star formation and evolution in spiral galaxies,' with W.J. Quirk, *Astrophys. J.* 179, 69-83, 1973.

'Dependence of the integrated background light on cosmology, galactic spectra, and galactic evolution,' *Astron. Astrophys.* 24, 89-98, 1973.

'A Critique of Hoyle and Narlikar's new cosmology,' with J.M. Barnothy, *Astrophys. J.* 182, 343-349, 1973.

'Photoionization by massive stars in protogalaxies,' *Astrophys. Letters* 14, 15-17, 1973.

'On the origin and evolution of the light elements,' in *Explosive Nucleosynthesis*, ed. D.N. Schramm and W.D. Arnett (University of Texas Press, Austin), 22-33, 1973.

'The cosmological significance of molecular band strengths in the infrared spectra of elliptical galaxies,' *Astrophys. J.* (Letters) 184, L41-L43, 1973.

'Analytical approximations to the evolution of galaxies,' *Astrophys. J.* 186, 35-49, 1973.

'Late stages of stellar evolution in the light of elliptical galaxies,' with W.K. Rose, *Astrophys. J.* 190, 243-251, 1974.

'Evolution of the nearby stellar population and its kinematics,' with P. Biermann, *Astron. Astrophys.* 30, 1-12, 1974.

'Galactic evolution and the formation of the light elements,' with J. Audouze, *Astrophys. J.* 192, 487-500, 1974.

'Photometric properties of model spherical galaxies,' with R.B. Larson, *Astrophys. J.* 192, 293-310, 1974.

'On stellar birthrates and age distributions,' *Astron. Astrophys.* 31, 463-465, 1974.

'Possible influence of comets on the chemical evolution of the galaxy,' with A.G.W. Cameron, *Astrophys. Space Sci.* 31, 31-35, 1974.

'On the correlation between M/L and colour for spiral galaxies,' with W.L.W. Sargent, Mon. Not. Roy. Astron. Soc. 168, 19P-22P, 1974.

'Necrology of the Hyades cluster,' Publ. Astron. Soc. Pacific 86, 554-557, 1974.

'Constraints on models for chemical evolution in the solar neighbourhood,' *Astrophys. J.* 192, 629-641, 1974.

'On the origin and evolution of s-process elements,' with D.N. Schramm, *Astrophys. J.* 193, 151-155, 1974.

'An unbound universe?' with J.R. Gott, J.E. Gunn, and D.N. Schramm, *Astrophys.J.* 194, 543-553, 1974.

'Predetonation lifetimes of Type II supernovae,' with P. Biermann, Publ. Astron. Soc. Pacific 86, 791-794, 1974.

'Galaxy counts as a cosmological test,' with G.S. Brown, *Astrophys. J.* 194, 555-558, 1974.

'Interpretation of the stellar metallicity distribution,' *Astrophys. J.* 197, 159-162, 1975.

'Nucleochronology and chemical evolution,' *Astrophys. J.* 198, 145-150, 1975.

'The evolution of galaxies and its significance for cosmology,' Ann. N.Y. Acad. Sci. 262, 436-446, 1975. (Invited review at Seventh Texas Symposium on Relativistic Astrophysics.)

'On the origin of S0 galaxies,' with P. Biermann, Astron. Astrophys. 41, 441-446, 1975.

'Is deuterium of cosmological or of galactic origin?' with J.P. Ostriker, Astrophys. J. (Letters) 201, L51-L54, 1975.

'Evolutionary synthesis of the stellar population in elliptical galaxies. I. Ingredients, broad band colors, and infrared features,' with J.E. Gunn, Astrophys. J. 203, 52-62, 1976.

'Effects of main-sequence brightening on the luminosity evolution of elliptical galaxies,' Astrophys. J. 203, 63-65, 1976.

'Composition gradients across spiral galaxies. II. The stellar mass limit,' with G.A. Shields, Astrophys. J. 203, 66-71, 1976.

'An accelerating universe?' with J.E. Gunn, Nature 257, 454-457, 1975.

'What stars become supernovae?' Publ. Astron. Soc. Pacific 87, 837-848, 1975.

'Luminosity functions and the evolution of low-mass Population I giants,' with J.E. Gunn, Astrophys. J. 206, 525-535, 1976.

'Rediscussion of the local space density of M dwarfs,'" with S.M. Faber, D. Burstein, and I.R. King, Astron. J. 81, 45-52, 1976.

'Chemical evolution of galaxies,' with J. Audouze, Ann. Rev. Astron. Astrophys. 14, 43-79, 1976.

'The indeterminacy of the age dependence of metallicities of nearby disk stars,' with R.D. McClure, Astrophys. J. 208, 480-486, 1976.

'Vital statistics of stars,' with J.P. Ostriker, Trans. IAU XVIA, Pt. 2 (Reidel, Dordrecht), 161-168, 1976.

'Chemical evolution in the solar neighbourhood. II. Statistical constraints, finite stellar lifetimes, and inhomogeneities,' Astrophys. J. 208, 797-811, 1976.

'Dynamical friction: the Hubble diagram as a cosmological test,' with J.E. Gunn, Astrophys. J. 210, 1-6, 1976.

'Distribution of evolved stars in Messier 67,' with I.R. King, Astron. J. 81, 835-839, 1976.

'Effects of metallicity on the mass-to-luminosity ratios of elliptical galaxies,' with H.A. Smith, Publ. Astron. Soc. Pacific 88, 370-373, 1976.

'Stellar production as a source of ^{3}He in the interstellar medium,' with R.T. Rood and G. Steigman, Astrophys. J. (Letters) 207, L57-L60, 1976.

'Evolution of chemical abundances and stellar populations,' in Galaxies, ed. L. Martinet and M. Mayor (Geneva Observatory), 155-251, 1976.

'Surface brightness parameters as tests of galactic evolution,' *Astrophys. J.* (Letters) 210, L49-L51, 1976.

'Galaxy counts, colour-redshift relations, and related quantities as probes of cosmology and galactic evolution,' *Astrophys. J.* 211, 621-637, 1977.

'Colors as indicators of the presence of spiral and elliptical components in N galaxies,' Publ. Astron. Soc. Pacific 89, 245-250, 1977.

'The cosmological constant and cosmological change,' *Physics Today* 30 (No. 6), 32-38, 1977.

'Chemical evolution in the solar neighbourhood. III. Time scales and nucleochronology,' *Astrophys. J.* 216, 548-559, 1977.

'Masses of supernova progenitors,' in *Supernovae*, e.d. D.N. Schramm (Reidel, Dordrecht), 117-129, 1977.

'Galactic evolution and the interpretation of cosmological tests,' in *Decalages ver le Rouge et Expansion de l'Univers; l'Evolution des Galaxies et ses Implications Cosmologiques*, ed. C. Balkowski and B.E. Westerlund (Meudon Observatory), 223-242, 1977.

'Final remarks: Connections between chemical and dynamical evolution,' in IAU Colloq. No. 45, *Chemical and Dynamical Evolution of Our Galaxy*, ed. E. Basinska-Grzesik and M. Mayor (Geneva Observatory), 309-319, 1977.

'The cosmological constant and cosmological change,' Ann. N.Y. Acad. Sci. 302, 423-436, 1978. (Invited review at Eighth Texas Symposium on Relativistic Astrophysics.)

'Star formation rates in normal and peculiar galaxies,' with R.B. Larson, *Astrophys. J.* 219, 46-59, 1978.

'Planetary nebulae and chemical evolution of the galaxy,' in IAU Symp. No. 76, *Planetary Nebulae, Observations and Theory*, e.d. Y. Terzian (Reidel, Dordrecht), 341-352, 1978.

'The past history of star formation in galaxies,' in IAU Symp. No. 77, *Structure and Properties of Nearby Galaxies*, ed. E.M. Berkhuijsen and R. Wielebinski (Reidel, Dordrecht), 15-21, 1978.

'The extragalactic background light and slow star formation in galaxies,' *Astrophys. J.*; 220, 816-821, 1978.

'Chemical evolution and the formation of galactic disks,' with R.B. Larson, *Astrophys.*, J., 221 554-561, 1978.

'Star formation rates and infrared radiation," with C. Struck-Marcell. *Astrophys. J.*, 221,. 562-566, 1978.

'HR diagrams of galaxies – ages and stages of evolution,' in IAU Symp. No. 80, *The HR Diagram*, ed. A.G.D. Philip and D.S. Hayes (Reidel, Dordrecht), 247-257, 1978.

'Evolutionary synthesis of the stellar population in elliptical galaxies. II.

Late M giants and composition effects,' *Astrophys. J.*, 222, 14-22, 1978.

'On the origin and evolution of isotopes of Carbon, Nitrogen, and Oxygen,' with D. Dearborn and D.N. Schramm, *Astrophys. J.* 223, 557-566, 1978.

'Accelerating universe revisited,' *Nature*, 273, 208-211, 1978.

'Observable properties of primeval giant elliptical galaxies,' with R.A. Sunyaev and D.L. Meier, *Comments on Astrophys.* 7, 183-195, 1978.

'The evolution of galaxies: evidence from optical observations,' in IAU Symp. No. 79, *Large-scale Structure of the Universe*, ed. M.S. Longair and J. Einasto, (Reidel, Dordrecht), 343-355, 1979.

'Stellar population explosions in proto-elliptical galaxies,' with R.B. Larson, Mon. Not. Roy. Astr. Soc. 186, 503-517, 1979.

'Stellar lifetimes and abundance ratios in chemical evolution,' *Astrophys. J.* 229, 1046-1056, 1979.

'Theoretical overview – Interactions among the Galaxy's components,' in IAU Symp. No. 84, *The Large-Scale Characteristics of the Galaxy*, ed. W.B. Burton (Reidel, Dordrecht), 431-440, 1979.

'Type I supernovae come from short-lived stars,' with A. Oemler, Jr., *Astron. J.* 84, 985-992, 1979.

'Evolution of the stars and gas in galaxies,' *Fundamentals of Cosmic Physics* 5, 287-388, 1980.

'Galactic evolution with the space telescope,' IAU Colloq. No. 54, Scientific *Research with the Space Telescope*, ed. M.S. Longair and J.Warner (U.S. Govt. Printing Office), 181-191, 1980.

'The detectability of young galaxies,' Phil. Trans. Roy. Soc. London A 296, 303-308, 1980.

'The ultraviolet continua of the nuclei of M31 and M81,' with C. - C. Wu, S.M. Faber, J.S. Gallagher, and M. Peck, *Astrophys. J.* 237, 290-302, 1980.

'The evolution of disk galaxies and the origin of S0 galaxies,' with R.B. Larson and C.N. Caldwell, *Astrophys. J.* 237, 692-707, 1980.

'On the interpretation of galaxy counts,' *Astrophys. J.* 241, 41-53, 1980.

'On the density of star formation in the universe,' with L. Danly, *Astrophys. J.* 242, 435-442, 1980.

'Relations between nucleosynthesis rates and the metal abundance,' Astron. Astrophys. 89, 246-248, 1980.

'Statistical evidence that Type I supernovae have short-lived progenitors,' with A. Oemler, in *Type I Supernovae*, ed. J.C. Wheeler (Univ. of Texas Press, Austin), 9-10, 1980.

'How much iron can each Type I supernova produce?' in *Type I Supernovae*, ed. J.C. Wheeler (Univ. of Texas Press, Austin), 196-198, 1980.

'Cosmology and galactic evolution,' in ? (North Holland, Amsterdam), 162-177, 1980.

'Correlation of the dark mass in galaxies with Hubble type,' Mon. Not. Roy. Astr. Soc. 194, 63-75, 1981.

'Evolutionary synthesis of the stellar population in elliptical galaxies. III. Detailed optical spectra,' with J.E. Gunn and L.L. Stryker, *Astrophys. J.* 249, 48-67, 1981.

'Chemical evolution in the solar neighbourhood. IV. Some revised general equations and a specific model,' *Astrophys. J.* 250, 758-768, 1981.

Book review on IAU Symp. No. 92, *Objects of High Redshift*, eds .G.O. Abell and P.J.E. Peebles, D. Reidel Publ. Co., 1980. Sky and Telescope 61, 336-339, 1981.

Bibliography

Publications cited or referred to in research

The letters of Beatrice Hill Tinsley to her family are in a private family archive.

Barrow, John D., *The Origin of the Universe*, HarperCollins, 1994
Catley, Christine Cole, *Springboard for Women*, Cape Catley, 1985
Chandrasekhar S., *Truth and Beauty*, University of Chicago Press, 1987
Cole, K.C., *Mind Over Matter*, Harcourt, 2003
Cole, K.C., *The Universe and the Teacup*, Harcourt Brace, 1998
Crosswell, Len, *The Alchemy of the Heavens*, Oxford University Press, 1995
Editors of Time-Life Books, *The Cosmos*, Series 1989
Ferris, Timothy, *Coming of Age in the Milky Way*, Morrow, 1988
Ferris, Timothy, *The Whole Shebang*, Touchstone, Simon & Schuster, 1997
Hamer, Dean, and Copeland, Peter, *Living with our Genes*, Doubleday, 1998
Hawking, Stephen W., *A Brief History of Time*, Bantam Press, 1988
Hawking, Stephen W., *Quest for a Theory of Everything*, Bantam, 1992
Hill, Edward, *My Daughter Beatrice*, American Physical Society of New York, 1986
Hoffleit, Dorrit, *Misfortunes Are Blessings in Disguise*, American Association of Variable Star Observers, 2002
Howe, Michael J. A., *Genius Explained*, Cambridge University Press, England, reprinted 2001.
Hoyle, Fred, *The Black Cloud*, Heinemann, 1957
Hoyle, Fred, *The Nature of the Universe*, Basil Blackwell, Oxford, 1950
Keller, Evelyn Fox, *A Feeling for the Organism*, W.H.Freeman, 1983
Levenson, Edgar, *A Perspective on Responsibility*, in Contemporary Psychoanalysis, 1978
Maddox, Brenda, *Rosalind Franklin, The Dark Lady of DNA*; HarperCollins, 2002
Mather, John C., and Boslough, John, *The Very First Light*, HarperCollins, 1996
Mountier, Neil, *Exponential Growth: Lincoln University's Computing Story to 1990*; Centre for Computing and Biometrics, Lincoln University
Munitz, Milton K. (ed), *Theories of the Universe*, The Free Press, 1957

Murdin, Paul, and Murdin, Lesley, *Supernovae*, Cambridge University Press, 1985
Neubauer, Peter B., and Alexander Neubauer, *Nature's Thumbprint*, Addison Wesley, 1990
Overbye, Dennis, *Lonely Hearts of the Cosmos*, HarperCollins, 1991
Plath, Aurelia Schober (ed. with a commentary), *Letters Home: Correspondence 1950-63*, London: Faber, 1975; New York, Harper & Row, 1975
Preston, Richard, *First Light, The Search for the Edge of the Universe*, Random House, 1996
Quinn, Susan, *Marie Curie, a Life*, William Heinemann, 1995
Stevenson, Anne, *Bitter Fame, A Life of Sylvia Plath*, Viking, England, 1989; Houghton Mifflin Co, USA, 1989
Thorne, Kip S., *Black Holes and Time Warps, Einstein's Outrageous Legacy*
Watson, James D, *Genes, Girls and Gamow*, Alfred A. Knopf, 2002
Watson, James D., *Double Helix*
Whineray, Scott, *Beatrice (Hill) Tinsley, 1941-1981, Astronomer*, Massey University and NZIPEC, 1985

Appendix I

EDWARD
Poem for my father
Rowena Hill

The form for telling of heroes is the epic;
this is a mini-epic, about an English gentleman.

Edward was born in the early years of the century,
his father had been an officer in the India army,
his beautiful mother studied the piano in Dresden
but gave up a career to be married and ride to hounds.
Edward was the second child, between two sisters
and early learnt the weight of 'ladies first'.
The only comforting arms and lap were Nanny's.

At prep school no one bothered to salve his chilblains,
and it was understood that little men aren't homesick.
Winchester was better, intelligence was allowed
and he was good at tennis (which his father said was for
    women).

At Oxford, after two terms with the smart set,
he joined a Christian crusade to make sense of life
and shocked the family by behaving like a socialist.

He got married and had no son to prove his manhood
(daughters didn't count and were not allowed to inherit);
he joined the army even before war started
and naturally took his place in the officer caste
but never saw battle because of fallen arches
while both his brothers were wounded and made much of.
When the war was over he said goodbye to all that –

or wanted to, but an English gent is born
and made and cannot be transformed so easily.
New Zealand acclaimed distinction only in sport
but Edward could never cut down his crests of attainment
or understand the force of sneering envy.

He rode his charisma, once, to power in a city
but the next election sent him back to limbo.

He was not to blame for the emptying out of an age –
for the babies of grace and honour going down the drain
with the bathwater of prejudice and social injustice.
It took a prophet's ear even at that late stage
to hear in Romantic chords from his wife's cello
soaring at night in the house full of English furniture
the bleeding to death of manhood as the West had known it.

The gorgeous pulses of man's heroic power
– hurled beyond measure, power for its own sake –
burst into abstract structures in space
where the human figure is lost, or dissolve
in morasses of effeminate emotion.
The heroic concept now is Hollywood
And sport and crude parallels in business;

but Edward had nothing to do with such travesties.
The steadfast ideal was part of his inheritance;
he had tried to strip it of its accreted lies
and found it fall apart from its rotten roots.
He wagered his life on the move against tradition,
but the prize of his courage was a diffident freedom
compromised by a never waning backlash.

Only his faith in God was constant, always.
Occasionally he preached, or advised commissions;
he wrote a history, rejected for lack of footnotes,
and librettos for a couple of short operas.
Through years of silence thickening around him
his talent for action transmuted into patience.
The spectre of wasted life he called God's will.

He would never consider going back to England
but stuck to his decision and his place
in the meltdown to the broth of a new culture,
only regretting when he was very old
he'd not had the chance to be more of a man,
but trusting in the Lord he'd always served
to raise him from the shadows after death.

Appendix II

OBITUARY
Beatrice Muriel Hill Tinsley

By Richard B. Larson and Linda Stryker, reprinted from the *Quarterly Journal of the Royal Astronomical Society* Vol. 23, 1982.

Beatrice M. Tinsley, one of the most widely known and respected theorists in modern astronomy, died on March 23 1981 at the age of 40, following a three-year battle with cancer. Her most untimely passing shocked and saddened all who had known her and who had been impressed by the extraordinary energy and verve with which she had pioneered many lines of study in cosmology and the evolution of galaxies.

Beatrice was born on January 27 1941 in Chester, England, the second of three daughters of Edward O.E. Hill and Jean Morton Hill. Her family moved in 1946 to New Zealand and eventually settled in the town of New Plymouth, where her father was an Anglican minister, and later mayor. She received her high-school education in New Plymouth and later attended the University of Canterbury in Christchurch, obtaining her BSc and MSc degrees in physics in 1961 and 1963. From an early age she had shown a special talent in mathematics and an abundance of industry and ambition, and as her interest in mathematics and science developed, so also did her determination to pursue her studies to the highest possible level and to excel and achieve distinction in them. Her exceptional ability and her enthusiastic pursuit of learning led, almost inevitably, to a brilliant scholastic record, and later to an equally outstanding scientific career.

In 1963, having recently married physicist Brian A. Tinsley, Beatrice moved to Dallas, Texas. Finding little opportunity there to pursue her scientific interests, she enrolled in 1964 as one of the first graduate students in the recently established Astronomy Department of the University of Texas in Austin. Despite commuting each week from Dallas and spending only half of the week in Austin, Beatrice mastered the graduate programme with awe-inspiring speed, and in 1967 completed a thesis which was a major scientific achievement, a pioneering study of the evolution of galaxies.

Her first interest in astronomy had been in cosmology, but she had quickly realised that further progress in this subject depended on understanding how galaxies evolve, so she devised the technique of synthesising models of galaxies by putting together the available empirical and theoretical information on the evolution of stars to deduce how galaxies evolve in colour and luminosity. This thesis marked the beginning of modern detailed studies of galactic evolution, and with it Beatrice Tinsley's career in the field was firmly launched. Her thesis contained two major conclusions: (1) the colours of galaxies of all Hubble types can be explained as a result of different histories of star formation without assuming different ages, and (2) galaxies fade rapidly enough that substantial corrections are required when they are used as 'standard candles' in cosmological studies. These conclusions were not immediately appreciated by all, and Beatrice felt at first that it was difficult for her to get her work accepted, but over the next several years her results became well established through further work on her part and confirmation or rediscovery by other investigators.

Family responsibilities brought a brief pause in her career after she adopted a son Alan in 1966 and a daughter Teresa in 1968, but her growing professional ambitions could not be set aside, and within a few years she had resumed her research career with the full force of her prodigious energy, beginning a prolific series of projects and publications that continued until her death. She rapidly established acquaintance with many prominent astronomers, and initiated numerous collaborative efforts combining the expertise of individuals working in diverse areas. All who worked with her were impressed with her broad knowledge and her ability to quickly grasp and synthesize information from many fields.

There is not space even to list all the projects on which she worked, but they expanded in many directions the study of galactic evolution and its relations to cosmology. To mention a few examples, several papers studied stellar populations in the solar neighbourhood and in star clusters, in order to obtain improved information about late stages of stellar evolution and the evolutionary history of galaxy. An important series of papers on the chemical evolution of galaxies combined concepts of galactic evolution with information on stellar chemical compositions and nucleosynthesis, and provided a careful study of the possible processes and models; these papers emphasised that the simplest models often fail and that realistic models must be more complex, involving features like gas flows and inhomogeneities. Several papers also addressed the evidence concerning the origins of different types of supernovae, and their relation to stellar evolution and nucleosynthesis.

In addition to the effect of luminosity dimming on the use of galaxies as standard candles in cosmology, Beatrice studied the use of the diameters and surface brightnesses of galaxies and the use of galaxy counts and the integrated background light as possible cosmological tests. In all cases similar conclusions were reached, namely that the observations are at least as sensitive to the evolution of galaxies as to the differences between the cosmological models being tested. As a result of this work, it eventually became clear that most 'cosmological' observations actually provide more information about galaxies than about cosmology, and that the evolution of galaxies is actually a very rich subject of study in itself.

For example, such 'cosmological' observations as counts of faint galaxies and measurements of the extragalactic background light are sensitive to whether galaxies experience an initial bright flash of star formation or whether star formation occurs more gradually; they are also sensitive to the mix of galaxy types and whether this changes with time. Tentative evidence for an early phase of rapid star formation in at least some galaxies has appeared in some of the data, but the identification of a *bona fide* 'primeval galaxy' undergoing an initial intense burst of star formation, a subject to which Beatrice also gave attention in several papers, has as yet proved elusive. A possible reason for this, discussed in yet another of her papers, is that star formation in young elliptical galaxies occurs not in a single large burst, but less spectacularly in a series of small ones.

Despite her burgeoning career and reputation, Beatrice was never able to obtain academic employment in Dallas, a fact which made her increasingly frustrated, even though she enjoyed a series of visiting appointments at the California Institute of Technology, the University of Maryland, the University of Texas at Austin, and the Lick Observatory. Finally she decided to leave Dallas and seek her fortunes elsewhere, and in 1975 she was appointed Associate Professor of Astronomy at Yale University, an appointment which pleased her greatly and marked the beginning of a very fruitful association with that institution. She interacted with and helped unify the existing activities at Yale in stellar evolution, star formation, and extragalactic astronomy, and contributed much to Yale's eminence in the areas of stellar and galactic evolution. In 1978 she was promoted to Professor of Astronomy, an achievement which finally amply fulfilled her once frustrated ambitions to find suitable academic employment. It was a tragic irony of fate that she learned of her promotion to professor at the same time that she learned that she had melanoma, a particularly intractable form of cancer that was to lead to her death three years later. Fortunately, her health held out remarkably well for most of that period, and she remained scientifically productive until very near her death in 1981.

Beatrice showed her talents in other ways than her prolific output of scientific papers, of which close to 100 were published. In 1977 she organised an outstandingly successful conference at Yale on 'The Evolution of Galaxies and Stellar Populations' at which many new trends in extragalactic research were exhibited. Her broad acquaintance with the field and her high standards were manifested in the content of the meeting and the published proceedings which she edited, and which became an important reference in the field. She herself attended a great many meetings as an invited speaker and was well known for her lucid and even inspiring review talks. She served with distinction on university and national committees and became known even in non-scientific circles for her vigour, astuteness, and uncompromising standards. In the Yale Astronomy Department, she served as Director of Graduate Studies, and devoted a generous share of her time and energy to the concerns of the graduate students. Mindful of the help she had received from more senior astronomers earlier in her career, she gave much help and encouragement to students and younger colleagues. She initiated a series of weekly student-faculty lunches, hosted gatherings with visitors, and took students to meetings and introduced them to other astronomers and promoted their work. Young astronomers anywhere in the world might receive a letter from Beatrice commending them for a piece of good work and urging them to continue their efforts; she did not regard them as rivals but rather as co-investigators in the greater enterprise of understanding the universe.

Perhaps one of her greatest contributions to science does not even appear in print, being alluded to only in the many acknowledgments at the end of papers where her name is mentioned. A great many people were stimulated and inspired by her vitality, her joy in the pursuit of knowledge, and the enthusiasm that she transmitted to others. All avenues of study were to be pursued with vigour, and many projects were launched or strongly influenced as a result of her initiatives. She provided the focus and, directly or indirectly, the driving force behind most of the work on galactic evolution that was done during her lifetime. Moreover, all those who worked in the field knew that their work would receive her close and critical attention, and her comments were often sought and sometimes feared, because she did not tolerate what she considered to be incorrect or inadequate arguments. Although her career was not long in years, it was long enough for her to have a greater impact on astronomy than most astronomers could hope to have over a much longer career.

As remarkable as the rest of her life was the way that Beatrice coped with her final catastrophic illness. As a person of strong ambition, she was initially devastated at the discovery of cancer, but quickly became determined to make the best possible use of whatever time remained. She underwent a series of operations, but her still abundant energy and drive helped her to recover quickly from them, and she soon reappeared in her office ready to carry on with all her normal activities. During this period she produced some of her most significant scientific contributions, including an extensive and widely quoted review on the evolution of the stars and gas in galaxies, the culmination of a long effort to model the stellar content of elliptical galaxies on the basis of detailed spectra, a study of the evolution of disc galaxies and the origin of S0 galaxies, a demonstration of evidence for different proportions of dark mass in galaxies of different Hubble type, and a number of other important works.

In 1980 September her health finally failed when it was found that her cancer had spread to vital organs, and in November a brain tumour left her partially paralysed and confined thereafter to the university infirmary. Although it had been expected that the subsequent course of her illness would be quick, she instead showed a remarkable improvement and surprised her doctors by how well and how long she held up. She was able to resume some of her scientific activities, such as reading, correspondence, and consulting with students and colleagues. Even the loss of the use of her right hand could not stop her, and her final paper, a detailed mathematical treatment of chemical evolution, conceived and completed while she was confined to the infirmary, was laboriously written out with her left hand. She died shortly after this paper was submitted for publication.

During her months in the infirmary, it became abundantly clear that she had gained vast numbers of friends and admirers throughout the scientific world, and mail, flowers, and visitors flowed into her room in a steady stream. All who visited her or spoke with her by telephone, even the doctors and nurses, were amazed at her cheerfulness and determination in a situation where she was expected to live only a matter of weeks. Inevitably she suffered moments of depression when her condition took a turn for the worse, but she always bounced right back. She greatly appreciated the attention and support she received from her many friends, insisting that it was the only thing that made it possible for her to appear courageous and cheerful. For her the pain of having to depart from the world at the height of her career was eased just a little by the realisation that she was doing so in the full glow of attention and recognition, rather than as a forgotten relic of a past era.

Index

Note: BH = Beatrice Hill in her early life; BHT = Beatrice Hill Tinsley after she married Brian Tinsley. Beatrice M. Tinsley when she publishes papers. Beatrice Hill Tinsley is the name by which she wished to be known.

Albee, Mary S., 365
Alcock, Charles, 383
Alderman, Leslie *see* Powell, Leslie
Allan, Doris Napier, 69
Allum, Rose, 54, 74-75
American Astronomical Society, 15, 182, 201, 237, 282, 305, 393
American Physical Society, 253
Andrews, Frank, 320
Annie J. Cannon Prize, 223, 228-29
Anthony-Twarog, Barbara, 275, 298, 304, 368, 394
Antigone, Ron, 150
Armatino family, 261, 271, 278, 285, 321
Arnett, Dave, 214
Arons, John, 237
Arizona State West University, 406
Ariyan, Stephan, 307
Asteroid (3087) Beatrice Tinsley, 398
Audouze, Jean,
  collaborates with BHT, 198, 200, 211, 214-15, 220, 252; friendship with BHT, 216, 220, 252, 264, 329; thoughts on BHT's impact on astronomy, 397; career in astronomy, 405

Barnothy, Jeno, 261, 278, 311
Barnothy, Madeline, 261, 278, 311
Bash, Frank N., 393, 394
Beatrice Tinsley Fund, 393
Beatrice M. Tinsley Prize, 393
Beatrice Tinsley Visiting Fellowship, 393, 394, 398
Beauchamp, Ruth, 94
Bell, Jocelyn, 287
Berkner, Lloyd, 129-130
Big Bang theory, 17, 18, 70, 120, 198-199
Black, Philippa, 81
black holes, 139-140, 286, 404
Blumenthal, George, 245

Bowen, Florence, 20
Bragg, Sir W. Lawrence, 100
Broadley, Colin, 78-80, 87
Bright Star (play), 398
Buchman, Frank, 21, 23, 24
Burbidge, Geoffrey, 189
Burbidge, Margaret, 189, 192
Burstein, David, 251, 252, 257, 296
Burstein, Gail, 257

Caldwell, Nelson,
  student of BHT, 272, 278, 284, 293, 295, 324, 327, 332; career in astronomy, 397
California Institute of Technology *see* Caltech
Caltech, 161, 193, 199-200, 211, 266, 288, Kellogg Radiation Laboratory, 198; Palomar Telescope, 193, 196
Cambridge Music School, 80
Campbell, Claire *see* Stewart, Claire
Canterbury University (Christchurch), 81ff, 207, Helen Connon Hall, 81; Mobil Computer Laboratory, 122; Rolleston House, 106
Carter Observatory (Wellington), 320
Cayrel, Guisa, 216
Cayrel, Roger, 216
Chandrasekhar, Subrahmanyan, 140
Christian, Carol, 346, 365-6
Charles Cook Prize, 127
Cohen, Marie,
  counsels Terry, 357, 371, 391
computers in astronomy, 121-123, 290
cosmological constant, 17, 254
cosmology, 18, 97, 141, 240, 399
Cowie, Len, 383
Crick, Francis, 325
Cudworth, Kyle, 251

DNA, 100, and Rosalind Franklin, 325
dark matter, 212

Davies, Alma, 75
Davies, Merlin, 37
Demarque, Pierre, 237, 238, 262, 266, 284, 288, 305, 338
da Costa, Gary S., 332
de Guider, Yvon, *see* Tinsley, Yvon
de Vaucouleurs, Gérard, 148, 188, 205, 216, 294, 296
Deeming, Terry, 149, 167
Dodd, Josephine, 65

Eastern Astronomers' Neighbourhood Meeting, 291
Easther, Richard, 240
Edmonds, Frank, 150
Einstein, Albert,
   cosmological constant, 17, 254; Theory of General Relativity, 17, 140, 153, 223, 273
Elizabeth II,
   visits New Zealand, 60, 62-63
Ellyett, Cliff, 125, 129, 130
Else, Anne, 113
Espeset, Bets, 215, 217, 219, 220, 223, 230
Evans, Margaret, 81
expanding universe, 17, 18, 120, 196, 200, 240, 254, 273
   and General Relativity Theory, 17, 153-54, 196, 223, 273; Hubble and, 17, 254
   see also Universe

Faber, Andy, 244, 257-58, 319;
   children of (Robin, Holly), 205, 244, 259, 318-19, 360-61
Faber, Sandra (Sandy),
   friendship with BHT, 201, 244, 257-59, 318-19, 350-52, 359, 376, 397; children of, (Robin, Holly)205, 244, 259, 318-19, 360-61; scientific work, 200, 205, 231, 252, 296, 404; collaborates with BHT, 252; last visit to BHT, 359; eulogy for BHT, 386-88; career in astronomy, 404-05
Fagan, Beatrice *see* Morton, Beatrice
Fagan, Patrick, 26
Fajardo, José, 124, 163, 170
   marries Rowena Hill, 124; children of (Sito, Lila), 163, 170, 176, 216, 230; marriage ends, 170

Family Planning Association, 118, 130
Faraday Centennial Medal, 25
Fardell, John, 41, 44, 108
Fardell, Teddy, 41, 42, 44, 65-66, 108, 112, 208
Fearnley, Mattie *see* Hill, Mattie
Ford, Jean *see* Ross, Jean
Fowler, Willie, 198
France, Guff,
   friendship with BH, 95, 107, 117, 164, 207, 377; visits BHT in Dallas, 171-173; visits BHT in New Haven, 344, last visit to BHT, 377
France, Necia,
   friendship with BH, 108, 207; visits BHT in Dallas, 171-173
Franklin, Rosalind, 325
Freeman, Ken, 284, 296, 320, 345, 394
Frogel, Jay, 213

galaxies,
   distances of, 139, 241; evolution of, 18, 166, 197, 213, 214, 252, 256, 297; chemical evolution of, 198-99, 252, 297; formation of, 254, 290, 296, 297; modelling of, 199, 290; observations of, 139, 198-99; Milky Way Galaxy, 294
Geddes, Dorothy, 56, 69
Gehret, Dolores Campbell, 298
Gilmore, Gerry, 115
Gilmour, Everard, 59
Geiringer, Erich, 128
Goddard Space Flight Center, 215, 291
Gott, Richard, 240, 264-5, 273, 278
Guenther, Dave, 365
Gullidge, Constance (Nanny), 34-35,37, 38, remains in England, 40; comes to New Zealand, 42, 43; returns to England, 44-45; correspondence with BHT, 194, 271
Guthrie, May, 372
Gunn, James (Jim),
   first meets BHT, 162; collaborates with BHT, 196-98, 211, 220, 223, 240-41, 259, 273, 330; personal relationship with BHT, 200-01, 220, 224-25, 243, 255-56, 270, 368, 369, 374; scientific work, 196-99, 203, 240-41, 322; death of father, 198, 256; marriage to Rosemary

Gunn ends, 282; relationship with Jill Knapp, 310-12; final visit to BHT, 374; writes eulogy for BHT read by Jerry Ostriker, 389-90; marries Jill Knapp, 405; career in astronomy, 404
Gunn, Rosemary, 201, 220, 224, 244, 264, relationship with BHT, 225, 242, 247-48, 251; asks Jim Gunn to leave, 282
Gunn, James Edward Sr, 198

Hale Observatories, 193
'Half a Life' (Edward Hill's unpublished autobiography), 20, 23, 224, 300-01
Hanlin, Jean, 124-25
Hanna, Dave, 90, 95, 102, 112
Hanna, (Nitschke) Nita, 81, 87, flats with BH, 88-90, 93; marries Dave Hanna, 112
Hanson, William B., 133
Harding, Pip *see* Jackson, Pip
Haydon Prize for Physics, 127
Hawking, Stephen, 286, black hole theory, 286
Head, Marilyn, 115
Heikkila, Walter, marries Bea Wolf, 190; friendship with BHT, 213; divorces Bea Wolf, 236
Heine, Jan, 94
Heward, Mary, 64
Hill, Beatrice Muriel *see* Tinsley, Beatrice Muriel Hill
Hill, Eustace, 19
Hill, Edward Owen Eustace (father), birth of and early life, 20-21; education, 21-22; involvement with Oxford Group, 21-24, 30-31, 248, 252, 300, 342, 400; first meets Jean Morton, 24; amateur acting, 23, 73; studies law, 24, 30; marries Jean Morton, 28-29; birth of Rowena, 30; call up for World War II, 31; birth of Beatrice, v, 19, 33; birth of Theodora, 35; decides to enter the clergy, 37; offered a curacy in Christchurch, 38; decides on immigration to New Zealand, 37-39; ordained in Christchurch, 41; as vicar in Merivale, Christchurch, 41; as vicar in Southbridge, 42; as vicar in New Plymouth, 45; resigns from clergy, 58; stands for mayoralty 59; elected as mayor, 60-61; duties as mayor, 67; mayor during official visit of Queen Elizabeth II, 62-63; buys house at Hatepe, 74; moves to Strathmore, Wellington, 97; tutor at Victoria University, Wellington, 109; marries Mattie Fearnley, 186; visits BHT in Dallas, 204; effect of BHT's divorce on, 229, 233, 235, 248; writes *My Daughter Beatrice*, 33, 45, 50, 83-84, 173, 177, 200, 252, 320, 391; writes 'Half a Life', 20, 23, 224, 300-01; death of his mother, 224; role as grandfather, 204, 252; visits BHT at the time of her cancer treatment, 314; death of, 404
Hill, Graham, 148-49
Hill, Jean (mother), early life, 27-28; first meets Edward Hill, 24; involvement with Oxford Group, 24, 28, 30-31; marries Edward Hill, 28-29; musical ability, 28, 46; amateur acting, 23, 28, 73; as a writer, 56, 97, 109; birth of Rowena, 30; birth of Beatrice, v, 19, 33; birth of Theodora, 35; relationship with children, 66, 77, 86, 208; ill-health, 27, 167; attitudes to divorce, 146; death of, 168
Hill, Matthew, 331
Hill, Mattie, marries Edward Hill, 186; visits BHT in Dallas, 204; suffers from multiple sclerosis, 207
Hill, Muriel, 19, 20
Hill, Rowena (sister), birth of, 30, childhood, 43-53; attends St Margaret's, 42; attends Southbridge Primary School, 43; attends New Plymouth Girls' High School, 54, 71; dux of New Plymouth Girls' High School, 68; university life, 72, 84 ; rebellion against parents, 66, 71, 77, 86; marries José Fajardo, 124; meets

BHT in Italy, 163, 176, 216, 228,
 marriage ends, 170; children of,
 (Sito, Lila) 163, 170, 176, 216,
 230, 246; meets BHT in Peru, 230;
 poetry of, 324, 378; retirement, 404
Hill, Theodora *see* Lee-Smith,
 Theodora
Hoffleit, Dorrit, 266, 299, 302
Holland, Sid, 62
Holt, Olav, 142, 163
Holt, Tordis, 142, 163
Honnor, Alf, 73-74
Hopkins, John, 94
Hoyle, Fred, 138, 139,
 Nature of the Universe 53, 70, 262;
 and Big Bang theory, 18, 70; and
 Steady State theory, 70, 121; 60th
 birthday of, 262, 263
Hubble, Edwin Powell,
 work on expanding universe, 17,
 399
Hubble constant, 296
 BHT and, 296; Sandage and, 296;
 de Vaucouleurs and, 296
Hubble Telescope, 197, 292, 404

Institute of Astrophysics (Paris), 405
International Astronomy Union
 Conference (Poland), 215, 216
International Astronomical Union
 Symposium (Grenoble), 282
International Conference on the
 Evolution of Galaxies and Stellar
 Populations (Yale, Connecticut),
 292-97
International Symposium on
 Gravitational Collapse (Dallas),
 137-141
Iowa State University, 396

Jackson, Pip, 80-81, 94, 183-184, 271,
 315, 377
Jarrold, Joyce, 69-71, 75
 reflections on BH's ability, 71
Johnson, Francis, 176
Johns Hopkins University, 291
Julian, Rae, 95

Kaplan, Moreson, 344, 382
Kardos, Thomas, 69
Kennedy, John F. (John Fitzgerald), 131

assassination of, 136-137
Kennicutt, Robert Jr, 166, 394
Kerr, Roy,
 solution to Einstein's gravitational
 field equations for rotating black
 holes, 140
King, Ivan, 166, 237, 252, 293
King, Martin Luther, Jr,
 assassination of, 168
Knapp, Jill,
 friendship with BHT, 212, 213-14,
 215, 224, 282, 310, 367, 369, 374;
 relationship with Jim Gunn, 310-
 11; final visit to BHT, 374; scientific
 work, 282, 406; marries Jim Gunn,
 405; career in astronomy, 394, 405
Knapp, Steve, 212, 224, 282
Komlos, Willi, 55

Larson, Richard,
 first meets BHT, 201; personal
 relationship with BHT, 261, 268,
 270, 274, 279, 309-10, 341, 356;
 music and BHT, 268, 274; scientific
 work, 166, 199; collaborates with
 BHT, 199, 267, 289, 294, 297, 346;
 relationship with Terry and Alan,
 285-86, 303, 316, 377, 385, 392;
 executor of BHT's will, 385; career
 in astronomy 237, 394, 404
Lee-Smith, David, 191, 207, 279, 315,
 marries Theodora Hill, 191;
 children of (Charlotte, Beatrice),
 315; BHT's last visit, 320
Lee-Smith, Theodora (sister)
 birth of, 35; childhood, 35-53;
 education, 43, 60, 71, 97, 109, 163;
 musical ability, 43, 46, 72; studies
 music in Florence, 163; studies music
 in London, 168; marries David Lee-
 Smith, 191; children of (Charlotte,
 Beatrice), 315; BHT's last visit, 320;
 career, 404
Leigh, Josephine *see* Dodd, Josephine
Lennon, John (Beatle),
 murder of, 373
Lester, Joan *see* Williman, Joan
Lick Observatory *see* University of
 California, Santa Cruz
Lobb, Julienne, 56, 61
Locke, Don, 99

Longair, Malcolm, 263

McClure, Bob, 237, 284, 305
McLellan, Alister, 97
McQueen, Harvey, 106, 107, 113
Marsh, Ngaio, 41-2
Matthews, Bill, 245, 247
Metcalf, Glen, 84, 90, 207, 337
Metcalf, Ruth *see* Beauchamp, Ruth
Metcalf, Wal, 84, 90, 207, 337
Milky Way Galaxy, 294
Moon, Bruce, 122
Moral Re-Armament, 22, 31, 301
  see also Oxford Group
Morgan, Bill, 314
Morton, Alexander, 24-25
Morton, Alistair, 26-27
Morton, Lady Beatrice, 26
Morton, James, 25-26,
  awarded Faraday Centennial Medal, 26
Morton, Jean *see* Hill, Jean
Morton, Jocelyn, 26, 27, 334
Muir, Margaret *see* Sparrow, Margaret
My Daughter Beatrice (Edward Hill), 33, 45, 50, 83-84, 173, 177 200, 252, 320, 391

National Aeronautics and Space Administration *see* NASA
NASA,
  Goddard Space Flight Center, 213;
  Jet Propulsion Laboratory, 115
NATO Advanced Study Unit (Cambridge, England), 222
National Youth Orchestra, 90, 94-95
Nature of the Universe (Fred Hoyle), 53, 70, 262
nepotism regulations, 127, 202-3, 207
New Plymouth Boys' High School, 61, 75, 104
New Plymouth Girls' High School, 45, 54-56, 64-65, 68-77
New Zealand Family Planning Association, 96
Niblock, Alec, 52-53
Niblock, Fanny, 52-53
Nitschke, Nita *see* Hanna, Nita
nucleosynthesis, 198-99

O'Dell, Tom, 205

O'Shea, John, 78
Oemler, Gus, 268, 285, 327,
  takes over BHT's students, 353, 375; scientific work, 325, 346
Oke, Beverley, 196
Olenik, Sandra, 348, 353, 391
Ostriker, Jeremiah (Jerry), 242, 269, 270, 322, 396,
  reads Jim Gunn's eulogy for BHT, 390; career in astronomy, 405
Oxford Group, 21-22, 23, 30-31, 38, 63, 66, 77, 156, 252, 300, 342, 400
  see also Moral Re-Armament

Papapetrou, Achilles,
  explains Roy Kerr's solution to Einstein's equations for rotating black holes, 140
Patchett, Bob, 107
Patterson, Tom, 176
Payne-Gaposchkin, Cecilia, 153
Penzias, Arno, 364
Persson, Jay, 213
Philip, Duke of Edinburgh,
  visits New Zealand, 60, 62-63
Pickering, William, 115
photometry, 252
Powell, Leslie, 61-62, 72, 74, 110
Princeton University, 322, 406
  Institute of Advanced Study, 254
pulsars, 287

quasars, 294

RNA, 100
redshift, 182, 297
Rees, Martin, 284, 336
Renzini, Alvio, 228, 279
Robinson, Ivor, 134, 137
Rindler, Wolfgang, 134,
Rose, Jim, 269
Ross, Archie, 110
Ross, Jean, 52
Ross, Vinnie, 46, 51, 55
Royal Society of New Zealand, 81, 100
Rubin, Vera, 212, 286, 316, 364, 369, 394, 397
Rutherford, Ernest, Lord Rutherford of Nelson, 82, 115
Ryle, Martin, 120-21

SCAS *see* Southwest Center for Advanced Studies
Saar, Enn, 343
Sachs, Rainer, 152, 158
Sadler, Mark, 118
Sandage, Allan,
    and BHT, 165-166, 238-40, 289; scientific work, 193, 239, 245, 296; and age of universe, 165, 239, 245
Sargent, Wallace, 288, 294, 295, 298
Saunders, Penny, 94, 164, 231, 271
Schild, A.E, 138
Schmidt, Maarten, 294,
    work on stellar photography, 138
Schombert, Jim, 325-26
Schramm, Dave, 214, 240, 273, 282
Schweizer, François, 365
Searle, Leonard, 288, 294
Seed, Tom, 97
Shields, Greg, 217
Sloan Foundation, 249
Smith, Harlan, 146, 223, 397
Smithsonian Astrophysical Observatory, 397
Socratic Society (Soc Soc), 84, 95-99, 117, 118-20, 125
Solheim, Jan-Erik, 182
Southbridge Primary School, 43
Southwest Center for Advanced Studies (Dallas, Texas), 129, 130, 133, 136-141, 151, 175 *see also* University of Texas at Dallas
Sparrow, Margaret, 76
spectroscopy, 106, 123, 198, 213
Spinrad, Hayron, 166
stars,
    blue, 289; formation of, 196, 198, 289-90; death of, 196; dwarf, 213, 251; giant, 215; infrared wavelengths, 199, 213; measurement of brightness, 252; red, 251, 289; temperature of, 198, 217
Steady State theory, 120, 121, 154
Steigman, Gary, 292, 305
stellar photography, 138
Stetson, Peter, 319
Steuart, Billie, 68, 69
Stewart, Claire, 56, 81
Struck, Curt,
    post-graduate student of BHT, 283, 286, 291, 293, 296, 303, 312, 379; remembers BHT as teacher, 283, 291; scientific work, 291; career in astronomy, 396
Struck-Marcell, Curtis *see* Struck, Curt
Stryker, Linda,
    post-graduate student of BHT, 301-02, 313, 333; friendship with BHT, 301, 303, 344; plays music with BHT and Richard Larson, 302; collaborates with BHT and Jim Gunn, 330; last days with BHT, 369ff; executor of BHT's will, 385; eulogy for BHT, 388-89; career in astronomy, 405

Tangiwai disaster, 62
Texas Christian University, 189
Texas International Symposium, Boston, 287; Dallas, 239; New York, 161
Theory of General Relativity 17, 140, 153, 223, 273
    and expanding universe, 17, 196, 223, 273
Theories of the Universe (Milton Munitz, ed.) 53, 240;
    impact on BHT, 53, 91, 299
Thomson, Helen, 68
Thorne, Kip, 140
Tinsley, Alan (son),
    birth of, (in New Zealand) 58; babyhood, 162, 167; childhood, 173, 178, 199, 203, 204, 219, 257, 270-71, 276-77, 281, 299, 302, 316, 363; custody arrangements, 233-4; relationship with Richard Larson, 285-86 ; thoughts on family relationships, 233, 328-29; career, 404
Tinsley, Beatrice Muriel Hill,
    CHILDHOOD AND EARLY LIFE
    birth of, v, 19, 33, 36; childhood, 43-53; known as 'Beetle', 43, 50, 61, 62, 65, 80, 86, 93, 95, 96, 104, 117, 120, 173, 193, 373, 381; voyage to NZ, 41; meets John and Teddy Fardell, 41; relationship with Nanny, 34-35, 40, 42, 44-45, 51, 57, 271; attitude to parents, 49, 54, 57, 60, 66, 77, 103, 146, 186, 300

EDUCATION
school days:
attends St Margaret's, 42; attends Southbridge Primary, 43; attends Central Primary School (New Plymouth), 43; attends New Plymouth Girls' High School, 54-56, 64-65, 68-77; dux of New Plymouth Girls' High School (1957), 76
university:
attends Canterbury University, 81ff university life, attends meetings of the Socratic Society, 84, 95-99, 117, 118-20; BSc, 110, 113; MSc, 110-111, 127, 130 decides on cosmology as a career, 97, 141

PERSONAL LIFE
Music, 35, 46, 49, 55, 72, 80, 85 124, 214, 268, 274,
joins National Youth Orchestra, 90, 94-95, joins Richardson Symphony Orchestra, 167, 170, 179; plays with Richard Larson, 268, 274; plays with Linda Stryker, 302; writes poetry, 71, 268, 358, 390; engagement to Colin Broadley 78-80, 87; makes début, 87; flats with Nita Nitschke, 88-90, 93; first meets Brian Tinsley, 95, 100; engagement to Brian, 102, 110; marries Brian, 113; honeymoon, 114-115; infertility, 143; adopts Alan, 156-158; adopts Terry, 169, 174; abortion, 174-175; relationship with children, 164, 171-173, 178, 199, 203, 208, 227-28, 241-42, 243, 254-55, 266, 271, 276, 280, 285-86, 321, 359-60, 363; visits New Zealand, 206-08; feminism, 145, 149, 161, 263, 265, 269, 370; works for Zero Population Growth, 181-184, 190, 194, 203, 207; meets Rowena in Italy, 163, 176, 216, 228; relationship with Jim Gunn, 211, 220, 224-5, 243, 254, 255-56, 256-57 264, 270, 321; friendship with Sandy Faber, 201, 205 244, 259-61, 318-19, 349-53, 359; friendship with Bea Wolf, 167, 190, 215, 274, 305, 353-55; friendship with Virginia Trimble, 199, 263; friendship with Jill Knapp, 211, 225, 282, 310-11; friendship with Jean Audouze, 218, 220, 252, 264, 329; friendship with Linda Stryker, 301, 303, 344; strains in marriage, 209, 211, 221-23, 225; meets Rowena in Peru, 230; divorces Brian, 233; custody of children, 225-27, 234, 244; life after divorce, 234, 236, 247; relationship with Richard Larson, 203, 261, 268-69, 270, 274, 279, 309-10, 323, 332, 356; discovery of melanoma, 306-07; cancer treatment, 308, 311, 324, 328, 337, 341, 363; final visit to New Zealand, 315, 320; writes 'living will', 309, 349, 382; Terry comes to live, 334 ff; diagnosis of terminal cancer, 347, 363; destroys diaries, 358; final days in Yale Infirmary, 364-385; makes funeral arrangements, 371; writes a paper from the infirmary, 381; death of, 383; memorial service of, 385-92; memorials to, 393.

SCIENTIFIC WORK IN THE UNITED STATES
PhD, 146-149, 158, 160; thesis, 152-154
collaborates with Jean Adouze, 200, 211, 214-15, 220, 252; disagrees with Allan Sandage, 165, 238-40, 289; works with Jim Gunn, 197, 196-199, 211, 220, 223, 240-41, 259, 273, 282, 330; works with Richard Larson, 199, 267, 289, 294, 297; works with Sandra Faber, 252; works with Virginia Trimble, 199, 211, 337; works with Greg Shields, 217-18; works with Jill Knapp, 212, 282; works with Scott Tremaine, 282; organises Yale's International Conference on the Evolution of Galaxies and Stellar Populations, 288, 292-97

BHT's scientific work on:
expanding universe, 153-54, 165, 196, 200, 223, 240-41, 273

galaxies,
  distances of, 152-154; evolution of, 17, 18, 166, 196-97, 213, 252; chemical evolution of, 198-99, 252; formation of, 290; modelling of, 254, 289-90; observations of, 196, 290;

CAREER IN ASTRONOMY
  teaches at Christchurch Girls' High School, 124, 126
  teaches at the South West Center, 179
  visiting scientist at the University of Texas at Dallas, 177, 187, 192, 204, 234
  appointment at Caltech, 193, 196, 210
  appointment to University of Maryland, 204, 207, 209-10, 211-14
  assistant professor at the University of Texas at Austin, 215, 216, 220, 222
  assistant research astronomer at University of California at Santa Cruz, Lick Observatory, 232, 244-45
  associate professor at Yale, 238-39, 265-66
  professor at Yale, 316
  post-graduate students, 274, 283-84, 312-13
  teaching, 266, 275, 279, 283, 290, 291, 313, 319, 332
  in Erice, Sicily, 228; at Saas-Fee, 279, at Goddard Space Flight Center, 291, at Johns Hopkins University, 291, at Les Houches, France, 329
  lasting contributions to astronomy, 18, 388, 396, 397, 399-400

Tinsley, Brian,
  childhood, 104; first meets BH, 95, 100; engagement to BH, 102, 110; marries BH, 113; honeymoon, 114-15; decides to study spectroscopy, 106; MSc, 106, 113, PhD, 106, 123, 125, 130, 135; goes to Texas, 131; adopts Alan, 156-158, adopts Terry, 169, 174; promoted to associate professor, 183; visits New Zealand, 206-08; strains in marriage, 209, 221-23, 225; divorces BHT, 233; children's custody, 225-27, 234, 244; feelings after divorce, 234, 281; marries Yvon de Guider, 302; birth of Winston Walter, 319; sends Terry to Dallas, 334-37; takes Terry back, 392; scientific work, 123-124, 143, 148, 175, 193, 195, 204, 219, 220; career, 404

Tinsley, Nola, 104

Tinsley, Teresa Jean (Terry) (daughter), birth of, 168; childhood, 173, 185, 199, 203, 204, 215, 219, 257, 259, 270-71, 281, 302; custody arrangements, 233-4; relationship with Richard Larson, 285-86, 303, 316, 377, 391; goes to live with BHT, 334 ff; attends Cheshire Academy, 358; returns to Dallas with Brian, 392; returns to Richard Larson, 392; career, 404

Tinsley, Terence, 104

Tinsley, Winston Walter, 319, 321, 335

Tinsley, Yvon,
  as Yvon de Guider, marries Brian Tinsley, 302; birth of Winston Walter, 319; relationship with Alan, 302; relationship with Terry, 302, 334-35, 392

Todd, Sir Alexander, 100

Toomre, Alar, 274, 296, 298

Tosi, Monica,
  post-graduate student of BHT, 341; remembers BHT as a teacher, 341, 372, 395; scientific work, 341, 373, 395; career in astronomy, 395

Tremaine, Scott, 256, 282-83, 383

Trimble, Virginia,
  first meets BHT, 199; friendship with BHT, 199, 211, 263, 336; career in astronomy, 394

Tsuruta, Sachiko, 161, 213

Twarog, Bruce,
  post-graduate student of BHT, 275, 368; career in astronomy, 394

UTD *see* University of Texas at Dallas

Unitarian Church, 171, 182, 203, 204, 218
universe, age of , 153 ; Hubble and, 17; open or closed, 165, 223, 240-41; size of, 153;
see also expanding universe
University of California, Santa Cruz, 244-45,
Lick Observatory, 200, 205, 231, 244-45
University of Kansas, 394
University of Maryland, 202, 205, 209-10, 211-14
University of Texas at Austin, 15, 146, 215, 393, 397, 404,
Astronomy Department, 146, 217, 220, 222, 398; McDonald Observatory, 146
University of Texas at Dallas, 176, 179, 187, 192, 192, 202-03, 205-06, 218, 234, 404

Van Altena, Bill, 265, 305, 344, 354, 370,
eulogy for BHT, 388
Van den Bergh, Sidney, 293, 319
Veale, Hilda, 55
Victoria University of Wellington, 109

Wardell, Phyl, 116, 129
Warwick House Prize, 127
Washington University (Saint Louis, Missouri),
McDonnell Center for Space Physics, 252

Watson, James D., 325
Weistrop, Donna, 251
West-Watson, Campbell, 37-38
Westphal, Jim, 197
Wheeldon, Robin, 187, 194, 204
Wheeldon, Ron, 187, 194, 204
Wheeler, John, 139,
black hole theory, 140
Wilkins, Maurice, 325
Williman, Alan, 122
Williman, Joan, 122
Wilson, Art, 182, 194
Wilson, Carol, 182, 194
Wilson, Chris, 268, 285
Wolf, Beatrice (Bea),
friendship with BHT, 167, 190, 213, 274, 305, 354-56; marries Walter Heikkila, 190; divorces Walter, 236
women in science, discrimination against, 69, 88, 127, 192, 204, 396
at Yale, 265, 269
World War, 1939-1945, v, 19, 32-33, 37
Woolf, Neville (Nick), 152

Yale University,
Astronomy Department, 238, 265-66, 303-04, 342, 388; Health Center, 307-08; Infirmary, 343, 346, 364-384; position of women, 265, 269, 388; student revues, 303-04, 338-40

Zero Population Growth, 128, 174, 181-184, 190, 194, 203, 206